THE SAUNDERS SERIES

Consulting Editor: John A. Thorpe, SUNY at Stony Brook

Howard Eves: AN INTRODUCTION TO THE HISTORY OF
MATHEMATICS, 5e   1983
Theodore W. Gamelin and Robert Everist Greene:
INTRODUCTION TO TOPOLOGY   1983
P. G. Kumpel and J. A. Thorpe: LINEAR ALGEBRA   1983

Forthcoming

Martin Guterman and Zbigniew Nitecki: DIFFERENTIAL
EQUATIONS: A FIRST COURSE   1984

 **SAUNDERS COLLEGE PUBLISHING**
Philadelphia   New York   Chicago
San Francisco   Montreal   Toronto
London   Sydney   Tokyo   Mexico City
Rio de Janeiro   Madrid

P. G. Kumpel
J. A. Thorpe

LINEAR ALGEBRA

with Applications to Differential Equations

THE SAUNDERS SERIES

Address orders to:
383 Madison Avenue
New York, NY 10017

Address editorial correspondence to:
West Washington Square
Philadelphia, PA 19105

Text Typeface: Times Roman
Compositor: York Graphic Services, Inc.
Acquisitions Editor: Leslie Hawke
Project Editor: Patrice L. Smith
Managing Editor & Art Director: Richard L. Moore
Design Assistant: Virginia A. Bollard
Text Design: Lawrence R. Didona
Cover Design: Lawrence R. Didona
Text Artwork: J & R Technical Services, Inc.
Production Manager: Tim Frelick
Assistant Production Manager: Maureen Read

Library of Congress Cataloging in Publication Data

Kumpel, P. G. (Paul G.)
  Linear algebra, with applications to differential
equations.

  (The Saunders series)
  Includes index.
  1. Algebras, Linear.   2. Differential equations.
I. Thorpe, John A.   II. Title
QA184.K85   1983   512′.5   82-60630
ISBN 0-03-060556-3

LINEAR ALGEBRA   ISBN 0-03-060556-3

345   038   987654321

CBS COLLEGE PUBLISHING
Saunders College Publishing
Holt, Rinehart and Winston
The Dryden Press

To our wives
Carol Graepp Kumpel
Marilyn Austin Thorpe

PREFACE

This text is designed for a one-semester, or two-quarter, sophomore level course in linear algebra. It is intended for students who have had a year of calculus.

A basic premise that has motivated the writing of this book is that students who have studied calculus are especially well prepared to understand what linear algebra is really about. The concepts of linear algebra permeate calculus, and calculus-based applications of linear algebra abound. The assumption that the reader of this book has studied calculus allows us to include many examples and applications that could otherwise not be presented. The student gains important insights into linear algebra from these examples and applications, as well as additional experience with the fundamentals of calculus.

But *we do not assume a mastery of calculus, only a familiarity with it.* We have taken great care to make the subject of linear algebra accessible to the average college sophomore.

Our most detailed calculus-based application of linear algebra is linear differential equations. Students easily recognize the importance of understanding differential equations and are quick to see the relevance of linear algebra for describing their solutions. A minimal treatment of differential equations is essential in any course based on this book. But *the required differential equations material can be taught in about two weeks,* or even less if the students have already had an introduction to the subject. This minimal investment of class time is more than justified by the large collection of nontrivial examples that then become available for use in motivation of linear algebra concepts and as a source of computational exercises. An instructor who wishes to make differential equations a major part of the course can do so by covering optional sections as described below.

We begin, in Chapter 1, with a study of linear systems of algebraic equations. Gaussian elimination is described here and the relevant vector and matrix ideas are introduced. Applications to the geometry of lines, planes, and hyperplanes in n-space are discussed.

In Chapter 2, we introduce the idea of a function space and use it, along with the various arithmetic and matrix spaces, to motivate and illustrate the study of vector spaces and their subspaces. The notion of a spanning set is discussed here. These ideas are used to describe completely the geometry of solution sets of systems of linear equations.

The introduction to linear differential equations comes in Chapter 3. The standard technique for solving homogeneous constant coefficient linear differential equations is described. Spanning sets play a central role here. The chapter concludes with an optional section on the method of undetermined coefficients.

For instructors who would like a more substantial treatment of differential equations than is provided in Chapter 3, we have included an optional chapter, Chapter 9, in which variation of parameters, power series methods, and the Laplace transform are discussed. The first three sections of that chapter can be taught at any time after Chapter 4 (linear independence).

Armed with a thorough understanding of systems of linear equations and homogeneous constant coefficient linear differential equations, the student is ready to study (in Chapter 4) linear independence, bases, and dimension. These concepts arise naturally within the context of the solution spaces discussed earlier. There is available at this point an ample supply of computational examples for illustration and motivation.

In Chapters 5 through 8, we discuss other standard topics in linear algebra: determinants (Chapter 5), orthogonality and orthonormal bases (Chapter 6), linear maps (Chapter 7), and eigenvector theory (Chapter 8). Applications are made to least squares approximations (Section 6.2), Fourier series (Section 6.3), systems of linear differential equations (Section 8.3), and linear difference equations (Section 8.4).

We strive throughout for a smooth development of the subject. We do not introduce a new concept until it is needed. We find that students have little patience with learning mathematics for its own sake. The promise "you will see later how important this material is" carries little weight. Pedagogically, it is preferable to build continuously on the student's prior knowledge so that the rewards of learning each new concept are immediately apparent. This approach does result in an occasional nonstandard ordering of topics. For example, matrix multiplication is not introduced until Chapter 7, at which point it arises naturally in connection with the study of the composition of linear maps. Similarly, matrix inverses first appear in connection with inverse maps in Section 7.4.

We have designed this text so that it can be used for two types of courses.

A. For a course with *minimal attention to differential equations,* the following chapters would ordinarily be covered: Chapters 1 through 8, omitting Sections 3.4 and 8.3.

B. For a course with an *emphasis on differential equations,* the following chapters would ordinarily be covered: Chapters 1 through 5 and 7 through 9.

Any of the following sections may be omitted in either type of course, if time is short, without loss of continuity: 3.4, 5.3, 6.2, 6.3, 8.3, 8.4, 9.2, 9.3, 9.4.

We have included in the text many examples and exercises. We are convinced that students at this level learn best when required to analyze and solve a large number of computational exercises. We have also included a variety of exercises that challenge the student's understanding of concepts and theory. We usually include one or two of these in each assignment when we teach this course.

We have included proofs for most of the theorems in the book. These may of course be omitted at the discretion of the instructor. We find, in teaching linear algebra to sophomores, that a clear explanation of the meaning of a theorem, followed by an example illustrating its use, is often much more effective as a teaching technique than a careful analysis of the proof.

Finally, we believe that students who have studied calculus should be encouraged to use it at every opportunity. It is our hope that students learning linear algebra from this text will gain not only an understanding of elementary linear algebra but also some appreciation of the interdependence between calculus and algebra.

We could not, of course, have completed this work without the help of others. We particularly wish to thank the many students at State University of New York at Stony Brook who studied linear algebra from a preliminary version of this text, our colleagues who taught these students from a less than perfect manuscript, and our typist, Wanda Mocarski, who patiently translated our handwriting into readable copy. Special thanks are due to Professor Raouf Doss of SUNY at Stony Brook, who read critically and commented thoughtfully on several sections of the manuscript, and to Professor Richard Koch of the University of Oregon, who provided us with an especially helpful prepublication review. Finally, we would like to thank the editorial staff at Saunders College Publishing for their expert help in preparing the manuscript for publication.

PAUL G. KUMPEL
JOHN A. THORPE

CONTENTS

Linear Equations

1.1 THE MATRIX OF A LINEAR SYSTEM

We begin by developing a systematic procedure, or algorithm, for solving a system of linear equations

(∗)

$$a_{11}x_1 + a_{12}x_2 + \cdots + a_{1n}x_n = b_1$$
$$a_{21}x_1 + a_{22}x_2 + \cdots + a_{2n}x_n = b_2$$
$$\cdots$$
$$a_{m1}x_1 + a_{m2}x_2 + \cdots + a_{mn}x_n = b_m.$$

Here, $a_{11}, \ldots, a_{mn}$ and $b_1, \ldots, b_m$ are real numbers and $x_1, \ldots, x_n$ are the "unknowns." Note that m is the number of equations and n is the number of "unknowns."

A **solution** of the system (∗) is an ordered n-tuple $(x_1, x_2, \ldots, x_n)$ of real numbers that satisfies (∗). Thus, for example, $(1, -1, 0)$ and $(3, -2, 3)$ are both solutions of the system

$$2x_1 + x_2 - x_3 = 1$$
$$x_1 - x_2 - x_3 = 2$$

whereas $(1, 2, 3)$ is not.

The number x_i is called the ith **entry** of $(x_1, \ldots, x_n)$. Thus, the ordered pair (ordered 2-tuple) $(1, -1)$ has 1 as its first entry and -1 as its second entry. Two ordered n-tuples $(x_1, \ldots, x_n)$ and $(y_1, \ldots, y_n)$ are **equal** if and only if all corresponding entries are the same. Thus, $(x_1, \ldots, x_n) = (y_1, \ldots, y_n)$ if and only if $x_1 = y_1, \ldots, x_n = y_n$.

The algorithm employs the familiar process of adding multiples of one equation to the others in order to eliminate one of the unknowns from all the other equations. The key to the success of the procedure lies in the fact that if we add a multiple of one equation to another, the new system of equations we get has the same set of solutions as the old. (Why?)

Before describing the algorithm explicitly, we shall work out a typical example.

1

Example 1. Consider the system

$$x_1 + x_2 + x_3 - x_4 \qquad = 2$$
(1)
$$2x_1 + 2x_2 + 3x_3 + x_4 \qquad = 5$$
$$x_1 - x_2 + 2x_3 + 3x_4 + x_5 = 4.$$

To solve this system, our first goal is to eliminate x_1 from all equations except the first. Thus, by adding -2 times the first equation to the second and then by adding -1 times the first equation to the third, we obtain

$$x_1 + x_2 + x_3 - x_4 \qquad = 2$$
$$x_3 + 3x_4 \qquad = 1$$
$$-2x_2 + x_3 + 4x_4 + x_5 = 2.$$

Observe that the operations we have done are reversible: by adding twice the first equation of our new system to the second and adding the first equation to the third we recover our original system. It follows that the set of solutions is unchanged.

The second step is to eliminate x_2 from all equations except the second. In order to do this, we interchange the second and third equations so as to bring an equation with nonzero x_2 coefficient into the second spot. It is helpful also to divide the equation by its leading coefficient, -2, so that the new coefficient of x_2 in this equation is 1. We then have

$$x_1 + x_2 + x_3 - x_4 \qquad = 2$$
$$x_2 - \tfrac{1}{2}x_3 - 2x_4 - \tfrac{1}{2}x_5 = -1$$
$$x_3 + 3x_4 \qquad = 1.$$

Finally, multiplying the second equation by -1 and adding to the first, we succeed in eliminating x_2 from all equations except the second:

$$x_1 \qquad + \tfrac{3}{2}x_3 + x_4 + \tfrac{1}{2}x_5 = 3$$
$$x_2 - \tfrac{1}{2}x_3 - 2x_4 - \tfrac{1}{2}x_5 = -1$$
$$x_3 + 3x_4 \qquad = 1.$$

The third step is to eliminate x_3 from all equations except the third. By adding appropriate multiples of the third equation to the first two, we obtain

$$x_1 \qquad - \tfrac{7}{2}x_4 + \tfrac{1}{2}x_5 = \tfrac{3}{2}$$
(2)
$$x_2 \qquad - \tfrac{1}{2}x_4 - \tfrac{1}{2}x_5 = -\tfrac{1}{2}$$
$$x_3 + 3x_4 \qquad = 1.$$

Now the solution set is evident. We can specify x_4 and x_5 arbitrarily and then x_1, x_2, and x_3 are completely determined. Thus if $x_4 = c_1$ and $x_5 = c_2$ then

$$x_1 = \tfrac{3}{2} + \tfrac{7}{2}c_1 - \tfrac{1}{2}c_2$$
$$x_2 = -\tfrac{1}{2} + \tfrac{1}{2}c_1 + \tfrac{1}{2}c_2$$
$$x_3 = 1 - 3c_1$$
$$x_4 = c_1$$
$$x_5 = c_2.$$

In other words, there are infinitely many solutions of our system, consisting of all 5-tuples of the form

(3) $$(\tfrac{3}{2} + \tfrac{7}{2}c_1 - \tfrac{1}{2}c_2, \ -\tfrac{1}{2} + \tfrac{1}{2}c_1 + \tfrac{1}{2}c_2, \ 1 - 3c_1, \ c_1, \ c_2)$$

where c_1 and c_2 are arbitrary real numbers. For every choice of c_1 and c_2 we get a solution, and every solution is of this form for some choice of c_1 and c_2. The 5-tuple (3) is called the **general solution** of (1), to distinguish it from a **particular solution** such as $(\tfrac{3}{2}, -\tfrac{1}{2}, 1, 0, 0)$, which is obtained by setting $c_1 = c_2 = 0$. ■

Our work can be made considerably easier by use of a shorthand notation. We can record all the basic information about our system by removing all the unknowns and all the plus and equal signs from our system, and arranging the numbers that remain in a rectangular array or **matrix**. Thus the matrix of our system (1) is

(1′)
$$\begin{pmatrix} 1 & 1 & 1 & -1 & 0 & 2 \\ 2 & 2 & 3 & 1 & 0 & 5 \\ 1 & -1 & 2 & 3 & 1 & 4 \end{pmatrix}.$$

Note that the last column of this matrix records the numbers on the right hand side of the system and the remaining columns represent the coefficients of the unknowns, the ith column listing the coefficients of x_i for $1 \le i \le 5$. The rows of the matrix represent the equations of the system, and these equations can clearly be recovered by filling in the x's, the $+$'s, and the $=$'s in the appropriate places. Observe that multiplying an equation of the system by a real number corresponds to multiplying a row of the matrix by that real number; that is, it corresponds to multiplying each entry in that row by the given number. Similarly, adding one equation to another corresponds to adding each entry in one row to the corresponding entry in another. Thus, the operations we performed on the system in order to solve it correspond to three **row operations** on the matrix of the system:

    (i)   multiply a row by a nonzero real number
    (ii) interchange two rows
    (iii) add a multiple of one row to another.

The operations we used to solve system (1) are conveniently represented as follows:

$$\begin{pmatrix} 1 & 1 & 1 & -1 & 0 & 2 \\ 2 & 2 & 3 & 1 & 0 & 5 \\ 1 & -1 & 2 & 3 & 1 & 4 \end{pmatrix} \xrightarrow{(iii)} \begin{pmatrix} 1 & 1 & 1 & -1 & 0 & 2 \\ 0 & 0 & 1 & 3 & 0 & 1 \\ 0 & -2 & 1 & 4 & 1 & 2 \end{pmatrix}$$

$$\xrightarrow{(ii)} \begin{pmatrix} 1 & 1 & 1 & -1 & 0 & 2 \\ 0 & -2 & 1 & 4 & 1 & 2 \\ 0 & 0 & 1 & 3 & 0 & 1 \end{pmatrix} \xrightarrow{(i)} \begin{pmatrix} 1 & 1 & 1 & -1 & 0 & 2 \\ 0 & 1 & -\frac{1}{2} & -2 & -\frac{1}{2} & -1 \\ 0 & 0 & 1 & 3 & 0 & 1 \end{pmatrix}$$

$$\xrightarrow{(iii)} \begin{pmatrix} 1 & 0 & \frac{3}{2} & 1 & \frac{1}{2} & 3 \\ 0 & 1 & -\frac{1}{2} & -2 & -\frac{1}{2} & -1 \\ 0 & 0 & 1 & 3 & 0 & 1 \end{pmatrix} \xrightarrow{(iii)} \begin{pmatrix} 1 & 0 & 0 & -\frac{7}{2} & \frac{1}{2} & \frac{3}{2} \\ 0 & 1 & 0 & -\frac{1}{2} & -\frac{1}{2} & -\frac{1}{2} \\ 0 & 0 & 1 & 3 & 0 & 1 \end{pmatrix}$$

Here the number above each arrow indicates which type of row operation was used in passing from each matrix to the next. Note that the last matrix obtained is just the matrix of the system (2) from which the set of solutions was evident.

A matrix is said to be in **_row echelon form_** provided a "staircase"

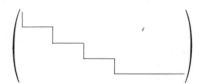

can be drawn in the matrix with the properties that

- (i)   each stair has a height of one row
- (ii)   below the staircase all entries are zeros
- (iii)   in each corner of the staircase there is a 1
- (iv)   in each column containing a corner 1, all other entries are zero.

The last matrix in the sequence above is in row echelon form:

$$\begin{pmatrix} 1 & 0 & 0 & -\frac{7}{2} & \frac{1}{2} & \frac{3}{2} \\ 0 & 1 & 0 & -\frac{1}{2} & -\frac{1}{2} & -\frac{1}{2} \\ 0 & 0 & 1 & 3 & 0 & 1 \end{pmatrix}.$$

Using this concept of row echelon form, it is easy to state the algorithm.

To find all solutions of a system of linear equations:

1. Write down the matrix of the system.

2. By a sequence of row operations, reduce the matrix to row echelon form. (The best procedure is to work on one column at a time, starting at the left.)

3. Write down the system of equations that has this row echelon form as its matrix, and read off the answer by solving for the unknowns corresponding to the corner 1's.

Remark. It can be shown that every matrix can be reduced to row echelon form, and that this row echelon matrix is uniquely determined by the original matrix.

Example 2. Solve the system

$$
\begin{aligned}
x_3 + x_4 &= 0 \\
-2x_1 - 4x_2 + x_3 &= -3 \\
3x_1 + 6x_2 - x_3 + x_4 &= 5
\end{aligned}
$$

Solution

$$
\begin{pmatrix} 0 & 0 & 1 & 1 & 0 \\ -2 & -4 & 1 & 0 & -3 \\ 3 & 6 & -1 & 1 & 5 \end{pmatrix}
\xrightarrow{\text{(ii)}}
\begin{pmatrix} 3 & 6 & -1 & 1 & 5 \\ -2 & -4 & 1 & 0 & -3 \\ 0 & 0 & 1 & 1 & 0 \end{pmatrix}
$$

$$
\xrightarrow{\text{(iii)}}
\begin{pmatrix} 1 & 2 & 0 & 1 & 2 \\ -2 & -4 & 1 & 0 & -3 \\ 0 & 0 & 1 & 1 & 0 \end{pmatrix}
\xrightarrow{\text{(iii)}}
\begin{pmatrix} 1 & 2 & 0 & 1 & 2 \\ 0 & 0 & 1 & 2 & 1 \\ 0 & 0 & 1 & 1 & 0 \end{pmatrix}
$$

$$
\xrightarrow{\text{(iii)}}
\begin{pmatrix} 1 & 2 & 0 & 1 & 2 \\ 0 & 0 & 1 & 2 & 1 \\ 0 & 0 & 0 & -1 & -1 \end{pmatrix}
\xrightarrow{\text{(i)}}
\begin{pmatrix} 1 & 2 & 0 & 1 & 2 \\ 0 & 0 & 1 & 2 & 1 \\ 0 & 0 & 0 & 1 & 1 \end{pmatrix}
$$

$$
\xrightarrow{\text{(iii)}}
\begin{pmatrix} 1 & 2 & 0 & 0 & 1 \\ 0 & 0 & 1 & 0 & -1 \\ 0 & 0 & 0 & 1 & 1 \end{pmatrix}
$$

Notice that in the second step we chose to introduce the first corner 1 by adding the second row to the first rather than by dividing the first row by 3. In this way we were able to avoid introducing fractions into our matrix. The system corresponding to the reduced matrix is

$$
\begin{aligned}
x_1 + 2x_2 \quad\quad &= \quad 1 \\
x_3 \quad &= -1 \\
x_4 &= \quad 1
\end{aligned}
$$

and hence all solutions are obtained by choosing x_2 arbitrarily (take $x_2 = c$) and solving for x_1, x_3 and x_4:

$$
\begin{aligned}
x_1 &= 1 - 2c \\
x_2 &= c \\
x_3 &= -1 \\
x_4 &= 1.
\end{aligned}
$$

Thus, the general solution is $(1 - 2c, c, -1, 1)$, c arbitrary. ■

EXERCISES

1. Find the general solution of each of the following systems using the method described in this section:

   (a) $\begin{aligned} x_1 + x_2 &= \quad 1 \\ x_1 - x_2 &= -1 \end{aligned}$ (b) $\begin{aligned} x_1 + x_2 &= \quad 1 \\ -x_1 - x_2 &= -1 \end{aligned}$ (c) $\begin{aligned} x_1 + x_2 &= 1 \\ -x_1 - x_2 &= 0 \end{aligned}$

2. Reduce each of the following matrices to row echelon form:

   (a) $\begin{pmatrix} 0 & 0 & 1 \\ 0 & 1 & 0 \\ 1 & 0 & 0 \end{pmatrix}$ (b) $\begin{pmatrix} 2 & -1 & 3 \\ 1 & 0 & -1 \\ 3 & -1 & 1 \\ -1 & 1 & -1 \end{pmatrix}$ (c) $\begin{pmatrix} 1 & 1 & -1 & -1 & 2 \\ 2 & 2 & 0 & 3 & 1 \\ -1 & -1 & 0 & 1 & 1 \end{pmatrix}$

3. Solve:

   $$
   \begin{aligned}
   2x_1 + x_2 - x_3 &= 6 \\
   x_1 - 2x_2 - 2x_3 &= 1 \\
   -x_1 + 12x_2 + 8x_3 &= 7
   \end{aligned}
   $$

4. Solve:

   $$
   \begin{aligned}
   x_1 + 2x_2 + 4x_3 &= 1 \\
   x_1 + x_2 + 3x_3 &= 2 \\
   2x_1 + 5x_2 + 9x_3 &= 1
   \end{aligned}
   $$

5. Solve:

   $$
   \begin{aligned}
   -x_1 + 2x_2 + x_3 - 2x_4 &= \quad 2 \\
   x_1 - x_2 - 2x_3 + 2x_4 &= \quad 0 \\
   -2x_1 + 4x_2 \quad\quad - 2x_4 &= -2
   \end{aligned}
   $$

6. Solve:

   $$
   \begin{aligned}
   3x_1 + 2x_2 - x_3 &= \quad 1 \\
   2x_1 - 2x_2 + 5x_3 &= -3 \\
   7x_1 - 2x_2 + 9x_3 &= -5
   \end{aligned}
   $$

7. Solve:

   $$
   \begin{aligned}
   3x_1 + x_2 - x_3 + 2x_4 &= \quad 7 \\
   2x_1 - 2x_2 + 5x_3 - 7x_4 &= \quad 1 \\
   -4x_1 - 4x_2 + 7x_3 - 11x_4 &= -13
   \end{aligned}
   $$

8. Solve:

$$\begin{array}{rcl} x_1 + x_2 - 3x_3 & = & 1 \\ x_2 + x_3 - x_4 & = & 2 \\ -x_1 + x_2 \qquad + x_4 & = & -1 \\ x_1 + 6x_2 - 8x_3 + x_4 & = & 3 \end{array}$$

9. Solve:

$$\begin{array}{rcl} 2x_2 - x_3 - 2x_4 & = & -1 \\ -2x_1 + 2x_2 - 3x_3 - 4x_4 - 8x_5 & = & 0 \\ x_3 + 3x_4 + 5x_5 & = & 1 \\ x_1 \qquad + x_3 + 2x_4 + 6x_5 & = & 0 \\ 2x_2 - x_3 - x_4 + 3x_5 & = & -1 \end{array}$$

10. Let a_{11}, a_{12}, a_{21}, and a_{22} be real numbers. Show that the system

$$a_{11}x_1 + a_{12}x_2 = 0$$
$$a_{21}x_1 + a_{22}x_2 = 0$$

has a nontrivial solution (one different from $(0, 0)$) if and only if $a_{11}a_{22} - a_{12}a_{21} = 0$.

11. Show that a system of linear equations has a solution if and only if the last column in its row echelon matrix does not contain a corner 1.

12. Show that a system of linear equations has a *unique* solution if and only if its row echelon matrix has a corner 1 in every column except the last.

1.2 SOLUTION VECTORS, ROW VECTORS, AND $\mathbb{R}^n$

The set of all ordered n-tuples of real numbers is commonly denoted by $\mathbb{R}^n$, and the n-tuples themselves are called **vectors.** Thus, a solution of a system of m equations in n unknowns is a vector in $\mathbb{R}^n$. It is called a **solution vector** of the system. Similarly, a row of an m by n matrix (a matrix with m rows and n columns) is a vector in $\mathbb{R}^n$ called a **row vector** of the given matrix. Since the matrix of a system of m linear equations in n unknowns is actually an m by $n + 1$ matrix, the rows of such a matrix are vectors in $\mathbb{R}^{n+1}$.

The row operations described in the previous section suggest a way of adding vectors in $\mathbb{R}^n$ and a way of multiplying vectors in $\mathbb{R}^n$ by real numbers. Thus, we define the **sum** of two vectors in $\mathbb{R}^n$ to be the vector obtained by adding corresponding entries:

$$(a_1, a_2, \ldots, a_n) + (b_1, b_2, \ldots, b_n) = (a_1 + b_1, a_2 + b_2, \ldots, a_n + b_n)$$

and we define the **scalar multiple** of a vector in $\mathbb{R}^n$ by a real number to be the vector obtained by multiplying each entry in the original vector by the given real number (scalar):

$$c(a_1, a_2, \ldots, a_n) = (ca_1, ca_2, \ldots, ca_n).$$

Here, the real number c is referred to as a **scalar** to emphasize that we are multiplying a vector by a real number and not by another vector. The set of all real numbers is usually denoted by $\mathbb{R}$. Thus, $\mathbb{R} = \mathbb{R}^1$.

Example 1. In $\mathbb{R}^3$,

$$(1, -1, 0) + (1, 1, 2) = (2, 0, 2)$$

$$3(-3, 2, 1) = (-9, 6, 3)$$

$$2(1, 3, 2) + (-1)(1, 1, 1) = (1, 5, 3). \quad \blacksquare$$

With these definitions of vector addition and scalar multiplication, the three row operations can be described as operations on the row vectors of the given matrix as follows:

  (i)   multiply a row vector by a nonzero scalar
  (ii) interchange two row vectors
  (iii) add a scalar multiple of one row vector to another.

When the entries of a vector $(a_1, a_2, \ldots, a_n)$ do not need to be displayed explicitly, we use the notation $\mathbf{a} = (a_1, a_2, \ldots, a_n)$. The sum of $\mathbf{a} = (a_1, \ldots, a_n)$ and $\mathbf{b} = (b_1, \ldots, b_n)$ is then $\mathbf{a} + \mathbf{b} = (a_1 + b_1, \ldots, a_n + b_n)$, and the scalar multiple of $\mathbf{a}$ by c is then $c\mathbf{a} = (ca_1, \ldots, ca_n)$.

The operation of vector addition in $\mathbb{R}^n$ has the following properties: for $\mathbf{a}$, $\mathbf{b}$, and $\mathbf{c}$ in $\mathbb{R}^n$,

(A_1) $\mathbf{a} + \mathbf{b} = \mathbf{b} + \mathbf{a}$

(A_2) $(\mathbf{a} + \mathbf{b}) + \mathbf{c} = \mathbf{a} + (\mathbf{b} + \mathbf{c})$

(A_3) $\mathbf{0} + \mathbf{a} = \mathbf{a} + \mathbf{0} = \mathbf{a}$, where $\mathbf{0}$ is the vector $(0, 0, \ldots, 0)$

(A_4) $\mathbf{a} + (-\mathbf{a}) = (-\mathbf{a}) + \mathbf{a} = \mathbf{0}$, where $-\mathbf{a}$ is the vector $(-1)\mathbf{a}$

These four properties follow easily from the corresponding properties of real numbers. Thus, for example, if $\mathbf{a} = (a_1, \ldots, a_n)$ and $\mathbf{b} = (b_1, \ldots, b_n)$ then

$$\mathbf{a} + \mathbf{b} = (a_1 + b_1, \ldots, a_n + b_n) = (b_1 + a_1, \ldots, b_n + a_n) = \mathbf{b} + \mathbf{a},$$

which proves property A_1.

The operation of scalar multiplication in $\mathbb{R}^n$ has the following properties: for c, c_1, c_2 in $\mathbb{R}$ and $\mathbf{a}$, $\mathbf{b}$ in $\mathbb{R}^n$,

(S_1) $c(\mathbf{a} + \mathbf{b}) = c\mathbf{a} + c\mathbf{b}$

(S_2) $(c_1 + c_2)\mathbf{a} = c_1\mathbf{a} + c_2\mathbf{a}$

(S_3) $c_1(c_2\mathbf{a}) = (c_1 c_2)\mathbf{a}$

(S_4) $1\mathbf{a} = \mathbf{a}$

These properties are also easy to check.

A third operation, **subtraction** of vectors, is defined as with real numbers: $\mathbf{a} - \mathbf{b} = \mathbf{a} + (-1)\mathbf{b}$.

Use of the operations of vector addition and scalar multiplication leads to a particularly nice way of expressing the general solution of a system of linear equations. Consider, for example, the solution to the system in Example 1 of the previous section. The general solution $\mathbf{x} = (x_1, \ldots, x_5)$ of this system was found to be

$$\mathbf{x} = (\tfrac{3}{2} + \tfrac{7}{2}c_1 - \tfrac{1}{2}c_2, -\tfrac{1}{2} + \tfrac{1}{2}c_1 + \tfrac{1}{2}c_2, 1 - 3c_1, c_1, c_2).$$

Using the definition of vector addition in $\mathbb{R}^5$, this can be written as

$$\mathbf{x} = (\tfrac{3}{2}, -\tfrac{1}{2}, 1, 0, 0) + (\tfrac{7}{2}c_1, \tfrac{1}{2}c_1, -3c_1, c_1, 0) + (-\tfrac{1}{2}c_2, \tfrac{1}{2}c_2, 0, 0, c_2).$$

Note that we have separated the vector $\mathbf{x}$ into the sum of three vectors, the first containing neither c_1 nor c_2, the second containing c_1 as a factor of each entry (note that $0 = 0c_1$), and the third containing c_2 as a factor of each entry. Now, using the definition of scalar multiplication in $\mathbb{R}^5$, $\mathbf{x}$ can be written as

$$\mathbf{x} = (\tfrac{3}{2}, -\tfrac{1}{2}, 1, 0, 0) + c_1(\tfrac{7}{2}, \tfrac{1}{2}, -3, 1, 0) + c_2(-\tfrac{1}{2}, \tfrac{1}{2}, 0, 0, 1).$$

Thus, the general solution is of the form

(1) $$\mathbf{x} = \mathbf{u} + c_1\mathbf{v}_1 + c_2\mathbf{v}_2$$

where $\mathbf{u}$, $\mathbf{v}_1$, and $\mathbf{v}_2$ are specific vectors in $\mathbb{R}^5$, and c_1 and c_2 are arbitrary real numbers. Observe that $\mathbf{u}$ is a *particular solution* of the given system (it is obtained by taking $c_1 = c_2 = 0$.) The vectors $\mathbf{v}_1$ and $\mathbf{v}_2$ are *not* solutions of the given system, but they do satisfy another system of equations: the one obtained from the given one by replacing the number on the right hand side of each equation by zero.

A system of linear equations is said to be ***homogeneous*** if the number on the right hand side of each equation is zero. Thus, the system

(2)
$$2x_1 + x_2 = 0$$
$$x_1 - x_2 = 0$$

is homogeneous, whereas the system

(3)
$$2x_1 + x_2 = 0$$
$$x_1 - x_2 = 1$$

is not. Given any system of linear equations, we can construct a homogeneous system from it simply by replacing each of the numbers on the right by zero. This new system of equations is called the homogeneous system ***associated*** with the given system. Thus, system (2) above is the homogeneous system associated with (3).

In Equation (1) above, the general solution of the system of Example 1 in Section 1.1 is expressed as $\mathbf{x} = \mathbf{u} + c_1\mathbf{v}_1 + c_2\mathbf{v}_2$ where $\mathbf{u}$ is a particular solution of the given system, v_1 and v_2 are particular solutions of the associated homogeneous system, and c_1 and c_2 are arbitrary real numbers.

In general, the algorithm given in Section 1.1 will always lead to a general solution of the form $x = u + c_1 v_1 + \cdots + c_k v_k$ where u is a particular solution of the system, $v_1, \ldots, v_k$ are particular solutions of the associated homogeneous system, and $c_1, \ldots, c_k$ are arbitrary real numbers. Of course, the algorithm may be applied in different ways by different people to arrive at the row echelon form. However, each matrix has a unique row echelon form associated with it. Therefore, the particular sequence of row operations you use to put the matrix into row echelon form does not affect the outcome. This means that the vectors $u, v_1, \ldots, v_n$ that result from this process are uniquely associated with the original system. (This does *not* mean that there is a unique way of choosing a particular solution u of the system and particular solutions $v_1, \ldots, v_k$ of the associated homogeneous system so that the general solution of the system is of the form $x = u + c_1 v_1 + \cdots + c_k v_k$. Indeed, as we shall see later, there are infinitely many ways of doing this. This *does* mean that if the general solution is obtained *using the algorithm*, then the vectors $u, v_1, \ldots, v_k$ are uniquely determined.)

Example 2.   Find the general solution of

$$x_1 + 2x_2 + 5x_3 + 5x_4 = 0$$
$$x_1 + 3x_2 + 8x_3 + 7x_4 = 0$$
$$2x_1 + 3x_2 + 7x_3 + 8x_4 = 0.$$

Solution.   Reducing the matrix of the system

$$
\begin{pmatrix} 1 & 2 & 5 & 5 & 0 \\ 1 & 3 & 8 & 7 & 0 \\ 2 & 3 & 7 & 8 & 0 \end{pmatrix}
\rightarrow
\begin{pmatrix} 1 & 2 & 5 & 5 & 0 \\ 0 & 1 & 3 & 2 & 0 \\ 0 & -1 & -3 & -2 & 0 \end{pmatrix}
$$

$$
\rightarrow
\begin{pmatrix} 1 & 0 & -1 & 1 & 0 \\ 0 & 1 & 3 & 2 & 0 \\ 0 & 0 & 0 & 0 & 0 \end{pmatrix}
$$

leads to the new system

$$x_1 \quad - x_3 + x_4 = 0$$
$$x_2 + 3x_3 + 2x_4 = 0$$
$$0 = 0$$

with general solution (obtained by setting $x_3 = c_1$, $x_4 = c_2$)

$$\begin{aligned}
\mathbf{x} &= (c_1 - c_2, -3c_1 - 2c_2, c_1, c_2) \\
&= (c_1, -3c_1, c_1, 0) + (-c_2, -2c_2, 0, c_2) \\
&= c_1(1, -3, 1, 0) + c_2(-1, -2, 0, 1).
\end{aligned}$$

Note that in this case $\mathbf{u} = \mathbf{0}$; $\mathbf{u}$ will always turn out to be the zero vector when the system is homogeneous. ∎

EXERCISES

1. Find
   (a) $(1, -1, 1, -1, 0) + (-1, 1, -1, 1, 0)$
   (b) $(2, 3, 5, 7) + (-1, 3, -5, 7)$
   (c) $2(1, 1, -1)$
   (d) $-1(1, 0, 0, -1)$
   (e) $3(1, 2, 3, 4) + 4(-1, -1, 0, 3)$
   (f) $(1, 3, -2, 1) - (1, -1, -3, 0)$

2. Solve the following vector equations:
   (a) $2\mathbf{x} = (-2, 4, 2, 6)$
   (b) $3\mathbf{x} + (-1, 1, 1) = (2, 0, 0)$
   (c) $\begin{cases} \mathbf{x} + \mathbf{y} = (1, 1) \\ \mathbf{x} - \mathbf{y} = (1, -1) \end{cases}$

3. Express each of the solutions to Exercises 3–9 of Section 1.1 in the form $\mathbf{u} + c_1\mathbf{v}_1 + \cdots + c_k\mathbf{v}_k$.

4. Solve: $\begin{aligned} x_1 + 2x_2 - x_3 &= 1 \\ 2x_1 + 4x_2 - 2x_3 &= 2 \\ -x_1 - 2x_2 + x_3 &= -1 \end{aligned}$

5. Solve: $\begin{aligned} 2x_1 + x_2 &= 2 \\ 7x_1 + 4x_2 &= 3 \\ 3x_1 + 2x_2 &= -1 \end{aligned}$

6. Solve: $\begin{aligned} x_1 - 2x_2 \quad\quad + x_4 \quad\quad - x_6 &= 0 \\ -x_1 + 2x_2 + x_3 \quad\quad - x_5 \quad\quad &= 1 \\ 2x_1 - 4x_2 + 3x_3 \quad\quad\quad - 3x_6 &= 4 \end{aligned}$

7. Verify properties A_2–A_4 for vector addition in $\mathbb{R}^n$.

8. Verify properties S_1–S_4 for scalar multiplication in $\mathbb{R}^n$.

9. Show that if $\mathbf{u}_1$ and $\mathbf{u}_2$ are any two solutions of a system of linear equations, then $\mathbf{u}_1 - \mathbf{u}_2$ is a solution of the associated homogeneous system. Conclude that, given any particular solution $\mathbf{u}$, every solution is of the form $\mathbf{u} + \mathbf{v}$ where $\mathbf{v}$ is a solution of the associated homogeneous system.

10. (a) Show that if both v_1 and v_2 are solutions of a given homogeneous system of linear equations, then so is $v_1 + v_2$.

(b) Show that if v is a solution of a homogeneous system of linear equations, then so is cv for every real number c.

(c) Conclude that if $v_1, \ldots, v_k$ are any k solutions of a homogeneous system of linear equations and $c_1, \ldots, c_k$ are any k real numbers, then $c_1 v_1 + \cdots + c_k v_k$ is also a solution.

11. Show that $(x_1, \ldots, x_n)$ is a solution of the system (∗) (p. 1) if and only if the numbers $x_1, \ldots, x_n$ satisfy the vector equation

$$x_1(a_{11}, a_{21}, \ldots, a_{m1}) + x_2(a_{12}, a_{22}, \ldots, a_{m2}) + \cdots + x_n(a_{1n}, a_{2n}, \ldots, a_{mn})$$
$$= (b_1, \ldots, b_n).$$

1.3 THE GEOMETRY OF $\mathbb{R}^n$

In the previous section we discussed the algebra of the space $\mathbb{R}^n$ of all ordered n-tuples of real numbers. In this section we shall discuss its geometry.

Let us first take a closer look at $\mathbb{R}^2$. We recall that a vector in $\mathbb{R}^2$ is an ordered pair of real numbers. But the reader has seen ordered pairs of real numbers before, as the Cartesian coordinates of points in the plane. Thus, it is natural to represent vectors in $\mathbb{R}^2$ as points of the Cartesian plane (see Figure 1.1). We may also represent vectors by arrows. Thus, the vector (a_1, a_2) in $\mathbb{R}^2$ can be viewed as an arrow whose tail is at the origin and whose head is at the point in the plane whose Cartesian coordinates are (a_1, a_2) (see Figure 1.2). Viewing vectors as arrows leads to a useful geometric interpretation of vector addition. If $a = (a_1, a_2)$ and $b = (b_1, b_2)$ are represented as arrows with tails at $(0, 0)$, then $a + b = (a_1 + b_1, a_2 + b_2)$ is the arrow along the diagonal of the parallelogram with sides a and b (see Figure 1.3). That the point $(a_1 + b_1, a_2 + b_2)$ is located at the fourth vertex of this parallelogram is a consequence of the fact that the shaded triangles in Figure 1.3 are congruent.

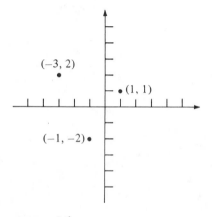

FIGURE 1.1

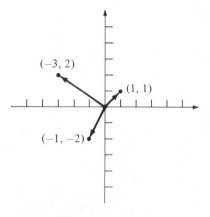

FIGURE 1.2

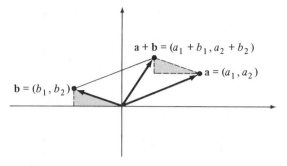

FIGURE 1.3

Scalar multiplication can also be interpreted using arrows. Given a vector $\mathbf{a} = (a_1, a_2)$ in $\mathbb{R}^2$ and a real number c, the vector $c\mathbf{a} = (ca_1, ca_2)$ lies on the same line through the origin as the vector $\mathbf{a}$. If $c > 0$, $c\mathbf{a}$ points in the same direction as $\mathbf{a}$ but has length c times that of $\mathbf{a}$ (see Figure 1.4a). If $c < 0$, $c\mathbf{a}$ points in the direction opposite to $\mathbf{a}$ and has length $|c|$ times that of $\mathbf{a}$ (see Figure 1.4b). To see this, use the fact that the triangles in Figures 1.4(a) and 1.4(b) are similar.

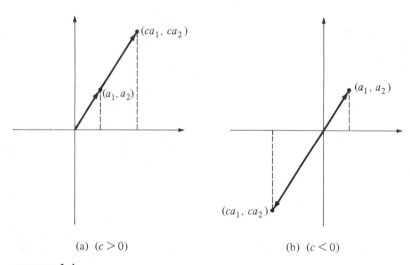

(a) $(c > 0)$ (b) $(c < 0)$

FIGURE 1.4

Another useful way to visualize vectors in $\mathbb{R}^2$ is to picture them as arrows with their tails at some point other than the origin. Thus the arrows in Figure 1.5 may *all* be considered to represent the vector $(-3, 2)$. In this visualization, two arrows are considered to represent the same vector if and only if they have the same length and same direction. Thus, the arrow whose tail is at (b_1, b_2) and whose head is at (c_1, c_2) represents the vector (a_1, a_2) if and only if $a, = c_1 - b_1$ and $a_2 = c_2 - b_2$ (see Figure 1.6).

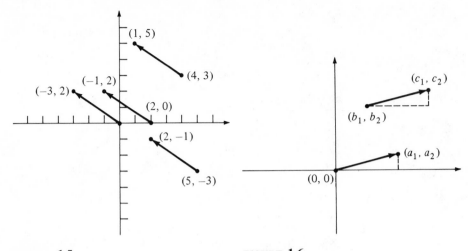

FIGURE 1.5   **FIGURE 1.6**

This geometric interpretation of vectors allows us to visualize vector addition in $\mathbb{R}^2$ another way. Given vectors **a** and **b** in $\mathbb{R}^2$, view **a** as an arrow with tail at the origin and view **b** as an arrow with tail at the head of **a**. Then **a** + **b** is the arrow that completes the triangle (see Figure 1.7).

Vector subtraction in $\mathbb{R}^2$ has a similar geometric interpretation. Let **a** and **b** be represented as arrows with tails at **0**. Since **b** − **a** is the vector that when added to **a** yields **b** (because **a** + (**b** − **a**) = **b**), the vector **b** − **a** can be represented by the arrow that goes from the head of the vector **a** to the head of the vector **b**. (See Figure 1.8.)

As an application of these ideas, we shall show hòw straight lines in the plane can be described by vector equations. Suppose **a** and **d** are vectors in $\mathbb{R}^2$ with **d** ≠ **0**. Viewing **a** as a point in the plane and **d** as an arrow with tail at **a**, let ℓ be the straight line passing through **a** in the direction of **d** (see Figure 1.9). Then a point **x** in the plane lies on ℓ if and only if the vector **x** − **a** is a scalar multiple

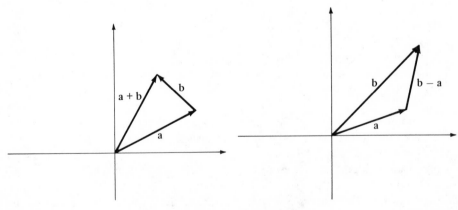

FIGURE 1.7   **FIGURE 1.8**

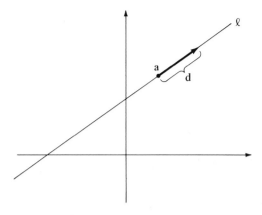

FIGURE 1.9

of **d** (see Figure 1.10). Thus **x** is on ℓ if and only if $\mathbf{x} - \mathbf{a} = t\mathbf{d}$ for some real number t. In other words, **x** is on ℓ if and only if

(1) $$\mathbf{x} = \mathbf{a} + t\mathbf{d}$$

for some real number t. Equation (1) is called a ***parametric representation*** for the line ℓ. Note that there are many parametric representations for the same line ℓ, one for each choice of point **a** on ℓ and vector **d** parallel to ℓ. It is often helpful to think of Equation (1) dynamically: as t runs through the real numbers, the point $\mathbf{x} = \mathbf{a} + t\mathbf{d}$ traces out the line ℓ (see Figure 1.11).

Example 1. Let ℓ be the line through $(-1, 1)$ in the direction of the vector $(2, 3)$. A parametric representation for ℓ is given by the equation

$$\mathbf{x} = (-1, 1) + t(2, 3).$$

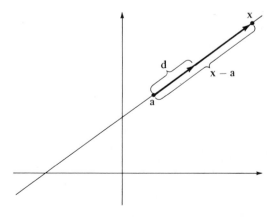

FIGURE 1.10

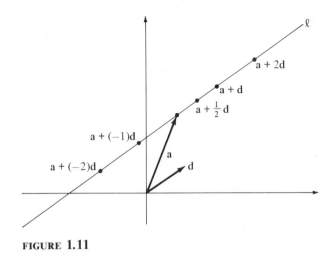

FIGURE 1.11

If we write the vector **x** as (x, y), this equation becomes

$$(x, y) = (-1, 1) + t(2, 3)$$

or

$$x = -1 + 2t$$
$$y = 1 + 3t.$$

Eliminating t from the pair of equations we see that if (x, y) is on ℓ then $3x - 2y = -5$ or $y = \frac{3}{2}x + \frac{5}{2}$. Thus we see that ℓ is the line with slope $\frac{3}{2}$ and y-intercept $\frac{5}{2}$. ■

Example 2.   Let **a** and **b** be points in the plane and let ℓ be the line passing through the points **a** and **b**. Then the vector $\mathbf{b} - \mathbf{a}$ points in the direction of ℓ, so ℓ is represented parametrically by the equation

$$\mathbf{x} = \mathbf{a} + t(\mathbf{b} - \mathbf{a})$$

(see Figure 1.12). ■

Now recall that the distance D between two points $\mathbf{a} = (a_1, a_2)$ and $\mathbf{b} = (b_1, b_2)$ in the plane is given by the distance formula

$$D = \sqrt{(b_1 - a_1)^2 + (b_2 - a_2)^2}.$$

In particular, the distance from the origin to $\mathbf{a} = (a_1, a_2)$ is $\sqrt{a_1^2 + a_2^2}$. Thus $\sqrt{a_1^2 + a_2^2}$ is the length of the arrow from the origin to **a**. We shall call this number the *length* of the vector **a** and denote it by the symbol $\|\mathbf{a}\|$:

$$\|\mathbf{a}\| = \sqrt{a_1^2 + a_2^2}.$$

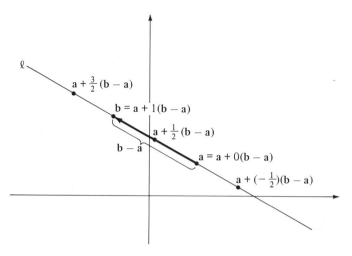

FIGURE **1.12**

For example, $\|(1, 2)\| = \sqrt{5}$, $\|(-1, 1)\| = \sqrt{2}$, and $\|(3, 4)\| = 5$.

Note that if **a** and **b** are two points in the plane, then the distance from **a** to **b** is $\|\mathbf{b} - \mathbf{a}\|$. If $\mathbf{x} = \mathbf{a} + t(\mathbf{b} - \mathbf{a})$ is any point on the line through **a** and **b** (see Example 2, above), then $\|\mathbf{x} - \mathbf{a}\| = |t| \ \|\mathbf{b} - \mathbf{a}\|$ and hence $|t|$ is the ratio of the distance from **a** to **x** to the distance from **a** to **b**. In particular, if $0 < t < 1$, then

$$\mathbf{x} = \mathbf{a} + t(\mathbf{b} - \mathbf{a}) = (1 - t)\mathbf{a} + t\mathbf{b}$$

lies between **a** and **b** on the line through **a** and **b**, and t is equal to the ratio (distance from **a** to **x**)/(distance from **a** to **b**). Thus, for example, $\mathbf{x} = \frac{1}{2}\mathbf{a} + \frac{1}{2}\mathbf{b}$ is the ***midpoint*** of the line segment from **a** to **b**, and $\mathbf{x} = \frac{2}{5}\mathbf{a} + \frac{3}{5}\mathbf{b}$ is the point which is three fifths of the way along the line segment from **a** to **b**.

We have seen that vectors in $\mathbb{R}^2$ can be viewed either as points in the plane or as arrows in the plane, and that these interpretations lead to nice descriptions of lines and of line segments in the plane. These ideas carry over directly to $\mathbb{R}^n$ for $n > 2$, although the figures are impossible to draw when $n > 3$.

Consider the case when $n = 3$. We may view the vector $\mathbf{a} = (a_1, a_2, a_3)$ in $\mathbb{R}^3$ either as a point in 3-space (see Figure 1.13a), as an arrow in 3-space with tail at the origin (see Figure 1.13b), or as an arrow in 3-space with tail at any point we choose (see Figure 1.13c). An arrow in 3-space with tail at the point (b_1, b_2, b_3) and head at the point (c_1, c_2, c_3) will represent **a** if and only if $a_1 = c_1 - b_1$, $a_2 = c_2 - b_2$, and $a_3 = c_3 - b_3$. If we view vectors **a** and **b** as arrows with tails at the same point (e.g., at the origin), then their sum $\mathbf{a} + \mathbf{b}$ is represented as the arrow along the diagonal of the parallelogram with sides along **a** and **b** (see Figure 1.14a). If we view **b** as an arrow whose tail is at the head of **a**, then $\mathbf{a} + \mathbf{b}$ is represented as the arrow from the tail of **a** to the head of **b** (see Figure 1.14b). Finally, if $c > 0$ then the vector $c\mathbf{a}$ is represented by an arrow in the same direction as **a** but c times as long, and if $c < 0$ then $c\mathbf{a}$ is represented as an arrow in the direction opposite to **a**, $|c|$ times as long as **a** (see Figure 1.14c).

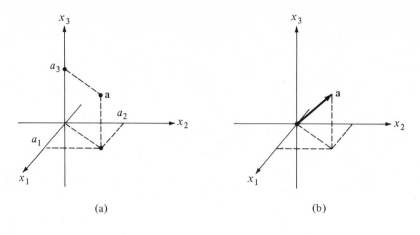

(a)   (b)

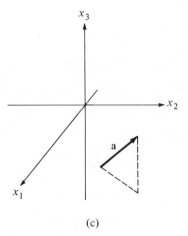

(c)

FIGURE 1.13

For the student who wants to check that vector addition in $\mathbb{R}^3$ is indeed given by the parallelogram construction, we have sketched Figure 1.15 from which this fact may be deduced. The figure shows that if **c** is the fourth vertex of the parallelogram with vertices **0**, **a**, and **b**, then $c_1 = a_1 + b_1$ and $c_2 = a_2 + b_2$. (Why, then, must $c_3 = a_3 + b_3$?)

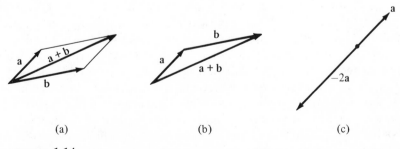

(a)   (b)   (c)

FIGURE 1.14

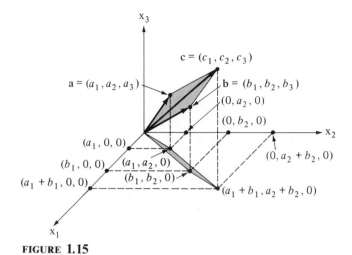

FIGURE 1.15

There is also a geometric interpretation of the sum $\mathbf{a} + \mathbf{b} + \mathbf{c}$ of three vectors in $\mathbb{R}^3$: viewing $\mathbf{a}$, $\mathbf{b}$, and $\mathbf{c}$ as arrows with tails at the origin, $\mathbf{a} + \mathbf{b} + \mathbf{c}$ is the arrow along the diagonal of the parallelepiped with edges $\mathbf{a}$, $\mathbf{b}$, and $\mathbf{c}$ (see Figure 1.16). To see this, simply draw $\mathbf{c}$ so that its tail is located at the head of $\mathbf{a} + \mathbf{b}$.

The vector equation for a line in $\mathbb{R}^3$ is derived in exactly the same way as the vector equation for a line in $\mathbb{R}^2$. Given vectors $\mathbf{a}$ and $\mathbf{d}$ in $\mathbb{R}^3$ with $\mathbf{d} \neq \mathbf{0}$, a point $\mathbf{x}$ will lie on the straight line ℓ through $\mathbf{a}$ in the direction of $\mathbf{d}$ if and only if $\mathbf{x} - \mathbf{a} = t\mathbf{d}$ for some real number t (see Figure 1.17). Thus $\mathbf{x}$ is on ℓ if and only if

$$\mathbf{x} = \mathbf{a} + t\mathbf{d}$$

for some real number t. The equation $\mathbf{x} = \mathbf{a} + t\mathbf{d}$ is a *parametric representation* for ℓ.

Example 3. Let ℓ be the line in $\mathbb{R}^3$ through $(1, 2, 1)$ in the direction of the vector $\mathbf{d} = (2, 1, 3)$. Then

$$\mathbf{x} = (1, 2, 1) + t(2, 1, 3)$$

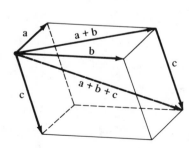

FIGURE 1.16

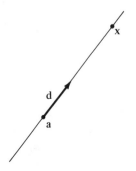

FIGURE 1.17

is a parametric representation for ℓ. If we write $\mathbf{x} = (x_1, x_2, x_3)$, this vector equation can be rewritten as the three scalar equations

$$x_1 = 1 + 2t$$
$$x_2 = 2 + t$$
$$x_3 = 1 + 3t.$$

There is no way to eliminate t from these equations to get a single scalar equation for ℓ as we can do for lines in $\mathbb{R}^2$, however. ∎

Example 4. Let ℓ be the line in $\mathbb{R}^3$ through the points $(2, 1, 2)$ and $(3, -1, 1)$. A direction vector $\mathbf{d}$ for ℓ may be obtained by subtraction:

$$\mathbf{d} = (3, -1, 1) - (2, 1, 2) = (1, -2, -1).$$

Hence ℓ is described by the vector equation

$$\mathbf{x} = (2, 1, 2) + t(1, -2, -1).$$

Notice that the point $(2, 1, 2)$ could be replaced by any other point on ℓ, $(3, -1, 1)$ for example, and the direction vector $(1, -2, -1)$ could be replaced by any other vector that points along ℓ, $(2, 1, 2) - (3, -1, 1) = (-1, 2, 1)$ for example, to obtain another parametric representation

$$\mathbf{x} = (3, -1, 1) + t(-1, 2, 1)$$

for the same line ℓ. ∎

Lines in $\mathbb{R}^n$ for $n > 3$ are described parametrically in the same way. Given $\mathbf{a}$ and $\mathbf{d}$ in $\mathbb{R}^n$ with $\mathbf{d} \neq \mathbf{0}$, *the line through* $\mathbf{a}$ *in the direction of* $\mathbf{d}$ is the set of all vectors $\mathbf{x}$ in $\mathbb{R}^n$ of the form

$$\mathbf{x} = \mathbf{a} + t\mathbf{d}.$$

Given two vectors $\mathbf{a}$ and $\mathbf{b}$ in $\mathbb{R}^n$, *the line through* $\mathbf{a}$ *and* $\mathbf{b}$ is the set of all vectors $\mathbf{x}$ in $\mathbb{R}^n$ of the form

$$\mathbf{x} = \mathbf{a} + t(\mathbf{b} - \mathbf{a}) = (1 - t)\mathbf{a} + t\mathbf{b}.$$

For $0 < t < 1$, the point $\mathbf{x} = (1 - t)\mathbf{a} + t\mathbf{b}$ lies *between* $\mathbf{a}$ and $\mathbf{b}$. When $t = \frac{1}{2}$, $\mathbf{x} = \frac{1}{2}\mathbf{a} + \frac{1}{2}\mathbf{b}$ is the *midpoint* of the line segment from $\mathbf{a}$ to $\mathbf{b}$.

Example 5. Let $\mathbf{a} = (1, 1, 0, 1)$ and $\mathbf{b} = (1, -1, 2, 1)$. Then the equation

$$\mathbf{x} = \mathbf{a} + t(\mathbf{b} - \mathbf{a})$$
$$= (1, 1, 0, 1) + t(0, -2, 2, 0)$$

describes the line in $\mathbb{R}^4$ through **a** and **b**. Setting $t = \frac{1}{2}$, we find the midpoint $(1, 0, 1, 1)$ of the line segment from **a** to **b**, and setting $t = \frac{9}{10}$ we find the point $(1, -\frac{4}{5}, \frac{9}{5}, 1)$, which is nine tenths of the way from **a** to **b**.   ■

EXERCISES

1. Let $\mathbf{a} = (1, 2)$ and $\mathbf{b} = (1, -3)$. Sketch each of the following vectors as arrows in the plane.

    (a) **a**   (e) **a** + **b**

    (b) **b**   (f) **b** − **a**

    (c) 2**a**   (g) 2**a** + 3**b**

    (d) −**b**

2. Let $\mathbf{a} = (-3, 3)$ and $\mathbf{b} = (2, 1)$. Sketch, and find parametric representations for, each of the following lines. Also, in each case, find the slope and y-intercept of the given line by eliminating the parameter t.

    (a) the line through **a** in the direction of **b**

    (b) the line through **a** in the direction of 2**b**

    (c) the line through 2**a** in the direction of **b**

    (d) the line through **a** and **b**

3. Find a parametric representation for the line $5x + 2y = 7$.
    [Hint: First locate two points on the line.]

4. Let $\mathbf{a} = (-1, 3)$ and $\mathbf{b} = (4, 9)$. Find

    (a) $\|\mathbf{a}\|$ (c) $\|\mathbf{a} + \mathbf{b}\|$

    (b) $\|\mathbf{b}\|$ (d) $\|\mathbf{a} - \mathbf{b}\|$

    (e) the midpoint of the line segment from **a** to **b**

    (f) the point on the line segment from **a** to **b** that lies two thirds of the way from **a** to **b**

    (g) the point on the line segment from **a** to **b** that lies three tenths of the way from **a** to **b**

5. Show that the diagonals of a parallelogram bisect each other. [Hint: Let the parallelogram have one vertex at **0**, choose **a** and **b** as in Figure 1.18, and compute the midpoints of the diagonals in terms of **a** and **b**.]

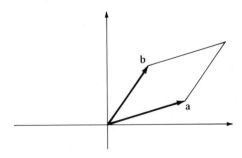

FIGURE 1.18

6. Show that the medians of a triangle meet at a point that is located on each median, two thirds of the way from the vertex to the midpoint of the opposite side. [Hint: Find a vector equation describing the point x_a in Figure 1.19. Then do the same for x_b and for x_c.]

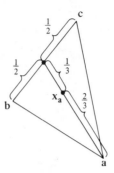

FIGURE 1.19

7. Show that the solution set of a system of n linear equations in two unknowns, viewed as a subset of the plane, must be one of the following: (a) the empty set, (b) a single point, (c) a line, or (d) the whole plane.

8. Sketch the following vectors as arrows in 3-space.
    (a) $(1, 0, 0)$ (d) $(1, 1, 0)$ (g) $(-1, 1, 0)$
    (b) $(0, 1, 0)$ (e) $(1, 0, 1)$ (h) $(1, 1, 2)$
    (c) $(0, 0, 1)$ (f) $(0, 1, 1)$ (i) $(1, 1, -1)$

9. Find a parametric representation for each of the following lines in $\mathbb{R}^3$.
    (a) the line through $(1, 2, 1)$ in the direction of $(-3, 1, 4)$
    (b) the line through $(-1, 1, -1)$ in the direction of $(2, 1, 2)$
    (c) the line through $(1, -2, 3)$ in the direction of $(-4, 1, 1)$
    (d) the line through the points $(1, 2, 1)$ and $(-1, 1, -1)$
    (e) the line through the points $(1, -2, 3)$ and $(-4, 1, 1)$

10. Find a parametric representation for each of the following lines:
    (a) the line in $\mathbb{R}^4$ through $(1, 1, -1, 0)$ in the direction of $(1, 2, 3, 4)$
    (b) the line in $\mathbb{R}^6$ through $(1, 0, 1, 0, 1, 0)$ in the direction of $(-2, -1, 0, 1, 2, 3)$
    (c) the line in $\mathbb{R}^4$ through the points $(1, 1, -1, 0)$ and $(1, 2, 3, 4)$
    (d) the line in $\mathbb{R}^5$ through the points $(1, 2, 3, 4, 5)$ and $(5, 4, 3, 2, 1)$
    (e) the line in $\mathbb{R}^n$ through the points $(1, 2, \dots, n)$ and $(0, 1, 2, \dots, n-1)$

11. Find the midpoint of the line segment from $\mathbf{a}$ to $\mathbf{b}$, where
    (a) $\mathbf{a} = (1, 3, 2)$, $\mathbf{b} = (-1, -3, -2)$
    (b) $\mathbf{a} = (-1, 0, 4)$, $\mathbf{b} = (1, 1, 1)$
    (c) $\mathbf{a} = (4, 3, 2, 1)$, $\mathbf{b} = (6, -1, -2, 7)$
    (d) $\mathbf{a} = (1, 2, \dots, n)$, $\mathbf{b} = (n, n-1, \dots, 1)$.

12. Let $\mathbf{a} = (-1, 2, -1)$ and $\mathbf{b} = (2, -7, 8)$.

    (a) Find the points on the line segment from $\mathbf{a}$ to $\mathbf{b}$ that divide this segment into three equal parts.

    (b) Find the point that lies one fourth of the way from $\mathbf{a}$ to $\mathbf{b}$.

    (c) Find the point that lies three fifths of the way from $\mathbf{a}$ to $\mathbf{b}$.

13. Find a parametric representation for the line connecting $(1, -1, 0)$ to the midpoint of the line segment from $(-2, 2, -2)$ to $(3, -1, 4)$.

14. (a) Find parametric representations for the medians of the triangle in $\mathbb{R}^3$ with vertices $(1, 0, 0)$, $(0, 1, 0)$, and $(0, 0, 1)$.

    (b) Find the point in $\mathbb{R}^3$ where the three lines of part (a) meet.

15. Find a parametric representation for the line in $\mathbb{R}^3$ passing through $(1, 3, 5)$ and parallel to the line $\mathbf{x} = (1, -3, 2) + t(-1, -1, 1)$. (Two lines in $\mathbb{R}^n$ are parallel if their direction vectors are scalar multiples of one another.)

16. Let ℓ be the line in $\mathbb{R}^3$ passing through the points $(\pi, 2\pi, 0)$ and $(0, \pi, 2\pi)$. Find a parametric representation for the line parallel to ℓ and passing through the origin.

17. Let ℓ be the line $\mathbf{x} = (1, -1, 1) + t(1, 0, 1)$. Which of the following points are on ℓ?

    (a) $(1, -1, 1)$ (c) $(3, -1, 1)$

    (b) $(1, 0, 1)$ (d) $(-1, -1, -1)$

18. For each of the following pairs of lines in $\mathbb{R}^3$, determine (i) if they are parallel, and (ii) if they intersect. Note that two lines $\mathbf{x} = \mathbf{a} + t\mathbf{d}$ and $\mathbf{x} = \mathbf{b} + t\mathbf{e}$ intersect if and only if $\mathbf{a} + t_1\mathbf{d} = \mathbf{b} + t_2\mathbf{e}$ for some $t_1, t_2 \in \mathbb{R}$.

    (a) $\mathbf{x} = (-2, -3, 0) + t(5, 3, 1)$, $\mathbf{x} = (5, -6, 8) + t(-1, -3, 2)$

    (b) $\mathbf{x} = (1, 0, 0) + t(0, 0, 1)$, $\mathbf{x} = t(0, 1, 0)$

    (c) $\mathbf{x} = (1, 1, 1) + t(-1, 0, 1)$, $\mathbf{x} = (0, 1, -1) + t(2, 0, -2)$

    (d) $\mathbf{x} = (1, 1, 1) + t(-1, 0, 1)$, $\mathbf{x} = (0, 1, 2) + t(-1, 0, 1)$

19. Suppose two particles are moving in 3-space according to the following equations:

    first particle: $\mathbf{x}(t) = (-2, -3, 0) + t(5, 3, 1)$

    second particle: $\mathbf{x}(t) = (5, -6, 8) + t(-1, -3, 2)$

    Here, t denotes time and $\mathbf{x}(t)$ denotes the position of the particle at time t. Do these particles collide?

1.4 THE DOT PRODUCT

An important problem in the geometry of $\mathbb{R}^n$ is that of finding the angle between two vectors. Let us look first at this problem when $n = 2$.

Let $\mathbf{a} = (a_1, a_2)$ and $\mathbf{b} = (b_1, b_2)$ be nonzero vectors in $\mathbb{R}^2$. The **angle between** $\mathbf{a}$ and $\mathbf{b}$ is defined to be the angle θ ($0 \leq \theta \leq \pi$) formed by the half-lines $\mathbf{x} = t\mathbf{a}$ ($t \geq 0$) and $\mathbf{x} = t\mathbf{b}$ ($t \geq 0$) (see Figure 1.20). Since $0 \leq \theta \leq \pi$, the angle θ is uniquely determined by its cosine. We can find a formula for $\cos \theta$ by applying the law of cosines to the triangle whose sides are $\mathbf{a}$, $\mathbf{b}$, and $\mathbf{b} - \mathbf{a}$ (see Figure 1.21).

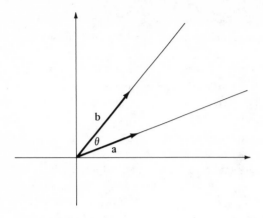

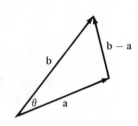

FIGURE **1.20**   FIGURE **1.21**

The law of cosines says that $\|\mathbf{b} - \mathbf{a}\|^2 = \|\mathbf{a}\|^2 + \|\mathbf{b}\|^2 - 2\|\mathbf{a}\| \; \|\mathbf{b}\| \cos \theta$. Since $\mathbf{a} = (a_1, a_2)$, $\mathbf{b} = (b_1, b_2)$, and $\mathbf{b} - \mathbf{a} = (b_1 - a_1, b_2 - a_2)$, we get $(b_1 - a_1)^2 + (b_2 - a_2)^2 = a_1^2 + a_2^2 + b_1^2 + b_2^2 - 2\|\mathbf{a}\| \; \|\mathbf{b}\| \cos \theta$. Expanding the left side, combining terms, and solving for $\cos \theta$, we get

$$\cos \theta = \frac{a_1 b_1 + a_2 b_2}{\|\mathbf{a}\| \; \|\mathbf{b}\|}.$$

The expression $a_1 b_1 + a_2 b_2$ in the numerator occurs frequently in vector computations and is therefore given a name of its own, the ***dot product*** of $\mathbf{a}$ and $\mathbf{b}$. We write

$$\mathbf{a} \cdot \mathbf{b} = a_1 b_1 + a_2 b_2$$

and have found that

$$\cos \theta = \frac{\mathbf{a} \cdot \mathbf{b}}{\|\mathbf{a}\| \; \|\mathbf{b}\|}.$$

Example 1.   To find the angle between the vectors $\mathbf{a} = (3, -2)$ and $\mathbf{b} = (1, 4)$, we compute

$$\|\mathbf{a}\|^2 = 3^2 + (-2)^2 = 9 + 4 = 13$$

$$\|\mathbf{b}\|^2 = 1^2 + 4^2 = 1 + 16 = 17,$$

and $\mathbf{a} \cdot \mathbf{b} = (3)(1) + (-2)(4) = 3 - 8 = -5$, to obtain

$$\cos \theta = \frac{-5}{\sqrt{13} \; \sqrt{17}}.$$

Thus θ is an obtuse angle (its cosine is negative). If we need some idea of how large θ is, a calculator computation shows $-\dfrac{5}{\sqrt{13}\,\sqrt{17}} \approx -0.336$. Therefore $\theta \approx 1.91$ radians, which seems reasonable if we sketch the vectors as in Figure 1.22. ∎

If two nonzero vectors **a** and **b** have $\mathbf{a} \cdot \mathbf{b} = 0$, then $\cos \theta = 0$, hence $\theta = \dfrac{\pi}{2}$. Conversely, if $\theta = \dfrac{\pi}{2}$ then $\cos \theta = 0$; hence $\dfrac{\mathbf{a} \cdot \mathbf{b}}{\|\mathbf{a}\|\,\|\mathbf{b}\|} = 0$ and therefore $\mathbf{a} \cdot \mathbf{b} = 0$. This shows that *two nonzero vectors are perpendicular if and only if their dot product is zero.*

Example 2. Let us verify that the diagonals of the square with vertices $(0, 0)$, $(4, 0)$, $(0, 4)$, $(4, 4)$ are perpendicular (see Figure 1.23). The diagonals are represented by the vectors $\mathbf{a} = (4, 4) - (0, 0) = (4, 4)$ and $\mathbf{b} = (0, 4) - (4, 0) = (-4, 4)$, and $\mathbf{a} \cdot \mathbf{b} = 4(-4) + 4(4) = 0$; hence **a** is perpendicular to **b**. ∎

The idea of dot product for vectors extends easily to vectors in $\mathbb{R}^n$. For $\mathbf{a} = (a_1, a_2, \ldots, a_n)$ and $\mathbf{b} = (b_1, b_2, \ldots, b_n)$, we define the *dot product* $\mathbf{a} \cdot \mathbf{b}$ by

$$\mathbf{a} \cdot \mathbf{b} = a_1 b_1 + a_2 b_2 + \cdots + a_n b_n.$$

Thus, for example,

$$(1, 2, 1) \cdot (-1, 0, 1) = (1)(-1) + (2)(0) + (1)(1) = 0$$

and

$$(1, 1, 3, 2) \cdot (-1, 0, 1, 2) = (1)(-1) + (1)(0) + (3)(1) + (2)(2) = 6.$$

We define the *length* $\|\mathbf{a}\|$ of a vector $\mathbf{a} = (a_1, a_2, \ldots, a_n)$ in $\mathbb{R}^n$ by the formula

$$\|\mathbf{a}\| = (a_1^2 + a_2^2 + \cdots + a_n^2)^{1/2}.$$

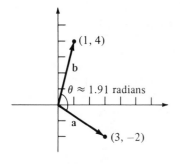

FIGURE 1.22

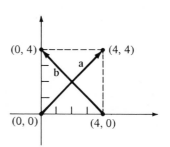

FIGURE 1.23

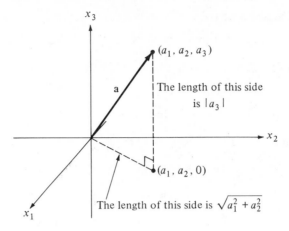

FIGURE 1.24

For a vector **a** in $\mathbb{R}^3$, it is clear from the Pythagorean theorem that $\|\mathbf{a}\|$ is indeed the length of the arrow representing **a** (see Figure 1.24).

For nonzero vectors **a** and **b** in $\mathbb{R}^n$ we define the *angle θ between **a** and **b*** by

$$\theta = \cos^{-1}\left(\frac{\mathbf{a} \cdot \mathbf{b}}{\|\mathbf{a}\| \, \|\mathbf{b}\|}\right), \; 0 \leq \theta \leq \pi,$$

so that the formula

$$\cos \theta = \frac{\mathbf{a} \cdot \mathbf{b}}{\|\mathbf{a}\| \, \|\mathbf{b}\|}$$

or

$$\mathbf{a} \cdot \mathbf{b} = \|\mathbf{a}\| \, \|\mathbf{b}\| \cos \theta$$

is valid in $\mathbb{R}^n$ for $n > 2$ as well as in $\mathbb{R}^2$. Notice that, once again, **a** is *perpendicular* to **b** $\left(\theta = \frac{\pi}{2}\right)$ if and only if $\mathbf{a} \cdot \mathbf{b} = 0$.

Remark 1. In order for the formula $\theta = \cos^{-1}(\mathbf{a} \cdot \mathbf{b}/\|\mathbf{a}\| \, \|\mathbf{b}\|)$ to actually define an angle θ, it is essential that the number $\mathbf{a} \cdot \mathbf{b}/\|\mathbf{a}\| \, \|\mathbf{b}\|$ lie between -1 and $+1$. It turns out (see Exercise 14) that this is always the case.

Remark 2. When $n = 3$, the formula $\theta = \cos^{-1}(\mathbf{a} \cdot \mathbf{b}/\|\mathbf{a}\| \, \|\mathbf{b}\|)$ does define the angle, as we know it, formed by the half-lines $\mathbf{x} = t\mathbf{a}$ ($t \geq 0$) and $\mathbf{x} = t\mathbf{b}$ ($t \geq 0$). This can be seen by applying the law of cosines as before to the triangle with sides **a**, **b**, and $\mathbf{b} - \mathbf{a}$ (see Exercise 15).

Example 3. In $\mathbb{R}^3$, let us find the angle between the vectors $\mathbf{a} = (1, -1, 2)$ and $\mathbf{b} = (-2, 1, 2)$. Since $\|\mathbf{a}\|^2 = (1)^2 + (-1)^2 + (2)^2 = 6$, $\|\mathbf{b}\|^2 = (-2)^2 + $

$(1)^2 + (2)^2 = 9$, and $\mathbf{a} \cdot \mathbf{b} = (1)(-2) + (-1)(1) + (2)(2) = 1$, we have $\cos \theta = \dfrac{1}{\sqrt{6}\sqrt{9}} = \dfrac{1}{3\sqrt{6}}$. Hence $\theta = \cos^{-1}\left(\dfrac{1}{3\sqrt{6}}\right) \approx 1.43$ radians. ◼

Example 4. A *parallelogram* in $\mathbb{R}^n$ is a quadrilateral whose vertices are points $\mathbf{a}$, $\mathbf{b}$, $\mathbf{c}$, $\mathbf{d}$ in $\mathbb{R}^n$ such that $\mathbf{b} - \mathbf{a} = \mathbf{c} - \mathbf{d}$ and $\mathbf{d} - \mathbf{a} = \mathbf{c} - \mathbf{b}$ (see Figure 1.25a). The parallelogram is a *rhombus* if, in addition, $\|\mathbf{b} - \mathbf{a}\| = \|\mathbf{d} - \mathbf{a}\|$. Let us show that the diagonals of any rhombus in $\mathbb{R}^n$ are perpendicular.

Let $\mathbf{u} = \mathbf{b} - \mathbf{a}$ and $\mathbf{v} = \mathbf{d} - \mathbf{a}$. Then the diagonals of the parallelogram are the vectors $\mathbf{u} + \mathbf{v}$ and $\mathbf{u} - \mathbf{v}$ (see Figure 1.25b). Let us compute the dot product of the diagonal vectors. Assuming that the parallelogram is a rhombus, that is, that $\|\mathbf{u}\| = \|\mathbf{v}\|$, we have

$$(\mathbf{u} + \mathbf{v}) \cdot (\mathbf{u} - \mathbf{v}) = \mathbf{u} \cdot \mathbf{u} - \mathbf{u} \cdot \mathbf{v} + \mathbf{v} \cdot \mathbf{u} - \mathbf{v} \cdot \mathbf{v}$$

$$= \|\mathbf{u}\|^2 - \|\mathbf{v}\|^2 \text{ (since } \mathbf{u} \cdot \mathbf{v} = \mathbf{v} \cdot \mathbf{u})$$

$$= 0 \qquad\qquad \text{(since } \|\mathbf{u}\| = \|\mathbf{v}\|).$$

Hence $\mathbf{u} + \mathbf{v}$ and $\mathbf{u} - \mathbf{v}$ are perpendicular, as asserted. ◼

In Example 4, we have used some of the algebraic properties of the dot product. Listed below are the properties of the dot product in $\mathbb{R}^n$ that should be remembered. For $\mathbf{a}$, $\mathbf{b}$, and $\mathbf{c}$ in $\mathbb{R}^n$ and t any real number,

(D_1) $\qquad\qquad\qquad\qquad\qquad \mathbf{a} \cdot \mathbf{b} = \mathbf{b} \cdot \mathbf{a}$

(D_2) $\qquad\qquad\qquad\qquad\qquad \mathbf{a} \cdot (\mathbf{b} + \mathbf{c}) = \mathbf{a} \cdot \mathbf{b} + \mathbf{a} \cdot \mathbf{c}$

(D_3) $\qquad\qquad\qquad\qquad\qquad (t\mathbf{a}) \cdot \mathbf{b} = \mathbf{a} \cdot (t\mathbf{b}) = t(\mathbf{a} \cdot \mathbf{b})$

(D_4) $\qquad\qquad \mathbf{a} \cdot \mathbf{a} = \|\mathbf{a}\|^2 \geq 0$, and $\mathbf{a} \cdot \mathbf{a} = 0$ if and only if $\mathbf{a} = \mathbf{0}$.

Verification of these properties is left to the exercises (see Exercise 16).

The dot product provides a useful tool for describing lines in $\mathbb{R}^2$ and planes in $\mathbb{R}^3$. Consider first the case of a plane P in 3-space. Suppose we are given a point $\mathbf{a}$ in P and a vector $\mathbf{d}$ perpendicular to P (see Figure 1.26). Then a point $\mathbf{x}$ in $\mathbb{R}^3$ lies on P if and only if $\mathbf{x} - \mathbf{a}$ is parallel to P, and this is the case if and only if

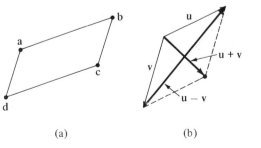

(a)

FIGURE 1.25

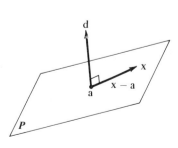

FIGURE 1.26

$\mathbf{x} - \mathbf{a}$ is perpendicular to $\mathbf{d}$; that is, if and only if $\mathbf{d} \cdot (\mathbf{x} - \mathbf{a}) = 0$. Thus, *the equation*

$$\mathbf{d} \cdot (\mathbf{x} - \mathbf{a}) = 0$$

describes the plane through $\mathbf{a}$ *perpendicular to* $\mathbf{d}$.

Example 5. Find an equation describing the plane in $\mathbb{R}^3$ through $\mathbf{a} = (1, 2, 3)$ perpendicular to $\mathbf{d} = (2, 2, 2)$.

Solution. The point $\mathbf{x} = (x_1, x_2, x_3)$ is in this plane if and only if $\mathbf{d} \cdot (\mathbf{x} - \mathbf{a}) = 0$, that is, if and only if $\mathbf{d} \cdot \mathbf{x} = \mathbf{d} \cdot \mathbf{a}$. Hence the equation is

$$(2, 2, 2) \cdot (x_1, x_2, x_3) = (2, 2, 2) \cdot (1, 2, 3)$$

or

$$2x_1 + 2x_2 + 2x_3 = 12$$

or

$$x_1 + x_2 + x_3 = 6. \quad \blacksquare$$

For $n = 2$, the equation $\mathbf{d} \cdot (\mathbf{x} - \mathbf{a}) = 0$ describes the *line* in $\mathbb{R}^2$ through $\mathbf{a}$ perpendicular to $\mathbf{d}$ (see Figure 1.27).

Example 6. Find an equation for the line through $\mathbf{a} = (-1, 1)$ perpendicular to $\mathbf{d} = (-3, 2)$.

Solution. The point $\mathbf{x} = (x, y)$ is on this line if and only if

$$(-3, 2) \cdot (\mathbf{x} - (-1, 1)) = 0.$$

That is, $(-3, 2) \cdot (x + 1, y - 1) = 0$

or $-3(x + 1) + 2(y - 1) = 0$

or $-3x + 2y = 5$

or $y = \tfrac{3}{2}x + \tfrac{5}{2}.$

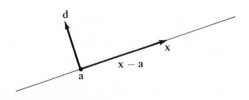

FIGURE 1.27

Compare this result with Example 1, Section 1.3. The answer is the same, but the data used to specify the line are different. ■

By analogy with the $n = 3$ case, we call the solution set of an equation $\mathbf{d} \cdot (\mathbf{x} - \mathbf{a}) = 0$, when $\mathbf{a}$ and $\mathbf{d}$ are vectors in $\mathbb{R}^n$ with $\mathbf{d} \neq \mathbf{0}$, a *hyperplane.* More specifically, we say that the equation $\mathbf{d} \cdot (\mathbf{x} - \mathbf{a}) = 0$ *describes the hyperplane in* $\mathbb{R}^n$ *through* $\mathbf{a}$ *perpendicular to* $\mathbf{d}$. Notice that a hyperplane in $\mathbb{R}^2$ is a line and that a hyperplane in $\mathbb{R}^3$ is a plane. A hyperplane in $\mathbb{R}^4$ looks like a copy of $\mathbb{R}^3$ sitting in $\mathbb{R}^4$.

Example 7. Find an equation for the hyperplane in $\mathbb{R}^4$ through $\mathbf{a} = (1, 3, 4, 2)$ perpendicular to the vector $\mathbf{d} = (-1, 1, 2, 5)$.

Solution. The equation is $\mathbf{d} \cdot (\mathbf{x} - \mathbf{a}) = 0$, or $\mathbf{d} \cdot \mathbf{x} = \mathbf{d} \cdot \mathbf{a}$, so we get

$$(-1, 1, 2, 5) \cdot (x_1, x_2, x_3, x_4) = (-1, 1, 2, 5) \cdot (1, 3, 4, 2)$$

or

$$-x_1 + x_2 + 2x_3 + 5x_4 = 20. \quad ■$$

Example 8. Show that the equation

$$d_1 x_1 + d_2 x_2 + \cdots + d_n x_n = c$$

describes a hyperplane H in $\mathbb{R}^n$, provided that at least one of the d_i's is not zero, and that the vector $\mathbf{d} = (d_1, \ldots, d_n)$ is perpendicular to H.

Solution. For simplicity, assume that $d_1 \neq 0$. Then the equation can be re-written as

$$d_1 x_1 + \cdots + d_n x_n = d_1 \cdot \frac{c}{d_1}$$

or

$$d_1 \left(x_1 - \frac{c}{d_1} \right) + \cdots + d_n x_n = 0$$

or

$$\mathbf{d} \cdot (\mathbf{x} - \mathbf{a}) = 0,$$

where $\mathbf{a} = \left(\dfrac{c}{d_1}, 0, \ldots, 0 \right)$. Thus H is the hyperplane through $\mathbf{a}$ perpendicular to $\mathbf{d} = (d_1, \ldots, d_n)$.

If $d_1 = 0$, then some other d_i is nonzero, say $d_k \neq 0$, and the same argument can be used to show H is the hyperplane through $\mathbf{a} = \left(0, \ldots, 0, \dfrac{c}{d_k}, 0, \ldots, 0 \right)$ perpendicular to $\mathbf{d}$. ■

We can use the ideas of this section to interpret geometrically the solution set of a system of linear equations. According to Example 8, the solution set of any equation of the form

$$a_{11}x_1 + a_{12}x_2 + \cdots + a_{1n}x_n = b_1$$

is a hyperplane in $\mathbb{R}^n$. The solution set of any system of linear equations,

$$a_{11}x_1 + \cdots + a_{1n}x_n = b_1$$

$$\cdots$$

$$a_{m1}x_1 + \cdots + a_{mn}x_n = b_m,$$

is therefore the intersection of m hyperplanes in $\mathbb{R}^n$. If $b_1 = \cdots = b_m = 0$, these equations can be rewritten as

$$\mathbf{a}_1 \cdot \mathbf{x} = 0$$

$$\cdots$$

$$\mathbf{a}_m \cdot \mathbf{x} = 0,$$

where $\mathbf{a}_1 = (a_{11}, \ldots, a_{1n}), \ldots, \mathbf{a}_m = (a_{m1}, \ldots, a_{mn})$. The solution set of this latter system therefore consists of all vectors $\mathbf{x}$ in $\mathbb{R}^n$ that are perpendicular to all of the vectors $\mathbf{a}_1, \mathbf{a}_2, \ldots, \mathbf{a}_m$.

Example 9. The planes $2x_1 - x_2 + 3x_3 = 1$ and $x_1 + 2x_2 - 5x_3 = -2$ in $\mathbb{R}^3$ intersect in a line ℓ. To find a parametric representation for ℓ, simply solve the system

$$2x_1 - x_2 + 3x_3 = 1$$
$$x_1 + 2x_2 - 5x_3 = -2.$$

Since the row echelon form of

$$\begin{pmatrix} 2 & -1 & 3 & 1 \\ 1 & 2 & -5 & -2 \end{pmatrix} \quad \text{is} \quad \begin{pmatrix} 1 & 0 & \frac{1}{5} & 0 \\ 0 & 1 & -\frac{13}{5} & -1 \end{pmatrix}$$

we see that $\mathbf{x} = (x_1, x_2, x_3)$ is on ℓ if and only if

$$\begin{array}{ll} x_1 \quad + \tfrac{1}{5}x_3 = \quad 0 & \quad x_1 = \quad -\tfrac{1}{5}x_3 \\ \quad x_2 - \tfrac{13}{5}x_3 = -1 & \quad x_2 = -1 + \tfrac{13}{5}x_3. \end{array} \quad \text{or}$$

Put $x_3 = t$. Then

$$\mathbf{x} = (-\tfrac{1}{5}t, -1 + \tfrac{13}{5}t, t)$$

or
$$\mathbf{x} = (0, -1, 0) + t(-\tfrac{1}{5}, \tfrac{13}{5}, 1)$$

is a parametric representation for ℓ. Notice that we could have set $x_3 = 5t$, obtaining
$$\mathbf{x} = (0, -1, 0) + t(-1, 13, 5)$$

as another parametric representation for ℓ.   ■

EXERCISES

1. Compute the dot product of the vectors $\mathbf{a}$ and $\mathbf{b}$ and find the cosine of the angle θ between them. Decide if θ is 0, $\pi/2$ or π, acute or obtuse.

   (a) $\mathbf{a} = (1, 2)$, $\mathbf{b} = (-2, 3)$
   (b) $\mathbf{a} = (-1, 1)$, $\mathbf{b} = (1, 1)$
   (c) $\mathbf{a} = (1, 1)$, $\mathbf{b} = (2, 2)$
   (d) $\mathbf{a} = (2, 1)$, $\mathbf{b} = (-4, -2)$
   (e) $\mathbf{a} = (3, 4)$, $\mathbf{b} = (1, -2)$
   (f) $\mathbf{a} = (1, 2, 1)$, $\mathbf{b} = (-1, 1, 1)$
   (g) $\mathbf{a} = (1, 2, -1)$, $\mathbf{b} = (3, 1, 1)$
   (h) $\mathbf{a} = (-1, 2, 3)$, $\mathbf{b} = (4, -2, -1)$
   (i) $\mathbf{a} = (-1, 1, 2, -1)$, $\mathbf{b} = (1, -1, 1, 0)$
   (j) $\mathbf{a} = (1, 1, 0, 1, 1)$, $\mathbf{b} = (2, 2, 3, 2, 2)$

2. (a)   Show that if $\mathbf{a} \neq 0$, then $\dfrac{1}{\|\mathbf{a}\|}\mathbf{a}$ is a unit vector in the direction of $\mathbf{a}$. (A *unit vector* is a vector of length 1).
   (b)   Find and sketch unit vectors in the directions of
      (i) $\mathbf{a} = (2, 3)$,   (ii) $\mathbf{a} = (-1, 1)$,   (iii) $\mathbf{a} = (2, 1, 2)$

3. Find an equation for the line in $\mathbb{R}^2$ through $(3, -2)$ perpendicular to the vector $(-\tfrac{1}{2}, \tfrac{1}{2})$. Also find a parametric representation for that line.

4. Find an equation for the plane in $\mathbb{R}^3$ through $\mathbf{a}$ and perpendicular to $\mathbf{d}$, where:
   (a) $\mathbf{a} = (1, 2, 1)$, $\mathbf{d} = (-3, 1, 4)$
   (b) $\mathbf{a} = (-1, 1, -1)$, $\mathbf{d} = (2, 1, 2)$
   (c) $\mathbf{a} = (1, -2, 3)$, $\mathbf{d} = (-4, 1, 1)$

5. Find an equation for the plane through $(1, 1, -1)$ perpendicular to the line $\mathbf{x} = (2, -1, 2) + t(2, 3, -1)$.

6. (a)   Find a parametric representation for the line through $(-2, 3, 1)$ perpendicular to the plane $3x_1 - 2x_2 + x_3 = 5$.
   (b)   Find the point where this line and this plane meet.

7. Find an equation for the hyperplane through $\mathbf{a}$ perpendicular to $\mathbf{d}$, where
   (a) $\mathbf{a} = (-1, 0, 1, -1)$, $\mathbf{d} = (2, -1, 3, 4)$
   (b) $\mathbf{a} = (0, 0, 0, 0, 0)$, $\mathbf{d} = (-2, -1, 0, 1, 2)$
   (c) $\mathbf{a} = \left(1, \dfrac{1}{2}, \ldots, \dfrac{1}{n}\right)$, $\mathbf{d} = (2, 4, \ldots, 2n)$

8. Find an equation for the hyperplane through $(1, 2, 3, 4)$ perpendicular to the line $\mathbf{x} = (\frac{1}{5}, \frac{2}{5}, \frac{3}{5}, \frac{4}{5}) + t(-1, 0, 0, -1)$.

9. Find a parametric representation for the line through the origin perpendicular to the hyperplane $-3x_1 + x_2 + 2x_3 - x_4 = 7$.

10. Find a parametric representation for the line obtained as the intersection of the indicated planes.

    (a) $x_1 + x_2 + x_3 = 5$, $x_1 - x_2 + x_3 = 1$

    (b) $2x_1 + x_3 = 0$, $3x_1 + 4x_2 = -1$

    (c) $x_1 = 0$, $x_2 = 0$

11. The three hyperplanes $x_1 + x_2 + x_3 + x_4 = 1$, $x_1 - x_2 + x_3 - x_4 = 0$, and $x_1 - x_3 + x_4 = 2$ intersect in a line ℓ. Find a parametric representation for ℓ.

12. Show that a parallelogram is a rhombus if its diagonals are perpendicular.

13. Use vector methods to show that if an altitude of a triangle intersects the opposite side in its midpoint, then the triangle is isosceles.

14. (a) Let $\mathbf{a}$ and $\mathbf{b}$ be nonzero vectors in $\mathbb{R}^n$ and let $f(t) = \|\mathbf{a} + t\mathbf{b}\|^2 = (\mathbf{a} + t\mathbf{b}) \cdot (\mathbf{a} + t\mathbf{b})$. Show that the function f attains its minimum when $t = -\mathbf{a} \cdot \mathbf{b}/\|\mathbf{b}\|^2$.

    (b) Calculate the minimum value of $f(t)$ and use the fact that $f(t) \geq 0$ for all t to show that $|\mathbf{a} \cdot \mathbf{b}| \leq \|\mathbf{a}\| \, \|\mathbf{b}\|$.

    Remark. The inequality $|\mathbf{a} \cdot \mathbf{b}| \leq \|\mathbf{a}\| \, \|\mathbf{b}\|$ is a famous expression called the **Cauchy-Schwarz inequality.** From this inequality follows immediately the fact that $-1 \leq \dfrac{\mathbf{a} \cdot \mathbf{b}}{\|\mathbf{a}\| \, \|\mathbf{b}\|} \leq 1$ and hence that the angle θ between $\mathbf{a}$ and $\mathbf{b}$ can be defined by the formula $\cos \theta = \mathbf{a} \cdot \mathbf{b}/\|\mathbf{a}\| \, \|\mathbf{b}\|$.

15. Let $\mathbf{a} = (a_1, a_2, a_3)$ and $\mathbf{b} = (b_1, b_2, b_3)$ be nonzero vectors in $\mathbb{R}^3$. Apply the law of cosines to the triangle with sides $\mathbf{a}$, $\mathbf{b}$, and $\mathbf{b} - \mathbf{a}$ to obtain the formula

$$\cos \theta = \frac{a_1 b_1 + a_2 b_2 + a_3 b_3}{\|\mathbf{a}\| \, \|\mathbf{b}\|}$$

for the angle θ formed by the half-lines $\mathbf{x} = t\mathbf{a}$ $(t \geq 0)$ and $\mathbf{x} = t\mathbf{b}$ $(t \geq 0)$.

16. Verify the properties D_1–D_4 for the dot product of vectors in $\mathbb{R}^n$.

17. Show that, for $\mathbf{a}$ and $\mathbf{b}$ any two vectors in $\mathbb{R}^n$,

$$\mathbf{a} \cdot \mathbf{b} = \tfrac{1}{4}(\|\mathbf{a} + \mathbf{b}\|^2 - \|\mathbf{a} - \mathbf{b}\|^2).$$

18. Show that if H is the hyperplane $\mathbf{d} \cdot \mathbf{x} = c$, or $\mathbf{d} \cdot \mathbf{x} - c = 0$, in $\mathbb{R}^n$ and $\mathbf{p}$ is any point in $\mathbb{R}^n$, then the perpendicular distance D from $\mathbf{p}$ to H is given by the formula

$$D = |\mathbf{d} \cdot \mathbf{p} - c|/\|\mathbf{d}\|.$$

[Hint: D is the distance from $\mathbf{p}$ to $\mathbf{p} + t\mathbf{d}$, where t is such that $\mathbf{p} + t\mathbf{d} \in H$ (see Figure 1.28).]

19. Use the formula of Exercise 18 to find

    (a) the distance in $\mathbb{R}^2$ from the origin to the line $3x + 2y = 7$

    (b) the distance from the point $(-1, 1)$ to the line $3x + 2y = 7$

    (c) the distance in $\mathbb{R}^3$ from the origin to the plane $-3x_1 + x_2 - x_3 = 10$

    (d) the distance from the point $(-2, 4, 0)$ to the plane $-3x_1 + x_2 - x_3 = 10$

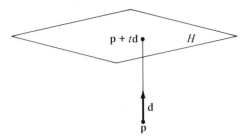

FIGURE 1.28

20. If **a** and **u** are vectors in $\mathbb{R}^n$ with $\|\mathbf{u}\| = 1$, the vector $\text{Proj}_{\mathbf{u}}\mathbf{a} = (\mathbf{a} \cdot \mathbf{u})\mathbf{u}$ is called the ***projection of* a *along* u** (see Figure 1.29). If **a** and **b** are any two vectors in $\mathbb{R}^n$ with $\mathbf{b} \neq \mathbf{0}$, the projection $\text{Proj}_{\mathbf{b}}\mathbf{a}$ of **a** along **b** is defined to be the same as the projection of **a** along the unit vector $\mathbf{b}/\|\mathbf{b}\|$.

   (a) Show that $\mathbf{a} - \text{Proj}_{\mathbf{u}}\mathbf{a}$ is perpendicular to **u**.

   (b) Show that $\|\text{Proj}_{\mathbf{u}}\mathbf{a}\| = \|\mathbf{a}\| \, |\cos \theta|$, where θ is the angle between **a** and **u**.

   (c) Show that the equation $\mathbf{d} \cdot \mathbf{x} = \mathbf{d} \cdot \mathbf{a}$ for a hyperplane says simply that **x** and **a** have the same projection along **d**.

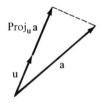

FIGURE 1.29

21. Show that the solution set of a system of n linear equations in three unknowns, viewed as a subset of 3-space, must be one of the following: (a) the empty set, (b) a single point, (c) a line, (d) a plane, or (e) all of 3-space.

1.5 THE CROSS PRODUCT

In the previous section, we learned that the equation $\mathbf{d} \cdot (\mathbf{x} - \mathbf{a}) = 0$ describes the plane in $\mathbb{R}^3$ through **a** perpendicular to **d**. Often the vector **d** is not given explicitly, however. Suppose, for example, that we wish to find an equation for the plane in $\mathbb{R}^3$ passing through the origin and the two points **a** and **b**. We assume that **0**, **a** and **b** are noncollinear, so that these points do determine a plane. The problem of finding a vector **d** perpendicular to this plane is the same as the problem of finding a vector perpendicular to the vectors **a** and **b** (see Figure 1.30). We shall now develop a formula for finding such a vector.

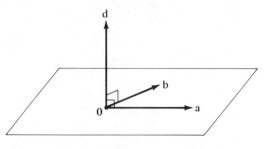

FIGURE 1.30

Let $\mathbf{a} = (a_1, a_2, a_3)$ and $\mathbf{b} = (b_1, b_2, b_3)$. Then we want to solve the equations $\mathbf{a} \cdot \mathbf{x} = 0$ and $\mathbf{b} \cdot \mathbf{x} = 0$; that is, we want to solve the linear system

$$a_1 x_1 + a_2 x_2 + a_3 x_3 = 0$$
$$b_1 x_1 + b_2 x_2 + b_3 x_3 = 0.$$

The general solution of this system is

$$\mathbf{x} = c(a_2 b_3 - a_3 b_2, a_3 b_1 - a_1 b_3, a_1 b_2 - a_2 b_1), \ c \in \mathbb{R}$$

(see Exercise 8). If we take $c = 1$, we obtain the particular solution

$$\mathbf{d} = (a_2 b_3 - a_3 b_2, a_3 b_1 - a_1 b_3, a_1 b_2 - a_2 b_1).$$

This vector $\mathbf{d}$ is perpendicular to both $\mathbf{a}$ and $\mathbf{b}$.

The vector $\mathbf{d} = (a_2 b_3 - a_3 b_2, a_3 b_1 - a_1 b_3, a_1 b_2 - a_2 b_1)$ is called the ***cross product*** of $\mathbf{a}$ and $\mathbf{b}$ and is denoted $\mathbf{a} \times \mathbf{b}$. To calculate $\mathbf{a} \times \mathbf{b}$, it is probably easiest to remember the formula in the following way. Given $a, b, c, d \in \mathbb{R}$, the number $ad - bc$ is called the ***determinant*** of the 2×2 matrix $\begin{pmatrix} a & b \\ c & d \end{pmatrix}$. Notice that

$$a_2 b_3 - a_3 b_2 = \det \begin{pmatrix} a_2 & a_3 \\ b_2 & b_3 \end{pmatrix}, \quad a_3 b_1 - a_1 b_3 = -\det \begin{pmatrix} a_1 & a_3 \\ b_1 & b_3 \end{pmatrix}, \text{ and}$$

$$a_1 b_2 - a_2 b_1 = \det \begin{pmatrix} a_1 & a_2 \\ b_1 & b_2 \end{pmatrix}.$$

Therefore

$$\mathbf{a} \times \mathbf{b} = \left(\det \begin{pmatrix} a_2 & a_3 \\ b_2 & b_3 \end{pmatrix}, \ -\det \begin{pmatrix} a_1 & a_3 \\ b_1 & b_3 \end{pmatrix}, \ \det \begin{pmatrix} a_1 & a_2 \\ b_1 & b_2 \end{pmatrix} \right).$$

Notice further that these three 2 × 2 matrices can be obtained simply by successively deleting columns of the 2 × 3 matrix

$$\begin{pmatrix} \mathbf{a} \\ \mathbf{b} \end{pmatrix} = \begin{pmatrix} a_1 & a_2 & a_3 \\ b_1 & b_2 & b_3 \end{pmatrix}.$$

Thus

$$\mathbf{a} \times \mathbf{b} = \left(\det \begin{vmatrix} a_1 & a_2 & a_3 \\ b_1 & b_2 & b_3 \end{vmatrix}, \; -\det \begin{pmatrix} a_1 & a_2 & a_3 \\ b_1 & b_2 & b_3 \end{pmatrix}, \; \det \begin{pmatrix} a_1 & a_2 & a_3 \\ b_1 & b_2 & b_3 \end{pmatrix}\right).$$

In this formula, and elsewhere in this book, the shading indicates the entries that are to be deleted. Don't forget the minus sign before the second entry!

Example 1. Find an equation for the plane through the points $(0, 0, 0)$, $(1, 2, 3)$, $(3, 2, 1)$.

Solution. Take

$$\mathbf{d} = (1, 2, 3) \times (3, 2, 1) = \left(\det \begin{pmatrix} 2 & 3 \\ 2 & 1 \end{pmatrix}, \; -\det \begin{pmatrix} 1 & 3 \\ 3 & 1 \end{pmatrix}, \; \det \begin{pmatrix} 1 & 2 \\ 3 & 2 \end{pmatrix}\right)$$

$$= (-4, 8, -4).$$

Thus the vector $(-4, 8, -4)$, and hence also the vector $(1, -2, 1)$, is perpendicular to both $(1, 2, 3)$ and $(3, 2, 1)$. So an equation for the plane in question is $x_1 - 2x_2 + x_3 = 0$. ∎

Example 2. Find an equation for the plane through the points $\mathbf{a} = (1, -1, 1)$, $\mathbf{b} = (2, 3, 1)$, and $\mathbf{c} = (-1, 2, 3)$.

Solution. Any vector perpendicular to this plane must be perpendicular to the two vectors $\mathbf{a} - \mathbf{c} = (2, -3, -2)$ and $\mathbf{b} - \mathbf{c} = (3, 1, -2)$ (see Figure 1.31). To

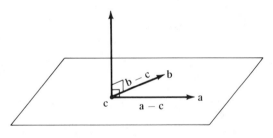

FIGURE 1.31

find such a vector, we compute

$$(\mathbf{a} - \mathbf{c}) \times (\mathbf{b} - \mathbf{c}) = (2, -3, -2) \times (3, 1, -2)$$

$$= (\det \begin{pmatrix} -3 & -2 \\ 1 & -2 \end{pmatrix}, \ -\det \begin{pmatrix} 2 & -2 \\ 3 & -2 \end{pmatrix}, \ \det \begin{pmatrix} 2 & -3 \\ 3 & 1 \end{pmatrix})$$

$$= (8, -2, 11).$$

Thus an equation for the plane in question is

$$8(x_1 - 1) - 2(x_2 + 1) + 11(x_3 - 1) = 0$$

or

$$8x_1 - 2x_2 + 11x_3 = 21. \quad \blacksquare$$

Remark. In Example 2 we had many choices, so the form of the answer can vary. For example, we could have taken the cross product of the vectors $\mathbf{a} - \mathbf{b}$ and $\mathbf{c} - \mathbf{b}$ to find $\mathbf{d}$, and we could have used the point $\mathbf{b}$ rather than $\mathbf{a}$ in the final computation. However,

$$(\mathbf{a} - \mathbf{b}) \times (\mathbf{c} - \mathbf{b}) = (-1, -4, 0) \times (-3, -1, 2)$$

$$= (\det \begin{pmatrix} -4 & 0 \\ -1 & 2 \end{pmatrix}, \ -\det \begin{pmatrix} -1 & 0 \\ -3 & 2 \end{pmatrix}, \ \det \begin{pmatrix} -1 & -4 \\ -3 & -1 \end{pmatrix})$$

$$= (-8, 2, -11)$$

so we would get

$$-8(x_1 - 2) + 2(x_2 - 3) - 11(x_3 - 1) = 0$$

or

$$-8x_1 + 2x_2 - 11x_3 = -21,$$

which, of course, is equivalent to our answer in Example 2.

There is a nice geometric description of the cross product of two vectors. We already know that $\mathbf{a} \times \mathbf{b}$ is perpendicular to the plane of $\mathbf{0}$, $\mathbf{a}$ and $\mathbf{b}$. Let us now compute a formula for the length $\|\mathbf{a} \times \mathbf{b}\|$ of $\mathbf{a} \times \mathbf{b}$.

Note that

$$\|\mathbf{a} \times \mathbf{b}\|^2 = \|(a_2 b_3 - a_3 b_2, \ -(a_1 b_3 - a_3 b_1), \ a_1 b_2 - a_2 b_1)\|^2$$

$$= (a_2 b_3 - a_3 b_2)^2 + (a_1 b_3 - a_3 b_1)^2 + (a_1 b_2 - a_2 b_1)^2$$

$$= a_2^2 b_3^2 + a_3^2 b_2^2 + a_1^2 b_3^2 + a_3^2 b_1^2 + a_1^2 b_2^2 + a_2^2 b_1^2$$

$$\quad -2(a_2 b_3 a_3 b_2 + a_1 b_3 a_3 b_1 + a_1 b_2 a_2 b_1).$$

If we compare this with

$$\|\mathbf{a}\|^2 \|\mathbf{b}\|^2 = (a_1^2 + a_2^2 + a_3^2)(b_1^2 + b_2^2 + b_3^2)$$
$$= a_1^2 b_1^2 + a_1^2 b_2^2 + a_1^2 b_3^2 + a_2^2 b_1^2 + a_2^2 b_2^2 + a_2^2 b_3^2$$
$$+ a_3^2 b_1^2 + a_3^2 b_2^2 + a_3^2 b_3^2$$

we see that

$$\|\mathbf{a}\|^2 \|\mathbf{b}\|^2 - \|\mathbf{a} \times \mathbf{b}\|^2 = a_1^2 b_1^2 + a_2^2 b_2^2 + a_3^2 b_3^2$$
$$+ 2(a_2 b_3 a_3 b_2 + a_1 b_3 a_3 b_1 + a_1 b_2 a_2 b_1)$$
$$= (a_1 b_1 + a_2 b_2 + a_3 b_3)^2$$
$$= (\mathbf{a} \cdot \mathbf{b})^2.$$

Solving for $\|\mathbf{a} \times \mathbf{b}\|^2$ thus yields the formula

$$\|\mathbf{a} \times \mathbf{b}\|^2 = \|\mathbf{a}\|^2 \|\mathbf{b}\|^2 - (\mathbf{a} \cdot \mathbf{b})^2, \text{ for all } \mathbf{a}, \mathbf{b} \in \mathbb{R}^3.$$

But $(\mathbf{a} \cdot \mathbf{b})^2 = \|\mathbf{a}\|^2 \|\mathbf{b}\|^2 \cos^2 \theta$, where θ is the angle between $\mathbf{a}$ and $\mathbf{b}$; hence

$$\|\mathbf{a}\|^2 \|\mathbf{b}\|^2 - (\mathbf{a} \cdot \mathbf{b})^2 = \|\mathbf{a}\|^2 \|\mathbf{b}\|^2 - \|\mathbf{a}\|^2 \|\mathbf{b}\|^2 \cos^2 \theta$$
$$= \|\mathbf{a}\|^2 \|\mathbf{b}\|^2 \sin^2 \theta$$

or

$$\boxed{\|\mathbf{a} \times \mathbf{b}\| = \|\mathbf{a}\| \|\mathbf{b}\| \sin \theta, \text{ where } \theta \text{ is the angle between } \mathbf{a} \text{ and } \mathbf{b}.}$$

(Note that $\sin \theta \geq 0$ since $0 \leq \theta \leq \pi$.)

This formula tells us that the *length of* $\mathbf{a} \times \mathbf{b}$ *is the same as the area of the parallelogram spanned by* $\mathbf{a}$ *and* $\mathbf{b}$ (see Figure 1.32). The cross product can therefore be used to compute areas of parallelograms and of triangles in $\mathbb{R}^3$, as in the following example.

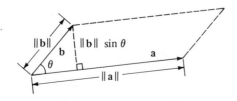

FIGURE 1.32

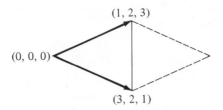

FIGURE 1.33

Example 3. The area of the parallelogram spanned by $\mathbf{a} = (1, 2, 3)$ and $\mathbf{b} = (3, 2, 1)$ is

$$\|\mathbf{a} \times \mathbf{b}\| = \|(-4, 8, -4)\| \text{ (see Example 1)}$$
$$= \sqrt{16 + 64 + 16} = \sqrt{96} \approx 9.8$$

The area of the triangle whose vertices are $(0, 0, 0)$, $(1, 2, 3)$ and $(3, 2, 1)$ is $\frac{1}{2}\sqrt{96} \approx 4.9$ (see Figure 1.33). ∎

We now know that $\mathbf{a} \times \mathbf{b}$ points in a direction perpendicular to the plane of $\mathbf{0}$, $\mathbf{a}$ and $\mathbf{b}$, and that its length is just the area of the parallelogram spanned by $\mathbf{a}$ and $\mathbf{b}$. This comes pretty close to specifying $\mathbf{a} \times \mathbf{b}$ uniquely. The only thing we don't yet know is which of the two vectors determined by this information is $\mathbf{a} \times \mathbf{b}$. It turns out that the direction of $\mathbf{a} \times \mathbf{b}$ along the line perpendicular to $\mathbf{a}$ and $\mathbf{b}$ is determined by the *right hand rule*. For example, if $\mathbf{e}_1 = (1, 0, 0)$, $\mathbf{e}_2 = (0, 1, 0)$, and $\mathbf{e}_3 = (0, 0, 1)$, then

$$\mathbf{e}_1 \times \mathbf{e}_2 = (1, 0, 0) \times (0, 1, 0) = (\det \begin{pmatrix} 0 & 0 \\ 1 & 0 \end{pmatrix}, -\det \begin{pmatrix} 1 & 0 \\ 0 & 0 \end{pmatrix}, \det \begin{pmatrix} 1 & 0 \\ 0 & 1 \end{pmatrix})$$

$$= (0, 0, 1) = \mathbf{e}_3.$$

If you curl the fingers of your right hand in the direction from $\mathbf{e}_1$ to $\mathbf{e}_2$, then your right thumb will point in the direction of $\mathbf{e}_3$ (see Figure 1.34a). In general, the

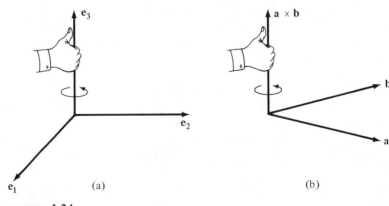

(a)   (b)

FIGURE 1.34

direction of **a** × **b** along the line perpendicular to **a** and **b** is the direction in which your right thumb points when the fingers of your right hand are curled in the direction from **a** to **b** (see Figure 1.34b). We shall not give a formal proof of this, but you can see intuitively why it is so. Move **a** and **b** continuously until **a** points in the direction of e_1 (**a** $= c_1 e_1$, $c_1 > 0$) and **b** points in the direction of e_2 (**b** $= c_2 e_2$, $c_2 > 0$), being careful to keep the angle between **a** and **b** always between 0 and π. At the end of this motion, the cross product

$$\mathbf{a} \times \mathbf{b} = (c_1 e_1) \times (c_2 e_2) = (c_1 c_2) e_1 \times e_2$$

must point in the direction of $e_1 \times e_2$; that is, it must point in the direction determined by the right hand rule. But, throughout the motion, the cross product **a** × **b** moved continuously so it must have always pointed to the side of the plane of **a** and **b** determined by the right hand rule.

We have come, then, to a complete geometric description of **a** × **b**.

The cross product **a** × **b** of two vectors **a** and **b** is the vector

(1)   that is perpendicular to the plane of **0**, **a** and **b**,

(2)   whose length is the area of the parallelogram spanned by **a** and **b**, and

(3)   whose direction is determined by the right hand rule.

Example 4. $e_1 \times e_2 = e_3$, $e_2 \times e_3 = e_1$, and $e_3 \times e_1 = e_2$. All three of these formulas are clear using the geometric characterization of the cross product (see Figure 1.35), since each pair of these vectors spans a square of area 1.   ■

The cross product has several useful algebraic properties. We list some of them in the following theorem. Most of these can be proved directly from the definition of the cross product.

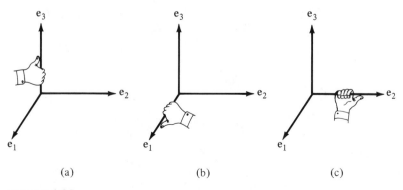

(a)   (b)   (c)

FIGURE 1.35

Theorem 1.   *For all vectors* **a**, **b**, **c** $\in \mathbb{R}^3$,

(a) $\mathbf{a} \times \mathbf{b} = -\mathbf{b} \times \mathbf{a}$

(b) $(\mathbf{a} \times \mathbf{b}) \cdot \mathbf{a} = 0$ *and* $(\mathbf{a} \times \mathbf{b}) \cdot \mathbf{b} = 0$

(c) $(\mathbf{a} + \mathbf{b}) \times \mathbf{c} = \mathbf{a} \times \mathbf{c} + \mathbf{b} \times \mathbf{c}$

(d) $\mathbf{a} \cdot (\mathbf{b} \times \mathbf{c}) = \mathbf{b} \cdot (\mathbf{c} \times \mathbf{a}) = \mathbf{c} \cdot (\mathbf{a} \times \mathbf{b})$

(e) $\mathbf{a} \times (\mathbf{b} \times \mathbf{c}) = (\mathbf{a} \cdot \mathbf{c})\mathbf{b} - (\mathbf{a} \cdot \mathbf{b})\mathbf{c}$

(f) $\|\mathbf{a} \times \mathbf{b}\|^2 = \|\mathbf{a}\|^2 \|\mathbf{b}\|^2 - (\mathbf{a} \cdot \mathbf{b})^2$.

Moreover,

(g)   *If* **a** *and* **b** *are nonzero, then*

$$\mathbf{a} \times \mathbf{b} = \mathbf{0} \Leftrightarrow \mathbf{a} \text{ and } \mathbf{b} \text{ are parallel.}$$

Remarks.   Property (a) shows that the cross product is not commutative. It is not associative either, since, for example, $\mathbf{e}_1 \times (\mathbf{e}_1 \times \mathbf{e}_2) = \mathbf{e}_1 \times \mathbf{e}_3 = -\mathbf{e}_3 \times \mathbf{e}_1 = -\mathbf{e}_2$, whereas $(\mathbf{e}_1 \times \mathbf{e}_1) \times \mathbf{e}_2 = \mathbf{0}$ from (g). Property (b) expresses the fact that $\mathbf{a} \times \mathbf{b}$ is perpendicular to both **a** and **b**, and (c) is a distributive law.

To prove (g), note that $\mathbf{a} \times \mathbf{b} = \mathbf{0} \Leftrightarrow \|\mathbf{a} \times \mathbf{b}\| = 0$. But $\|\mathbf{a} \times \mathbf{b}\| = \|\mathbf{a}\| \|\mathbf{b}\| \sin \theta$ so, if $\mathbf{a} \neq 0$ and $\mathbf{b} \neq 0$, then $\mathbf{a} \times \mathbf{b} = \mathbf{0} \Leftrightarrow \sin \theta = 0 \Leftrightarrow \theta = 0$ or π. Property (f) was proved above, in the derivation of the formula $\|\mathbf{a} \times \mathbf{b}\| = \|\mathbf{a}\| \|\mathbf{b}\| \sin \theta$. You are asked to verify the remaining properties in the exercises.

The product $\mathbf{a} \cdot (\mathbf{b} \times \mathbf{c})$ that appears in (d) of Theorem 1 is called the ***triple scalar product*** of **a**, **b** and **c**. It is, of course, a real number, and it has an interesting geometric interpretation. We know that

$$\mathbf{a} \cdot (\mathbf{b} \times \mathbf{c}) = \|\mathbf{a}\| \|\mathbf{b} \times \mathbf{c}\| \cos \theta,$$

where θ is the angle between **a** and $\mathbf{b} \times \mathbf{c}$. Also, we know that $\|\mathbf{b} \times \mathbf{c}\|$ is the area of the parallelogram spanned by **b** and **c**. That parallelogram is the base of the parallelepiped spanned by **a**, **b** and **c** (see Figure 1.36). Clearly, $\|\mathbf{a}\| \, |\cos \theta|$ is the altitude of this parallelepiped; hence $\|\mathbf{b} \times \mathbf{c}\| \, \|\mathbf{a}\| \, |\cos \theta| = |\mathbf{a} \cdot (\mathbf{b} \times \mathbf{c})|$ is its volume. Thus, we have the following:

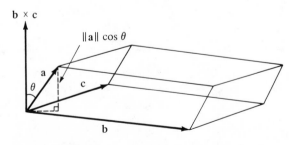

FIGURE 1.36

Theorem 2.   *The volume V of the parallelepiped in* $\mathbb{R}^3$ *spanned by vectors* **a**, **b** *and* **c** *is equal to the absolute value of the triple scalar product of* **a**, **b**, *and* **c**:

$$V = |\mathbf{a} \cdot (\mathbf{b} \times \mathbf{c})|.$$

Example 6.   Let us compute the volume V of the parallelepiped spanned by the vectors $\mathbf{a} = (3, 0, 1)$, $\mathbf{b} = (3, 3, 0)$, and $\mathbf{c} = (1, 2, 3)$. Since

$$\mathbf{b} \times \mathbf{c} = (3, 3, 0) \times (1, 2, 3) = \left(\det\begin{pmatrix} 3 & 0 \\ 2 & 3 \end{pmatrix}, -\det\begin{pmatrix} 3 & 0 \\ 1 & 3 \end{pmatrix}, \det\begin{pmatrix} 3 & 3 \\ 1 & 2 \end{pmatrix} \right)$$
$$= (9, -9, 3),$$

we find that $V = |\mathbf{a} \cdot (\mathbf{b} \times \mathbf{c})| = |(3, 0, 1) \cdot (9, -9, 3)| = 30.$ ■

We can also use Theorem 2 to compute volumes of tetrahedra. Since the volume of the tetrahedron spanned by three vectors **a**, **b** and **c** is one sixth the volume of the parallelepiped spanned by these same vectors (see Exercise 11), the volume of the tetrahedron is $V = \frac{1}{6}|\mathbf{a} \cdot (\mathbf{b} \times \mathbf{c})|$.

Example 7.   The tetrahedron with vertices $\mathbf{a} = (1, 0, 0), \mathbf{b} = (0, 1, 0), \mathbf{c} = (1, 1, 0)$ and $\mathbf{d} = (1, 1, 1)$ has the same volume as the one spanned by the vectors $\mathbf{b} - \mathbf{a} = (-1, 1, 0)$, $\mathbf{c} - \mathbf{a} = (0, 1, 0)$, and $\mathbf{d} - \mathbf{a} = (0, 1, 1)$ (see Figure 1.37). Now

$$(\mathbf{c} - \mathbf{a}) \times (\mathbf{d} - \mathbf{a}) = \left(\det\begin{pmatrix} 1 & 0 \\ 1 & 1 \end{pmatrix}, -\det\begin{pmatrix} 0 & 0 \\ 0 & 1 \end{pmatrix}, \det\begin{pmatrix} 0 & 1 \\ 0 & 1 \end{pmatrix} \right)$$
$$= (1, 0, 0),$$

so the volume of the tetrahedron is

$$V = \tfrac{1}{6}|(\mathbf{b} - \mathbf{a}) \cdot ((\mathbf{c} - \mathbf{a}) \times (\mathbf{d} - \mathbf{a}))| = \tfrac{1}{6}|(-1, 1, 0) \cdot (1, 0, 0)| = \tfrac{1}{6}.$$ ■

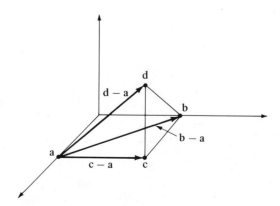

FIGURE 1.37

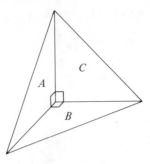

FIGURE 1.38

Theorem 3. *(3-dimensional Pythagorean theorem). Let A, B, C, and D be the areas of the four faces of a right tetrahedron, with D the area of the face opposite the vertex where the edges meet at right angles (see Figure 1.38). Then*

$$D^2 = A^2 + B^2 + C^2.$$

Proof. Move the tetrahedron so that the vertex where the edges meet at right angles is at the origin **0**. Let **a**, **b**, and **c** be the other three vertices, with **a** opposite the face with area A, **b** opposite the face with area B, and **c** opposite the face with area C (see Figure 1.39). By the formula for the area of a triangle in terms of the cross product we see that

$$D = \tfrac{1}{2}\|(\mathbf{b} - \mathbf{a}) \times (\mathbf{c} - \mathbf{a})\|$$

and hence

$$
\begin{aligned}
D^2 &= \tfrac{1}{4}\|(\mathbf{b} - \mathbf{a}) \times (\mathbf{c} - \mathbf{a})\|^2 \\
&= \tfrac{1}{4}\|\mathbf{b} \times \mathbf{c} - \mathbf{a} \times \mathbf{c} - \mathbf{b} \times \mathbf{a}\|^2 \\
&= \tfrac{1}{4}\|\mathbf{b} \times \mathbf{c} + \mathbf{c} \times \mathbf{a} + \mathbf{a} \times \mathbf{b}\|^2.
\end{aligned}
$$

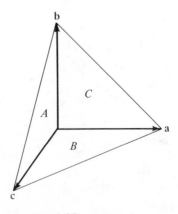

FIGURE 1.39

But, since **a**, **b**, and **c** are mutually perpendicular, **b** × **c** is a scalar multiple of **a**, **c** × **a** is a scalar multiple of **b**, and **a** × **b** is a scalar multiple of **c**. In particular, the three vectors **b** × **c**, **c** × **a**, and **a** × **b** are mutually perpendicular. It follows that

$$D^2 = \tfrac{1}{4}(\mathbf{b} \times \mathbf{c} + \mathbf{c} \times \mathbf{a} + \mathbf{a} \times \mathbf{b}) \cdot (\mathbf{b} \times \mathbf{c} + \mathbf{c} \times \mathbf{a} + \mathbf{a} \times \mathbf{b})$$

$$= \tfrac{1}{4}\|\mathbf{b} \times \mathbf{c}\|^2 + \tfrac{1}{4}\|\mathbf{c} \times \mathbf{a}\|^2 + \tfrac{1}{4}\|\mathbf{a} \times \mathbf{b}\|^2$$

$$= A^2 + B^2 + C^2. \quad \blacksquare$$

EXERCISES

1. Compute the following cross products.
   (a) $(2, -1, 1) \times (1, 2, 1)$
   (b) $(-3, 4, 2) \times (4, 3, -2)$
   (c) $(1, 2, 1) \times (-1, -2, -1)$
   (d) $(3, 2, 1) \times (-1, 2, 3)$
   (e) $(1, -1, 1) \times (-1, 1, 1)$
   (f) $(2, 2, 1) \times (-1, 2, 2)$

2. Find an equation for the plane in $\mathbb{R}^3$ through the points
   (a) $(0, 0, 0), (1, 1, 2), (2, -1, 1)$
   (b) $(1, 1, 1), (1, -1, 1), (1, 2, 2)$
   (c) $(1, 2, 1), (-1, 5, 3), (2, 0, 1)$
   (d) $(1, 0, 0), (0, 1, 0), (0, 0, 1)$
   (e) $(1, 1, 1), (-1, -1, -1), (2, 3, 0)$
   (f) $(3, 2, 3), (-4, 1, 2), (-1, 3, 2)$

3. Find the area of the parallelogram spanned by the vectors
   (a) $(2, -1, 1)$ and $(1, 1, 1)$
   (b) $(-1, 3, 1)$ and $(2, 2, 2)$
   (c) $(0, 1, 0)$ and $(0, 1, 1)$
   (d) $(1, 2, -1)$ and $(4, 1, 3)$
   (e) $(-1, 2, 2)$ and $(3, 0, 1)$
   (f) $(2, 5, 1)$ and $(1, 1, 4)$

4. Find the area of the parallelogram whose vertices are
   (a) $(1, 2, 1), (3, 1, 2), (2, 3, 3), (4, 2, 4)$
   (b) $(3, 5, 7), (4, 6, 8), (3, 4, 5), (4, 5, 6)$
   (c) $(2, -1, 3), (4, 6, -2), (5, 7, 3), (7, 14, -2)$
   (d) $(0, 0, 0), (1, 1, 0), (1, 0, 0), (2, 1, 0)$
   (e) $(0, 0, 0), (1, 2, 0), (2, 1, 0), (3, 3, 0)$

5. Find the area of the triangle whose vertices are
   (a) $(1, 0, 0), (0, 1, 0), (0, 0, 1)$
   (b) $(0, 1, -1), (5, 4, 3), (-1, 1, -1)$
   (c) $(2, 2, -1), (3, 1, 0), (-1, 1, 2)$
   (d) $(0, 0, 0), (1, 2, 0), (2, 1, 0)$
   (e) $(1, 3, 0), (-3, 1, 0), (1, 1, 1)$

6. Find the volume of the parallelepiped spanned by the three vectors
   (a) $(-3, 1, 4)$, $(3, 2, -1)$, $(1, 0, 1)$
   (b) $(2, 1, 1)$, $(3, 1, 1)$, $(1, 3, 2)$
   (c) $(5, 4, 3)$, $(2, 1, 0)$, $(0, 1, -1)$
   (d) $(-1, 1, 2)$, $(0, 1, 1)$, $(0, 1, -1)$
   (e) $(2, 1, 0)$, $(1, 2, 0)$, $(0, 0, 1)$

7. Find the volume of the tetrahedron whose vertices are
   (a) $(0, 0, 0)$, $(1, 0, 0)$, $(0, 1, 0)$, $(0, 0, 1)$
   (b) $(0, 0, 0)$, $(1, 0, 0)$, $(0, 2, 0)$, $(1, 1, 3)$
   (c) $(1, 1, 0)$, $(-1, 2, 0)$, $(1, -3, 0)$, $(0, 0, 6)$
   (d) $(1, 3, -4)$, $(-2, 1, 1)$, $(1, -1, 0)$, $(3, -2, -1)$

8. Let (a_1, a_2, a_3) and (b_1, b_2, b_3) be nonzero vectors in $\mathbb{R}^n$ that are not multiples of one another. Verify that the general solution of the equations

$$a_1 x_1 + a_2 x_2 + a_3 x_3 = 0$$
$$b_1 x_1 + b_2 x_2 + b_3 x_3 = 0$$

   is $\mathbf{x} = c(a_2 b_3 - a_3 b_2, a_3 b_1 - a_1 b_3, a_1 b_2 - a_2 b_1)$, $c \in \mathbb{R}$. (You will need to consider several special cases. Start by assuming that $a_1 \neq 0$ and that $a_1 b_2 - a_2 b_1 \neq 0$.)

9. Show, by direct calculation, that $(\mathbf{a} \times \mathbf{b}) \cdot \mathbf{a} = 0$ and $(\mathbf{a} \times \mathbf{b}) \cdot \mathbf{b} = 0$ wherever $\mathbf{a}$ and $\mathbf{b}$ are vectors in $\mathbb{R}^n$.

10. Prove parts (a), (c), (d) and (e) of Theorem 1.

11. Show that the volume of the tetrahedron spanned by three vectors in $\mathbb{R}^3$ is one-sixth the volume of the parallelepiped spanned by those vectors. [Hint: See Figure 1.40. The plane $ABDF$ cuts the volume of the parallelepiped in

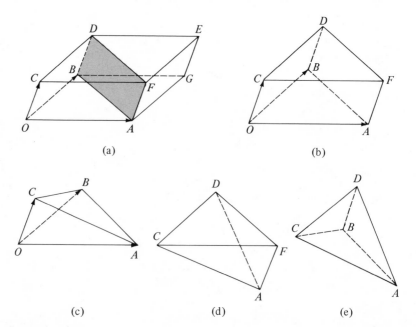

(a)   (b)

(c)   (d)   (e)

FIGURE 1.40

half. The prism with vertices O, A, B, C, D, F can be subdivided into three tetrahedra as shown in Figures 1.40 (b)–(e). Use the fact that tetrahedra with bases of equal areas and with equal altitudes have equal volumes to show that those three tetrahedra have equal volumes.]

12. Prove the 3-dimensional law of cosines: Let A, B, C, and D be the areas of the four faces of a tetrahedron. Then

$$D^2 = A^2 + B^2 + C^2 - 2BC \cos \alpha - 2AC \cos \beta - 2AB \cos \gamma$$

where α is the angle between the faces with areas B and C, β is the angle between the faces with areas A and C, and γ is the angle between the faces with areas A and B. [Hint: Imitate the proof of Theorem 3. Use the fact that $\|\mathbf{v}\|^2 = \mathbf{v} \cdot \mathbf{v}$ to evaluate $\|\mathbf{b} \times \mathbf{c} - \mathbf{a} \times \mathbf{c} - \mathbf{b} \times \mathbf{a}\|^2$. Note that the angle between two faces is the same as the angle between vectors perpendicular to these faces (see Figure 1.41).]

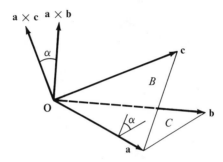

FIGURE 1.41

Vector Spaces

2.1 FUNCTION SPACES AND OTHER VECTOR SPACES

In Chapter 1 we studied the spaces $\mathbb{R}^n$ of ordered n-tuples of real numbers. We now turn our attention to spaces that at first glance seem very different from the spaces $\mathbb{R}^n$, but that turn out to have many of the same properties. These spaces are called function spaces. You have seen them before in calculus, but probably have not looked at them in quite the same way as we will here. You will see in Chapter 3 that these spaces are very useful in the study of differential equations.

Let a and b be real numbers with $a < b$. We shall denote by $I_{(a,b)}$ the set of all real numbers x such that $a < x < b$. $I_{(a,b)}$ is called the **open interval** from a to b. When a and b are fixed throughout a given discussion we shall often drop the subscripts and denote the interval simply by I. We shall allow the possibilities that $a = -\infty$ and $b = \infty$, so, for example, $I_{(-\infty,\infty)} = \mathbb{R}$.

Given such an interval $I = I_{(a,b)}$, we shall denote by $\mathcal{F}(I)$ the set of all real valued functions with domain I. Thus an element of $\mathcal{F}(I)$ is a function $f: I \to \mathbb{R}$ (read "f from I into $\mathbb{R}$") that assigns to each x with $a < x < b$ a real number $f(x)$. As examples of elements of $\mathcal{F}(I_{(0,\pi)})$ we list the four functions f_1, f_2, f_3, and f_4 defined by $f_1(x) = x^2$, $f_2(x) = 1/x$, $f_3(x) = \csc x$, and $f_4(x) = 1$.

$\mathcal{F}(I)$ is called a space of functions, or a **function space.** Note that, given any two functions f and g in $\mathcal{F}(I)$, we can add them: their sum is the function $f + g$ in $\mathcal{F}(I)$ defined by

$$(f + g)(x) = f(x) + g(x) \quad \text{for all } x \text{ in } I.$$

It is easy to check that this operation of addition in $\mathcal{F}(I)$ has the following properties: for f, g, and h in $\mathcal{F}(I)$,

(A_1) $\qquad f + g = g + f$

(A_2) $\qquad (f + g) + h = f + (g + h)$

(A_3) $\qquad 0 + f = f + 0 = f$, where 0 is the constant function that assigns the value zero to each x in I

(A_4) $\qquad f + (-f) = (-f) + f = 0$, where $-f$ is the function in in $\mathcal{F}(I)$ defined by $(-f)(x) = -[f(x)]$ for all x in I

To prove A_1, we recall that two functions in $\mathcal{F}(I)$ are equal if and only if they have the same value at every x in I; that is, $f = g$ if and only if $f(x) = g(x)$ for all x in I. But for any x in I,

$$(f + g)(x) = f(x) + g(x) \quad \text{by definition of } f + g$$
$$= g(x) + f(x) \quad \text{by a property of real numbers}$$
$$= (g + f)(x) \quad \text{by definition of } g + f.$$

Hence $(f + g)(x) = (g + f)(x)$ for all x in I and so $f + g = g + f$. Properties A_2-A_4 are proved similarly.

There is another familiar operation on $\mathcal{F}(I)$ that we call **scalar multiplication.** Given any function f in $\mathcal{F}(I)$ and any real number c we define cf to be the function in $\mathcal{F}(I)$ given by

$$(cf)(x) = cf(x) \quad \text{for all } x \text{ in } I.$$

This operation of scalar multiplication has the following properties: for c, c_1, c_2 in $\mathbb{R}$ and f, g in $\mathcal{F}(I)$,

(S$_1$) $\qquad\qquad\qquad\qquad c(f + g) = cf + cg$

(S$_2$) $\qquad\qquad\qquad\qquad (c_1 + c_2)f = c_1 f + c_2 f$

(S$_3$) $\qquad\qquad\qquad\qquad c_1(c_2 f) = (c_1 c_2)f$

(S$_4$) $\qquad\qquad\qquad\qquad 1f = f.$

Verification of these properties is easy and is left for the exercises.

Observe that these are the same properties that we verified for addition and scalar multiplication in $\mathbb{R}^n$. Hence anything we can do in $\mathbb{R}^n$ using only addition and scalar multiplication we can also do in $\mathcal{F}(I)$. For example, given functions f and g in $\mathcal{F}(I)$, we can consider the "line through f and g" to be the set of all functions h in $\mathcal{F}(I)$ that can be expressed in the form $h = f + t(g - f) = (1 - t)f + tg$, where $t \in \mathbb{R}$. Note that, by analogy with our previous work on lines in $\mathbb{R}^n$, the function $g - f$ serves as a sort of "direction vector" for the line $h = f + t(g - f)$, so we speak of this line as "the line through f in the direction of $g - f$."

Example 1.   The trigonometric identity

$$\cos 2x = \cos^2 x + (-1) \sin^2 x$$

shows that the function $\cos 2x$ lies on the line in $\mathcal{F}(\mathbb{R})$ through $\cos^2 x$ in the direction of $\sin^2 x$. ■

Example 2.   The identity

$$\cosh x = \tfrac{1}{2}e^x + \tfrac{1}{2}e^{-x}$$

shows that the function $\cosh x$ is the midpoint of the line segment in $\mathcal{F}(\mathbb{R})$ from e^x to e^{-x}. ∎

Example 3. Let $f \in \mathcal{F}(I)$ be continuous. Then the set of all functions in $\mathcal{F}(I)$ that are differentiable and have derivative equal to f is a line in $\mathcal{F}(I)$. Indeed, from elementary calculus we know that if g is any function on I whose derivative is f, then every function h on I whose derivative is f is of the form

$$h = g + c = g + c \cdot 1,$$

for some real number c. But this just says that h is on the line in $\mathcal{F}(I)$ through g in the direction of the constant function 1. ∎

Example 4. Consider the differential equation

$(*)$
$$xy' + y = 2x.$$

A **solution** of $(*)$ on the interval I is any function f in $\mathcal{F}(I)$ that is differentiable on I and that satisfies

$$xf'(x) + f(x) = 2x$$

for all x in I. The left side of $(*)$ is just the derivative of xy, so we can rewrite $(*)$ in the form

$$(xy)' = 2x.$$

Integrating both sides of this equation, we get

$$xy = x^2 + c$$

for some real number c. Dividing by x yields

$$y = x + c \cdot \frac{1}{x}$$

so we see that, on any interval I not containing 0, the set of all solutions of $(*)$ is precisely the line in $\mathcal{F}(I)$ through the function $f(x) = x$ in the direction of the function $g(x) = 1/x$. ∎

In later sections of this book we shall focus our attention on $\mathbb{R}^n$ in order to study some of its subsets (e.g., solution sets of systems of linear equations) and in order to study functions with domains that are subsets of $\mathbb{R}^n$ (e.g., the function f of two variables defined by $f(x, y) = x + y$, which is a function with domain $\mathbb{R}^2$). In other sections, we shall focus our attention on $\mathcal{F}(I_{(a,b)})$ in order to study some of its subsets (e.g., solution sets of differential equations) and in order to study functions with domains that are subsets of $\mathcal{F}(I)$ (e.g., the function G defined by $G(f) = \int_a^b f(t)dt$, which is a function with domain a subset of $\mathcal{F}(I_{(a,b)})$.) Since $\mathbb{R}^n$ and $\mathcal{F}(I_{(a,b)})$ share many properties, it makes sense to study them together. By doing this we not only gain economy of presentation, but we also gain additional insight. To this end, we develop the unifying concept of vector space.

A *vector space* is a set V, whose elements will be called *vectors,* together with two operations, (i) *addition:* to each pair $\mathbf{u}, \mathbf{v}$ of vectors in V there is associated a vector $\mathbf{u} + \mathbf{v}$ in V, called "$\mathbf{u}$ plus $\mathbf{v}$," and (ii) *scalar multiplication:* to each pair c and $\mathbf{v}$, where c is a real number and $\mathbf{v}$ is a vector in V, there is associated a vector $c\mathbf{v}$ in V, called "c times $\mathbf{v}$." These operations must have the following properties:

(A_1) $\mathbf{u} + \mathbf{v} = \mathbf{v} + \mathbf{u}$ for all $\mathbf{u}, \mathbf{v}$ in V.

(A_2) $(\mathbf{u} + \mathbf{v}) + \mathbf{w} = \mathbf{u} + (\mathbf{v} + \mathbf{w})$ for all $\mathbf{u}, \mathbf{v}, \mathbf{w}$ in V.

(A_3) There exists an element $\mathbf{0}$ in V, called zero, with the property that $\mathbf{0} + \mathbf{v} = \mathbf{v} + \mathbf{0} = \mathbf{v}$ for all $\mathbf{v}$ in V.

(A_4) For each $\mathbf{v}$ in V there exists an element, denoted $-\mathbf{v}$, in V, such that

$$\mathbf{v} + (-\mathbf{v}) = (-\mathbf{v}) + \mathbf{v} = \mathbf{0}.$$

(S_1) $c(\mathbf{u} + \mathbf{v}) = c\mathbf{u} + c\mathbf{v}$ for all $\mathbf{u}, \mathbf{v}$ in V and c in $\mathbb{R}$.

(S_2) $(c_1 + c_2)\mathbf{v} = c_1\mathbf{v} + c_2\mathbf{v}$ for all c_1, c_2 in $\mathbb{R}$ and $\mathbf{v}$ in V.

(S_3) $(c_1 c_2)\mathbf{v} = c_1(c_2\mathbf{v})$ for all c_1, c_2 in $\mathbb{R}$ and $\mathbf{v}$ in V.

(S_4) $1\mathbf{v} = \mathbf{v}$ for all $\mathbf{v}$ in V.

Example 5. We have already seen that $\mathbb{R}^n$ satisfies the above axioms, so $\mathbb{R}^n$ is a vector space. ■

Example 6. We have also seen that $\mathcal{F}(I_{(a,b)})$ satisfies the above axioms, so $\mathcal{F}(I_{(a,b)})$ is a vector space. ■

Example 7. Let $\mathcal{P}$ denote the set of all polynomials with real coefficients. An element of $\mathcal{P}$ is a function $p \colon \mathbb{R} \to \mathbb{R}$ of the form

$$p(x) = a_0 + a_1 x + \cdots + a_n x^n$$

where n is a nonnegative integer and $a_0, a_1, \ldots, a_n$ are real numbers. Clearly the sum of two polynomials is a polynomial, as is the product cp of any real number c with any polynomial p. It is easy to check that properties A_1–A_4 and S_1–S_4 are satisfied, so $\mathcal{P}$ is a vector space. ■

Example 8. Let $\mathbb{R}^\infty$ denote the set of all infinite sequences of real numbers. For $\mathbf{a} = (a_1, a_2, a_3, \ldots)$ and $\mathbf{b} = (b_1, b_2, b_3, \ldots)$ in $\mathbb{R}^\infty$, define $\mathbf{a} + \mathbf{b} = (a_1 + b_1, a_2 + b_2, a_3 + b_3, \ldots)$. For $\mathbf{a} = (a_1, a_2, a_3, \ldots)$ in $\mathbb{R}^\infty$ and c in $\mathbb{R}$, define $c\mathbf{a} = (ca_1, ca_2, ca_3, \ldots)$. With these operations of addition and scalar multiplication, $\mathbb{R}^\infty$ is a vector space. ∎

Example 9. Let $\mathfrak{M}_{2\times2}$ denote the set of all 2 by 2 matrices; that is, $\mathfrak{M}_{2\times2}$ is the set of all matrices with two rows and two columns. For

$$A = \begin{pmatrix} a_{11} & a_{12} \\ a_{21} & a_{22} \end{pmatrix} \quad \text{and} \quad B = \begin{pmatrix} b_{11} & b_{12} \\ b_{21} & b_{22} \end{pmatrix}$$

and for c in $\mathbb{R}$, define

$$A + B = \begin{pmatrix} a_{11} + b_{11} & a_{12} + b_{12} \\ a_{21} + b_{21} & a_{22} + b_{22} \end{pmatrix} \quad \text{and} \quad cA = \begin{pmatrix} ca_{11} & ca_{12} \\ ca_{21} & ca_{22} \end{pmatrix}.$$

With these operations of addition and scalar multiplication, $\mathfrak{M}_{2\times2}$ is a vector space. ∎

Example 10. For m and n positive integers, let $\mathfrak{M}_{m\times n}$ denote the set of all m by n (m rows, n columns) matrices. Addition and scalar multiplication in $\mathfrak{M}_{m\times n}$ can be defined as in Example 9: for A and B in $\mathfrak{M}_{m\times n}$ and c in $\mathbb{R}$, $A + B$ is the matrix in $\mathfrak{M}_{m\times n}$ whose (i, j)-entry (the entry in the ith row and jth column) is equal to the (i, j)-entry of A plus the (i, j)-entry of B, and cA is the matrix in $\mathfrak{M}_{m\times n}$ whose (i, j)-entry is equal to c times the (i, j)-entry in A, for each i and j with $1 \leq i \leq m$ and $1 \leq j \leq n$. Thus, for example,

$$\begin{pmatrix} 1 & 2 & -1 \\ 0 & -2 & 1 \end{pmatrix} + \begin{pmatrix} 1 & 3 & 2 \\ 2 & -1 & 0 \end{pmatrix} = \begin{pmatrix} 2 & 5 & 1 \\ 2 & -3 & 1 \end{pmatrix}$$

and

$$-3 \begin{pmatrix} 1 & 2 & -1 \\ 0 & -2 & 1 \end{pmatrix} = \begin{pmatrix} -3 & -6 & 3 \\ 0 & 6 & -3 \end{pmatrix}.$$

$\mathfrak{M}_{m\times n}$ is another example of a vector space, for each m and n. ∎

There are many more examples of vector spaces. Some of these will appear as exercises at the end of this section; others will appear in later sections.

The axioms A_1–A_4 and S_1–S_4 have as consequences many of the rules of algebra that are familiar for $\mathbb{R}^1 = \mathbb{R}$ and for $\mathcal{F}(I_{(a,b)})$. For example, given vectors $\mathbf{v}_1, \mathbf{v}_2, \ldots, \mathbf{v}_k$ in V and real numbers $c_1, c_2, \ldots, c_k$, we would like to be able to talk about the vector

$$\mathbf{v} = c_1\mathbf{v}_1 + c_2\mathbf{v}_2 + \cdots + c_k\mathbf{v}_k.$$

Such a vector is called a ***linear combination*** of the vectors $v_1, v_2, \ldots, v_k$. Axiom A_2, applied repeatedly, tells us that we do not need any parentheses in this sum. The order in which the additions are performed doesn't matter.

Example 11. In $\mathbb{R}^3$, the vector $(-1, -9, 4)$ is a linear combination of the vectors $(1, -1, 2)$ and $(2, 3, 1)$. To show this, we must find real numbers c_1 and c_2 such that

$$c_1(1, -1, 2) + c_2(2, 3, 1) = (-1, -9, 4).$$

Thus we seek real numbers c_1 and c_2 satisfying

$$c_1 + 2c_2 = -1$$
$$-c_1 + 3c_2 = -9$$
$$2c_1 + c_2 = 4.$$

Since the solution of this linear system is $(c_1, c_2) = (3, -2)$, we see that

$$(-1, -9, 4) = 3(1, -1, 2) - 2(2, 3, 1)$$

so $(-1, -9, 4)$ is a linear combination of $(1, -1, 2)$ and $(2, 3, 1)$, as claimed. ∎

Other algebraic properties of vector addition and scalar multiplication that will surprise no one, but that are not included in the properties A_1–A_4 and S_1–S_4, can be derived from A_1–A_4 and S_1–S_4.

Example 12. The vector $\mathbf{0}$ is unique; that is, there is only one vector $\mathbf{0}$ in V with the property that

(1) $$\mathbf{0} + \mathbf{v} = \mathbf{v} + \mathbf{0} = \mathbf{v}$$

for all $\mathbf{v}$ in V.

Proof Suppose there were another vector, say $\tilde{\mathbf{0}}$, with this property:

(2) $$\tilde{\mathbf{0}} + \mathbf{v} = \mathbf{v} + \tilde{\mathbf{0}} = \mathbf{v}$$

for all $\mathbf{v}$ in V. Then

$$\tilde{\mathbf{0}} = \mathbf{0} + \tilde{\mathbf{0}} = \mathbf{0},$$

where the first equality follows from (1) and the second from (2). Thus $\mathbf{0} = \tilde{\mathbf{0}}$, as was to be shown. ∎

Example 13. $0\mathbf{v} = \mathbf{0}$ for all $\mathbf{v}$ in V.

Proof $0v = (0 + 0)v = 0v + 0v$ by S_2. Addition of $-0v$ to both sides of this equation yields

$$0 = 0v + (-0v) = (0v + 0v) + (-0v)$$
$$= 0v + (0v + (-0v)) \text{ by } A_2$$
$$= 0v + 0 \text{ by } A_4$$
$$= 0v \quad \text{ by } A_3,$$

as was to be shown. ∎

EXERCISES

1. Verify the properties A_2–A_4 of addition in $\mathfrak{F}(I)$.

2. Verify the properties S_1–S_4 of scalar multiplication in $\mathfrak{F}(I)$.

3. Show that the function $(x + 2)^2$ lies on the line in $\mathfrak{F}(\mathbb{R})$ through x^2 in the direction of $x + 1$.

4. Show that the constant function 1 is on the line in $\mathfrak{F}(\mathbb{R})$ through $\sec^2 x$ in the direction of $\tan^2 x$.

5. Show that $\cos^2 x$ is the midpoint of the line segment in $\mathfrak{F}(\mathbb{R})$ joining the constant function 1 to the function $\cos 2x$.

6. Show that the solution set of each of the following differential equations is a line in $\mathfrak{F}(\mathbb{R})$.
   (a) $(1 + x^2)y' + 2xy = x^2$
   (b) $e^x y' + e^x y = 1$
   (c) $e^{-x} y' - e^{-x} y = x$
   (d) $e^{p(x)} y' + p'(x)e^{p(x)} y = q(x)$, where p is any differentiable function on $\mathbb{R}$ and q is any continuous function on $\mathbb{R}$.

7. Let $\mathbf{a} = (1, 2, 3, 4, 5, \ldots)$ and $\mathbf{b} = (0, 1, 2, 3, 4, \ldots)$. Using the operations of addition and scalar multiplication in $\mathbb{R}^\infty$, find
   (a) $\mathbf{a} + \mathbf{b}$ (d) $\frac{1}{2}\mathbf{a} - \frac{3}{2}\mathbf{b}$
   (b) $\mathbf{a} - \mathbf{b}$ (e) $\pi\mathbf{b}$
   (c) $2\mathbf{a}$

8. Let $A = \begin{pmatrix} -1 & 0 \\ 1 & 3 \\ 0 & 2 \end{pmatrix}$ and $B = \begin{pmatrix} 1 & 1 \\ -1 & 1 \\ 2 & 4 \end{pmatrix}$. Using the operations of addition and scalar multiplication in $\mathfrak{M}_{3\times 2}$, find
   (a) $A + B$ (d) $5A + 2B$
   (b) $A - B$ (e) $2\pi A$
   (c) $\frac{1}{2}B$

9. In $\mathbb{R}^2$, show that the vector $(1, 0)$ is a linear combination of the vectors $(1, -1)$ and $(2, 1)$.

10. In $\mathbb{R}^3$, which of the following vectors are linear combinations of $(1, -1, 1)$ and $(2, 0, 1)$?

    (a) $(1, 1, 0)$ (c) $(-1, 3, 2)$

    (b) $(1, 1, 1)$ (d) $(1, -3, 2)$

11. (a) Show that every vector in $\mathbb{R}^2$ can be expressed as a linear combination of the vectors $(1, 0)$ and $(0, 1)$.

    (b) Show that every vector in $\mathbb{R}^3$ can be expressed as a linear combination of $(1, 0, 0)$, $(0, 1, 0)$, and $(0, 0, 1)$.

12. In $\mathcal{F}(\mathbb{R})$, show that $\cos 2x$ is a linear combination of $\cos^2 x$ and $\sin^2 x$.

13. In the space $\mathcal{P}$ of polynomials, show that $2x^2 + 3x - 1$ is a linear combination of 1, $x - 1$, and $(x - 1)^2$.

14. In $\mathcal{F}(\mathbb{R})$, show that e^{-x} is a linear combination of e^x and $\cosh x$.

15. Consider the differential equation

    $(*)$ $y'' + ay' + by = 0$

    where a and b are any two real numbers. Show that if f_1 and f_2 in $\mathcal{F}(I)$ are any two solutions of $(*)$ and c_1 and c_2 are any two real numbers, then $c_1 f_1 + c_2 f_2$ is also a solution of $(*)$.

16. Verify that the operations of addition and scalar multiplication in $\mathcal{P}$, $\mathbb{R}^\infty$, and $\mathfrak{M}_{m \times n}$ do indeed have properties A_1–A_4 and S_1–S_4.

17. Let V be a set consisting of a single element a. Define $a + a = a$ and define $ca = a$ for all real numbers c. Show that, with these operations, V is a vector space.

18. Let V and W be vector spaces and let X be the set of all ordered pairs $(\mathbf{v}, \mathbf{w})$ where $\mathbf{v} \in V$ and $\mathbf{w} \in W$. For $(\mathbf{v}, \mathbf{w})$ and $(\mathbf{v}', \mathbf{w}')$ in X and c in $\mathbb{R}$, define

$$(\mathbf{v}, \mathbf{w}) + (\mathbf{v}', \mathbf{w}') = (\mathbf{v} + \mathbf{v}', \mathbf{w} + \mathbf{w}')$$

$$c(\mathbf{v}, \mathbf{w}) = (c\mathbf{v}, c\mathbf{w}).$$

Show that, with these operations, X is a vector space.

19. Let V be a vector space and let S be any set. Let $\mathcal{F}(S, V)$ denote the set of all functions $f \colon S \to V$ with domain S and with values in V. For f and g in $\mathcal{F}(S, V)$, define $f + g$ in $\mathcal{F}(S, V)$ by

$$(f + g)(s) = f(s) + g(s) \quad \text{for all } s \in S$$

and for c in $\mathbb{R}$ and f in $\mathcal{F}(S, V)$ define cf in $\mathcal{F}(S, V)$ by

$$(cf)(s) = c(f(s)).$$

Show that, with these operations, $\mathcal{F}(S, V)$ is a vector space.

20. Let S be a set and let V be the collection of all subsets of S. For A and B in V, define $A + B = A \cup B$. For $A \in V$ and $c \in \mathbb{R}$, define $cA = A$. Which of the properties A_1–A_4 and S_1–S_4 are satisfied by these operations? Is V, with these operations, a vector space?

21. Define a funny addition $\oplus$ on the set $\mathbb{R}_+$ of positive real numbers by $a \oplus b = ab$ for $a, b \in \mathbb{R}_+$ and define a funny scalar multiplication $\odot$ on this set by $c \odot a = a^c$ for $a \in \mathbb{R}_+$ and $c \in \mathbb{R}$. With these operations, is $\mathbb{R}_+$ a vector space?

22. Prove that, in a vector space, negatives are unique. That is, show that if V is a vector space and $\mathbf{v}$ is a vector in V, then there is only one vector $\mathbf{w}$ in V such that $\mathbf{v} + \mathbf{w} = \mathbf{w} + \mathbf{v} = \mathbf{0}$.

23.   Prove that, in a vector space, $c\mathbf{0} = \mathbf{0}$ for all $c \in \mathbb{R}$.

24.   Let V be a vector space and let $\mathbf{v}_1, \ldots, \mathbf{v}_k$ be vectors in V. Show that if $\mathbf{v}$ and $\mathbf{w}$ are both linear combinations of $\mathbf{v}_1, \ldots, \mathbf{v}_k$, then so is $a\mathbf{v} + b\mathbf{w}$ for all a, $b \in \mathbb{R}$.

2.2 SUBSPACES

A **subspace** of a vector space V is a subset that contains the zero vector and is closed under addition and scalar multiplication. In other words, a subset W of V is a subspace if

(i) $\mathbf{0} \in W$,

(ii) $\mathbf{u} + \mathbf{v} \in W$ whenever $\mathbf{u} \in W$ and $\mathbf{v} \in W$, and

(iii) $c\mathbf{v} \in W$ whenever $\mathbf{v} \in W$ and $c \in \mathbb{R}$.

Subspaces are important because each subspace of a vector space is itself a vector space. Indeed, we can add any two vectors in W because we know how to add them in V, and the result of the addition gives us an element of W. Similarly, we can multiply vectors in W by real numbers and get vectors in W. The properties A_1–A_4 and S_1–S_4 of vector addition and of scalar multiplication are automatically satisfied because they are satisfied in V. Thus we can expand our list of vector spaces by listing subspaces of vector spaces we already know.

Example 1.   Let $V = \mathbb{R}^n$ and let ℓ be a line through the origin in $\mathbb{R}^n$ (see Figure 2.1). Take $\mathbf{d}$ to be any nonzero vector along ℓ. We know that $\mathbf{x} \in \ell$ if and only if $\mathbf{x} = t\mathbf{d}$ for some $t \in \mathbb{R}$. We shall show that ℓ is a subspace of $\mathbb{R}^n$ by verifying conditions (i)–(iii). (i) Taking $t = 0$, we see that $\mathbf{0} \in \ell$. (ii) If $\mathbf{u} = t_1\mathbf{d}$ and $\mathbf{v} = t_2\mathbf{d}$ are any two points on ℓ, then $\mathbf{u} + \mathbf{v} = (t_1 + t_2)\mathbf{d}$ is also on ℓ. Finally, (iii) if $\mathbf{v} = t\mathbf{d}$ is on ℓ and $c \in \mathbb{R}$, then $c\mathbf{v} = c(t\mathbf{d}) = (ct)\mathbf{d}$ is on ℓ. Thus ℓ is a subspace of $\mathbb{R}^n$. ∎

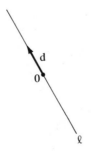

FIGURE 2.1

Example 2.   Let $V = \mathbb{R}^3$ and let S be the solution set of the equation $x_1 + x_2 + x_3 = 0$. Let us show that S is a subspace of $\mathbb{R}^3$.

(i) $\mathbf{0} = (0, 0, 0) \in S$ since $0 + 0 + 0 = 0$.

(ii)   If $\mathbf{u} = (u_1, u_2, u_3) \in S$ and $\mathbf{v} = (v_1, v_2, v_3) \in S$, then

$$\mathbf{u} + \mathbf{v} = (u_1 + v_1, u_2 + v_2, u_3 + v_3)$$

and

$$(u_1 + v_1) + (u_2 + v_2) + (u_3 + v_3) = (u_1 + u_2 + u_3) + (v_1 + v_2 + v_3)$$
$$= 0 + 0 = 0$$

so $\mathbf{u} + \mathbf{v} \in S$.

(iii)   If $\mathbf{v} = (v_1, v_2, v_3) \in S$, then $c\mathbf{v} = (cv_1, cv_2, cv_3)$ and

$$cv_1 + cv_2 + cv_3 = c(v_1 + v_2 + v_3) = c \cdot 0 = 0$$

so $c\mathbf{v} \in S$.
Thus S is a subspace of $\mathbb{R}^3$.   ■

Example 3.   Let $V = \mathbb{R}^3$ and let P be any plane through the origin in $\mathbb{R}^3$ (see Figure 2.2). Then P is described by an equation of the form

$$\mathbf{d} \cdot \mathbf{x} = 0$$

where $\mathbf{d}$ is a nonzero vector perpendicular to P. Certainly,

(i) $\mathbf{0} \in P$.

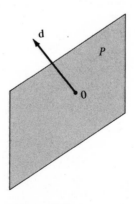

FIGURE 2.2

Also,

(ii)   if $\mathbf{u} \in P$ and $\mathbf{v} \in P$, then $\mathbf{d} \cdot \mathbf{u} = 0$ and $\mathbf{d} \cdot \mathbf{v} = 0$,

hence $\mathbf{d} \cdot (\mathbf{u} + \mathbf{v}) = \mathbf{d} \cdot \mathbf{u} + \mathbf{d} \cdot \mathbf{v} = 0$, and so $\mathbf{u} + \mathbf{v} \in P$.
Finally,

(iii)   if $\mathbf{v} \in P$ and $c \in \mathbb{R}$, then $\mathbf{d} \cdot c\mathbf{v} = c\mathbf{d} \cdot \mathbf{v} = 0$,

so $c\mathbf{v} \in P$. Thus P is a subspace of $\mathbb{R}^3$. ■

Remark. The same argument shows that hyperplanes through $\mathbf{0}$ in $\mathbb{R}^n$ are subspaces of $\mathbb{R}^n$.

Example 4. Let I be an open interval in $\mathbb{R}$ and let $V = \mathcal{F}(I)$. Let $\mathcal{C}(I)$ denote the set of all continuous real-valued functions with domain I. Then $\mathcal{C}(I)$ is a subset of $\mathcal{F}(I)$. $\mathcal{C}(I)$ is in fact a subspace of $\mathcal{F}(I)$ because (i) the zero function (the constant function zero) is continuous, (ii) the sum of continuous functions is continuous, and (iii) constant multiples of continuous functions are continuous. ■

Example 5. As in Example 4, let $V = \mathcal{F}(I)$. Let $\mathcal{D}(I)$ denote the set of all differentiable functions with domain I. Then $\mathcal{D}(I)$ is a subspace of $\mathcal{F}(I)$ because (i) the zero function is differentiable, (ii) the sum of differentiable functions is differentiable, and (iii) constant multiples of differentiable functions are differentiable. ■

Example 6. Let $V = \mathcal{P}$, the vector space of all polynomials with real coefficients, and let $\mathcal{P}^n$ denote those polynomials with degree $<n$. The zero polynomial is in $\mathcal{P}^n$ (it has degree zero), the sum of two polynomials of degree $<n$ has degree $<n$, and any constant multiple of a polynomial of degree $<n$ also has degree $<n$. Thus $\mathcal{P}^n$ is a subspace of $\mathcal{P}$. ■

Example 7. Let us find *all* the subspaces of $\mathbb{R}^2$. Suppose W is a subspace of $\mathbb{R}^2$. Then $\mathbf{0} \in W$. Possibly $W = \{\mathbf{0}\}$, since $\{\mathbf{0}\}$ is a subspace of $\mathbb{R}^2$ (Why?). If W contains a nonzero vector $\mathbf{v}$, then W must contain $c\mathbf{v}$ for each $c \in \mathbb{R}$; that is, W must contain the line $\mathbf{x} = t\mathbf{v}$ through $\mathbf{0}$ in the direction of $\mathbf{v}$. Since this line is a subspace, possibly W is this line.

Suppose W contains a nonzero vector $\mathbf{v}$ and a nonzero vector $\mathbf{w}$ that is not on the line $\mathbf{x} = t\mathbf{v}$. Then W must contain all linear combinations $c_1\mathbf{v} + c_2\mathbf{w}$ of $\mathbf{v}$ and $\mathbf{w}$. But every vector in $\mathbb{R}^2$ can be expressed as a linear combination of $\mathbf{v}$ and $\mathbf{w}$. This can be seen geometrically as follows. Let $\mathbf{u}$ be any vector in $\mathbb{R}^2$, and consider the lines $\mathbf{x} = \mathbf{u} + t\mathbf{v}$ and $\mathbf{x} = t\mathbf{w}$ (see Figure 2.3). Since these lines are not parallel ($\mathbf{w}$ is not a multiple of $\mathbf{v}$) they must intersect. Hence there exist real numbers c_1 and c_2 such that $\mathbf{u} + c_1\mathbf{v} = c_2\mathbf{w}$. Solving for $\mathbf{u}$ we find that

$$\mathbf{u} = (-c_1)\mathbf{v} + c_2\mathbf{w}.$$

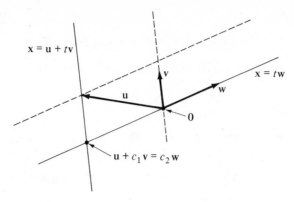

FIGURE 2.3

So **u** is a linear combination of **v** and **w** as claimed. Thus W contains every vector in $\mathbb{R}^2$; that is $W = \mathbb{R}^2$.

We conclude, then, that *the only subspaces of* $\mathbb{R}^2$ *are* $\{0\}$, *lines through* **0**, *and* $\mathbb{R}^2$ *itself.* ∎

Remark. In a similar way it can be shown that the only subspaces of $\mathbb{R}^3$ are $\{0\}$, lines through **0**, planes through **0**, and $\mathbb{R}^3$ itself.

EXERCISES

1. Which of the following are subspaces of $\mathbb{R}^4$?
   (a) the hyperplane $x_1 + x_2 + x_3 + x_4 = 0$
   (b) the hyperplane $2x_1 + 3x_4 = 0$
   (c) the hyperplane $2x_1 + 3x_4 = 1$
   (d) the subset $x_1^2 - x_4^2 = 0$
   (e) the solution set of the system
$$x_1 + x_2 + x_3 \qquad\quad = 0$$
$$2x_1 - x_2 \qquad + x_4 = 0$$
   (f) the solution set of the system
$$x_1 + x_2 + x_3 \qquad\quad = \quad 3$$
$$2x_1 - x_2 \qquad + x_4 = -1$$
   (g) all vectors of the form $(a, a, -a, 2a)$
   (h) all vectors of the form $(a, b, a + b, 0)$
   (i) all vectors of the form $(a, b, a^2 + b^2, 0)$

2. Which of the following are subspaces of the vector space $\mathcal{D} = \mathcal{D}(\mathbb{R})$? [Note: "$\{x \in \mathcal{D} \mid A\}$" means "the set of all x in $\mathcal{D}$ such that A."]
   (a) $\{f \in \mathcal{D} \mid f' = 0\}$ (c) $\{f \in \mathcal{D} \mid f' = f\}$
   (b) $\{f \in \mathcal{D} \mid f' = 1\}$ (d) $\{f \in \mathcal{D} \mid f' = f^2\}$

3. Which of the following are subspaces of the vector space $\mathcal{F}(\mathbb{R})$?
   (a) $\{f \in \mathcal{F}(\mathbb{R}) \mid f(1) = 0\}$ (c) $\{f \in \mathcal{F}(\mathbb{R}) \mid f(x) \geq 0 \text{ for all } x\}$
   (b) $\{f \in \mathcal{F}(\mathbb{R}) \mid f(1) = 1\}$ (d) $\{f \in \mathcal{F}(\mathbb{R}) \mid f(x) = f(-x) \text{ for all } x\}$

4. Which of the following are subspaces of the vector space $\mathcal{C}(I_{[0,1]})$?

    (a) $\{f \in \mathcal{C}(I_{[0,1]}) \mid \int_0^1 f = 0\}$ (c) $\{f \in \mathcal{C}(I_{[0,1]}) \mid \int_0^{\frac{1}{2}} f = \int_{\frac{1}{2}}^1 f\}$

    (b) $\{f \in \mathcal{C}(I_{[0,1]}) \mid \int_0^1 f = 1\}$ (d) $\{f \in \mathcal{C}(I_{[0,1]}) \mid \int_0^1 xf(x)dx = 0\}$

5. Which of the following is a subspace of the given vector space?

    (a) $\{\begin{pmatrix} a & b \\ c & d \end{pmatrix} \in \mathfrak{M}_{2\times2} \mid ad - bc = 0\}$, in $\mathfrak{M}_{2\times2}$

    (b) $\{\begin{pmatrix} a & b \\ c & d \end{pmatrix} \in \mathfrak{M}_{2\times2} \mid a + d = 0\}$, in $\mathfrak{M}_{2\times2}$

    (c) $\{(a_1, a_2, a_3, \ldots) \in \mathbb{R}^\infty \mid a_1 = a_3 = a_5 = \ldots\}$, in $\mathbb{R}^\infty$

    (d) $\{a_0 + a_1 x + a_2 x^2 \mid a_0 = 0\}$, in $\mathcal{P}^3$

    (e) the hyperplane $\mathbf{d} \cdot \mathbf{x} = 0$ in $\mathbb{R}^n$, where $\mathbf{d} \in \mathbb{R}^n$, $\mathbf{d} \neq \mathbf{0}$

    (f) the hyperplane $\mathbf{d} \cdot \mathbf{x} = c$ in $\mathbb{R}^n$, where $\mathbf{d} \in \mathbb{R}^n$, $\mathbf{d} \neq \mathbf{0}$, and $c \in \mathbb{R}$, $c \neq 0$

6. Describe all subspaces of $\mathbb{R}^1$.

7. Let W be a subspace of $\mathbb{R}^n$ and let $W^\perp = \{\mathbf{v} \in \mathbb{R}^n \mid \mathbf{v} \cdot \mathbf{w} = 0 \text{ for all } \mathbf{w} \in W\}$. Show that $W^\perp$ is a subspace of $\mathbb{R}^n$. ($W^\perp$ is called the **orthogonal complement** of W.)

8. Show that the intersection of any two subspaces of a vector space is a subspace.

9. Show analytically that if $\mathbf{v}$ and $\mathbf{w}$ are nonzero vectors in $\mathbb{R}^2$ and $\mathbf{w}$ is not a scalar multiple of $\mathbf{v}$, then every vector in $\mathbb{R}^2$ can be expressed as a linear combination of $\mathbf{v}$ and $\mathbf{w}$. [Hint: If $\mathbf{v} = (v_1, v_2)$ and $\mathbf{w} = (w_1, w_2)$, then $\mathbf{u} = (u_1, u_2)$ is a linear combination of $\mathbf{v}$ and $\mathbf{w}$ if and only if there exist x_1 and x_2 such that

$$x_1(v_1, v_2) + x_2(w_1, w_2) = (u_1, u_2).$$

Convert this vector equation into a system of two scalar equations and use the fact that $\mathbf{w}$ is not a multiple of $\mathbf{v}$ to show that this system is consistent and hence has a solution.]

10. Show that if W is a *nonempty* subset of a vector space V that is closed under vector addition and under scalar multiplication, then the zero vector must be in W.

11. Show that a subset W of a vector space V is a subspace if and only if

      (i) W is nonempty

   and

      (ii) $c_1\mathbf{v}_1 + c_2\mathbf{v}_2 \in W$ whenever $\mathbf{v}_1, \mathbf{v}_2 \in W$ and $c_1, c_2 \in \mathbb{R}$.

2.3 SPANNING SETS

    In this section we shall discuss subspaces that are determined by finite sets of vectors. Such subspaces are particularly easy to describe. We shall see in the next section of this chapter that such subspaces arise naturally as solution spaces of homogeneous linear systems of equations, and we shall see in the next chapter that they also arise naturally as solution spaces of homogeneous linear differential equations.

Let $\{\mathbf{v}_1, \ldots, \mathbf{v}_k\}$ be a finite nonempty subset of a vector space V and consider the set $\mathfrak{L}(\mathbf{v}_1, \ldots, \mathbf{v}_k)$ of all linear combinations

$$\mathbf{v} = c_1\mathbf{v}_1 + c_2\mathbf{v}_2 + \cdots + c_k\mathbf{v}_k$$

of $\{\mathbf{v}_1, \ldots, \mathbf{v}_k\}$. Note that $\mathfrak{L}(\mathbf{v}_1, \ldots, \mathbf{v}_k)$ is a subspace of V. Indeed,

(i) $\mathbf{0} = 0\mathbf{v}_1 + \cdots + 0\mathbf{v}_k \in \mathfrak{L}(\mathbf{v}_1, \ldots, \mathbf{v}_k)$,

(ii) if $\mathbf{v} = c_1\mathbf{v}_1 + \cdots + c_k\mathbf{v}_k$ and $\mathbf{w} = d_1\mathbf{v}_1 + \cdots + d_k\mathbf{v}_k$ are in $\mathfrak{L}(\mathbf{v}_1, \ldots, \mathbf{v}_k)$, then $\mathbf{v} + \mathbf{w} = (c_1 + d_1)\mathbf{v}_1 + \cdots + (c_k + d_k)\mathbf{v}_k$ is in $\mathfrak{L}(\mathbf{v}_1, \ldots, \mathbf{v}_k)$, and

(iii) if $\mathbf{v} = c_1\mathbf{v}_1 + \cdots + c_k\mathbf{v}_k \in \mathfrak{L}(\mathbf{v}_1, \ldots, \mathbf{v}_k)$ and $c \in \mathbb{R}$, then $c\mathbf{v} = (cc_1)\mathbf{v}_1 + \cdots + (cc_k)\mathbf{v}_k \in \mathfrak{L}(\mathbf{v}_1, \ldots, \mathbf{v}_k)$.

The subspace $\mathfrak{L}(\mathbf{v}_1, \ldots, \mathbf{v}_k)$ is called the **space spanned by** $\{\mathbf{v}_1, \ldots, \mathbf{v}_k\}$. It is the smallest subspace of V that contains the set $\{\mathbf{v}_1, \ldots, \mathbf{v}_k\}$. For if W is any subspace of V containing $\{\mathbf{v}_1, \ldots, \mathbf{v}_k\}$ then, since W is closed under addition and scalar multiplication, W must contain every linear combination of $\{\mathbf{v}_1, \ldots, \mathbf{v}_k\}$; that is, W must contain $\mathfrak{L}(\mathbf{v}_1, \ldots, \mathbf{v}_k)$.

Example 1. In $\mathbb{R}^n$, the space spanned by a single nonzero vector $\mathbf{v}$ is the line $\mathbf{x} = t\mathbf{v}$ through the origin consisting of all scalar multiples of $\mathbf{v}$ (see Figure 2.4). ∎

Example 2. In $\mathbb{R}^3$, the space $\mathfrak{L}(\mathbf{a}, \mathbf{b})$ spanned by a pair $\{\mathbf{a}, \mathbf{b}\}$ of vectors, neither of which is a multiple of the other, is the plane $\mathbf{d} \cdot \mathbf{x} = 0$ where $\mathbf{d} = \mathbf{a} \times \mathbf{b}$ (see Figure 2.5). ∎

Example 3. Let A be an $m \times n$ matrix. The row vectors of A are vectors in $\mathbb{R}^n$. The space spanned by these row vectors is then a subspace of $\mathbb{R}^n$, called the **row space of** A. Thus if

$$A = \begin{pmatrix} 1 & 0 & 0 \\ 0 & 1 & 0 \end{pmatrix}$$

then the row space of A is the subspace $\mathfrak{L}((1, 0, 0), (0, 1, 0))$ of $\mathbb{R}^3$ consisting of all vectors of the form $a(1, 0, 0) + b(0, 1, 0) = (a, b, 0)$. ∎

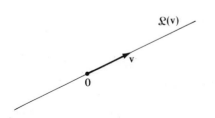

FIGURE 2.4

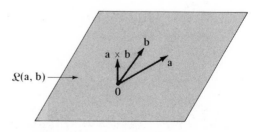

FIGURE 2.5

Example 4. The columns of an $m \times n$ matrix are vectors in the vector space $\mathfrak{M}_{m \times 1}$. Thus if

$$A = \begin{pmatrix} a_{11} & a_{12} & \vdots & a_{1n} \\ \vdots & \vdots & \vdots & \vdots \\ a_{m1} & a_{m2} & & a_{mn} \end{pmatrix}$$

then the **column vectors** of A are the $m \times 1$ matrices

$$\begin{pmatrix} a_{11} \\ \vdots \\ a_{m1} \end{pmatrix}, \begin{pmatrix} a_{12} \\ \vdots \\ a_{m2} \end{pmatrix}, \ldots, \begin{pmatrix} a_{1n} \\ \vdots \\ a_{mn} \end{pmatrix}.$$

The space spanned by these column vectors is called the **column space of** A.   ■

As an application of the concept of column space we prove:

Theorem 1.   *The system of linear equations*

$$a_{11}x_1 + a_{12}x_2 + \cdots + a_{1n}x_n = b_1$$
$$\vdots$$
$$a_{m1}x_1 + a_{m2}x_2 + \cdots + a_{mn}x_n = b_m$$

is consistent (i.e., has a solution) if and only if $\begin{pmatrix} b_1 \\ \vdots \\ b_m \end{pmatrix}$ *is in the column space of the coefficient matrix*

$$A = \begin{pmatrix} a_{11} & a_{12} & \vdots & a_{1n} \\ \vdots & \vdots & \vdots & \vdots \\ a_{m1} & a_{m2} & & a_{mn} \end{pmatrix}.$$

Proof   The system is consistent if and only if there exist real numbers $c_1, \ldots, c_n$ such that

$$a_{11}c_1 + a_{12}c_2 + \cdots + a_{1n}c_n = b_1$$
$$\vdots$$
$$a_{m1}c_1 + a_{m2}c_2 + \cdots + a_{mn}c_n = b_m.$$

This system of scalar equations is equivalent to the single matrix equation

$$c_1 \begin{pmatrix} a_{11} \\ \vdots \\ a_{m1} \end{pmatrix} + c_2 \begin{pmatrix} a_{12} \\ \vdots \\ a_{m2} \end{pmatrix} + \cdots + c_n \begin{pmatrix} a_{1n} \\ \vdots \\ a_{mn} \end{pmatrix} = \begin{pmatrix} b_1 \\ \vdots \\ b_m \end{pmatrix}.$$

But to say that there exist real numbers $c_1, \ldots, c_n$ satisfying this matrix equation

is the same as saying that $\begin{pmatrix} b_1 \\ \vdots \\ b_m \end{pmatrix}$ is a linear combination of the column vectors of A;

that is, that $\begin{pmatrix} b_1 \\ \vdots \\ b_m \end{pmatrix}$ is in the column space of A. ■

The row space of a matrix has an interesting property that will be important later. We say that two $m \times n$ matrices A and B are **row equivalent** if B can be obtained from A by a finite sequence of row operations. Thus, for example, each matrix A is row equivalent to its row echelon matrix. Note that row equivalence is a symmetric relation; that is, if B can be obtained from A by a finite sequence of row operations, then A can also be obtained from B by a finite sequence of row operations. This is because each row operation is reversible: we can undo the effect of any row operation by applying another row operation of the same type.

Theorem 2.   *Suppose A and B are row equivalent $m \times n$ matrices. Then A and B have the same row space. In particular, each matrix has the same row space as its row echelon matrix.*

Proof   It suffices to check that if B can be obtained from A by a row operation, then A and B have the same row space. Suppose $\mathbf{a}_1, \ldots, \mathbf{a}_m$ are the row vectors of A. If B is obtained from A by multiplying the ith row vector $\mathbf{a}_i$ by a nonzero scalar c, then

   (i)   each vector $\mathbf{b}$ in the row space of B is of the form

$$\mathbf{b} = c_1\mathbf{a}_1 + \cdots + c_i(c\mathbf{a}_i) + \cdots + c_m\mathbf{a}_m$$
$$= c_1\mathbf{a}_1 + \cdots + (cc_i)\mathbf{a}_i + \cdots + c_m\mathbf{a}_m$$

so $\mathbf{b}$ is also in the row space of A, and

   (ii)   each vector $\mathbf{a}$ in the row space of A is of the form

$$\mathbf{a} = c_1\mathbf{a}_1 + \cdots + c_i\mathbf{a}_i + \cdots + c_m\mathbf{a}_m$$
$$= c_1\mathbf{a}_1 + \cdots + \left(\frac{c_i}{c}\right)(c\mathbf{a}_i) + \cdots + c_m\mathbf{a}_m$$

so $\mathbf{a}$ is also in the row space of B. Thus if B is obtained from A by this type of row operation, then A and B have the same row space.

The proofs for the other two types of row operations are similar and are left for the exercises (see Exercise 15). ■

Since a subspace $\mathcal{L}(\mathbf{v}_1, \ldots, \mathbf{v}_k)$ spanned by a finite set of vectors is relatively easy to understand, it is reasonable to ask if a subspace W of a vector space V

can be described as $\mathcal{L}(v_1, \ldots, v_k)$ for some choice of vectors $v_1, \ldots, v_k$. In other words, do there exist vectors $v_1, \ldots, v_k$ in W such that every vector in W is a linear combination of $\{v_1, \ldots, v_k\}$? If such vectors $v_1, \ldots, v_k$ do exist, then we say that $\{v_1, \ldots, v_k\}$ *spans* W, or that $\{v_1, \ldots, v_k\}$ is a *spanning set* for the subspace W.

Example 5. $\{(1, -1, 1), (1, 1, -1), (1, 1, 1)\}$ spans $\mathbb{R}^3$. To verify this we must show that, given any vector (a_1, a_2, a_3) in $\mathbb{R}^3$, we can find real numbers c_1, c_2, c_3 such that

$$(a_1, a_2, a_3) = c_1(1, -1, 1) + c_2(1, 1, -1) + c_3(1, 1, 1),$$

i.e., such that

$$c_1 + c_2 + c_3 = a_1$$
$$-c_1 + c_2 + c_3 = a_2$$
$$c_1 - c_2 + c_3 = a_3.$$

Reducing the matrix for this system we find that

$$\begin{pmatrix} 1 & 1 & 1 & a_1 \\ -1 & 1 & 1 & a_2 \\ 1 & -1 & 1 & a_3 \end{pmatrix} \rightarrow \begin{pmatrix} 1 & 0 & 0 & \dfrac{a_1 - a_2}{2} \\ 0 & 1 & 0 & \dfrac{a_1 - a_3}{2} \\ 0 & 0 & 1 & \dfrac{a_2 + a_3}{2} \end{pmatrix}$$

and hence the system is satisfied if

$$c_1 = \frac{a_1 - a_2}{2}, \ c_2 = \frac{a_1 - a_3}{2}, \ c_3 = \frac{a_2 + a_3}{2}.$$

Thus

$$(a_1, a_2, a_3) = \frac{a_1 - a_2}{2}(1, -1, 1) + \frac{a_1 - a_3}{2}(1, 1, -1) + \frac{a_2 + a_3}{2}(1, 1, 1)$$

is a linear combination of the given vectors, as asserted.   ■

Example 6. $\{1, x, x^2\}$ is a spanning set for the space $\mathcal{P}^3$ of polynomials of degree less than 3. This is because every polynomial $p(x)$ of degree less than 3 is of the form

$$p(x) = a_0 \cdot 1 + a_1 x + a_2 x^2.$$

Notice, however, that $\{1, x, x^2\}$ is not the only spanning set for $\mathcal{P}^3$. For example $\{x - 1, x + 1, x^2\}$ also spans $\mathcal{P}^3$ since each polynomial $a_0 + a_1 x + a_2 x^2$ in $\mathcal{P}^3$ can be expressed in the form

$$a_0 + a_1 x + a_2 x^2 = \frac{a_1 - a_0}{2}(x - 1) + \frac{a_1 + a_0}{2}(x + 1) + a_2 x^2.$$

Furthermore, notice that not every spanning set for $\mathcal{P}^3$ has three elements, as do the sets $\{1, x, x^2\}$ and $\{x - 1, x + 1, x^2\}$. In fact, we can construct arbitrarily large spanning sets for $\mathcal{P}^3$ just by adjoining to a given spanning set, say $\{1, x, x^2\}$, as many additional elements of $\mathcal{P}^3$ as we wish. Thus, for example, $\{1, x, x^2, x^2 - x, x^2 + x\}$ is a five-element spanning set for $\mathcal{P}^3$, since each element of $\mathcal{P}^3$ can be expressed in the form

$$a_0 + a_1 x + a_2 x^2 = a_0 + a_1 x + a_2 x^2 + 0(x^2 - x) + 0(x^2 + x). \quad \blacksquare$$

It is important to be aware of the fact that not every vector space has a finite spanning set. Consider, for example, the space $\mathcal{P}$ of all polynomials. We shall show that no finite set of elements of $\mathcal{P}$ can possibly span $\mathcal{P}$. Indeed, given $\{p_1(x), \ldots, p_k(x)\}$ in $\mathcal{P}$, let d be the degree of a polynomial of largest degree in $\{p_1(x), \ldots, p_k(x)\}$. Then every linear combination of the polynomials $\{p_1(x), \ldots, p_k(x)\}$ will necessarily have degree less than or equal to d. Thus no polynomial of degree greater than d can be in the set $\mathcal{L}(p_1(x), \ldots, p_k(x))$, and so $\{p_1(x), \ldots, p_k(x)\}$ cannot span $\mathcal{P}$.

EXERCISES

1. Find an equation $\mathbf{d} \cdot \mathbf{x} = 0$ describing the given plane.
    (a) $\mathcal{L}((1, 0, 0), (0, 1, 0))$
    (b) $\mathcal{L}((-1, 1, 0), (0, -1, 1))$
    (c) $\mathcal{L}((-2, 1, 3), (-1, 0, 3))$
    (d) $\mathcal{L}((1, 1, 1), (-1, 0, 1))$
    (e) $\mathcal{L}((-1, 5, 3), (2, 3, 1))$

2. In $\mathbb{R}^3$, which of the following vectors are in $\mathcal{L}((1, 1, 2), (1, -1, 3))$?
    (a) $(1, 1, 2)$ (c) $(0, 2, 1)$
    (b) $(2, 0, 5)$ (d) $(1, 3, 1)$

3. In $\mathcal{P}$, which of the following polynomials are in $\mathcal{L}(x^2 + x + 1, x^2 - 1, x + 2)$?
    (a) x^3
    (b) $x^2 + 3x + 5$
    (c) $x + 1$

4. In $\mathfrak{M}_{2 \times 2}$, which of the following matrices are in $\mathcal{L}\left(\begin{pmatrix} 1 & 0 \\ 0 & 1 \end{pmatrix}, \begin{pmatrix} 1 & 0 \\ 0 & -1 \end{pmatrix}\right)$?
    (a) $\begin{pmatrix} 1 & 0 \\ 0 & 0 \end{pmatrix}$ (b) $\begin{pmatrix} 0 & 1 \\ 0 & 0 \end{pmatrix}$ (c) $\begin{pmatrix} 0 & 0 \\ 0 & 1 \end{pmatrix}$

5. In $\mathcal{F}(\mathbb{R})$, which of the following functions are in $\mathcal{L}(1, \cos x, \cos 2x, \cos 3x)$?
    [Hint: Notice that any linear combination of $\{1, \cos x, \cos 2x, \cos 3x\}$ must
    satisfy the relation $f(-x) = f(x)$.]
    (a) $\sin x$ (d) $\cos^2 x$
    (b) $\cos x$ (e) $\sin^3 x$
    (c) $\sin^2 x$ (f) $\cos^3 x$

6. Show that $\{(a_1, a_2), (b_1, b_2)\}$ spans $\mathbb{R}^2$ if and only if $a_1 b_2 - a_2 b_1 \neq 0$.

7. Which of the following are spanning sets for $\mathbb{R}^3$?
    (a) $\{(1, 0, 0), (0, 1, 0), (0, 0, 1)\}$
    (b) $\{(1, 0, 0), (1, 1, 0), (1, 1, 1)\}$
    (c) $\{(1, -1, 1), (-1, 2, 0), (-1, 3, 1)\}$

8. Find a spanning set for the space $\mathcal{P}^4$ of all polynomials of degree <4.

9. Find a spanning set for the space $\mathcal{P}^n$ of all polynomials of degree $<n$.

10. Find a spanning set for $\mathbb{R}^n$.

11. Let W be the subset of $\mathbb{R}^\infty$ consisting of those $(a_1, a_2, a_3, \ldots)$ with only finitely
    many $a_i \neq 0$.
    (a) Show that W is a subspace of $\mathbb{R}^\infty$.
    (b) Show that W has no finite spanning set.

12. Find a spanning set for the space of all solutions of the differential equation
    $y'' = 0$.

13. Show that if $\{v_1, \ldots, v_k\}$ spans V, then so does $\{v_1, \ldots, v_k, v_{k+1}\}$ for any vector
    v_{k+1} in V.

14. Show that if $\{v_1, \ldots, v_k\}$ spans V and v_k is a linear combination of
    $\{v_1, \ldots, v_{k-1}\}$, then $\{v_1, \ldots, v_{k-1}\}$ also spans V.

15. Complete the proof of Theorem 2 by showing that if B is obtained from A
    either by interchanging two rows or by adding a scalar multiple of one row to
    another, then A and B have the same row space.

16. Let V be a vector space and let $\{v_1, \ldots, v_k\}$ be vectors in V. Show that
    $\mathcal{L}(v_1, \ldots, v_k)$ is equal to the intersection of the collection of all subspaces of V
    that contain $\{v_1, \ldots, v_k\}$.

2.4 SOLUTION SETS OF LINEAR EQUATIONS

In this section we apply the ideas of the previous sections to study the
structure of solution sets of systems of linear equations. We consider first ***homogeneous systems***—that is, systems of equations of the form

$$a_{11}x_1 + a_{12}x_2 + \cdots + a_{1n}x_n = 0$$

(1)
$$\cdots$$

$$a_{m1}x_1 + a_{m2}x_2 + \cdots + a_{mn}x_n = 0$$

(all zeros on the right hand side). Using the dot product, we can write this system more compactly as

$$\mathbf{a}_1 \cdot \mathbf{x} = 0$$
$$\vdots$$
$$\mathbf{a}_m \cdot \mathbf{x} = 0$$

(1')

where $\mathbf{a}_1, \ldots, \mathbf{a}_m$ are the row vectors of the coefficient matrix

$$\begin{pmatrix} a_{11} & \cdots & a_{1n} \\ & \cdots & \\ a_{m1} & \cdots & a_{mn} \end{pmatrix}$$

and $\mathbf{x} = (x_1, \ldots, x_n)$.

Our first observation is that *the solution set S of (1) is a subspace of* $\mathbb{R}^n$. Indeed:

   (i)   The zero vector $\mathbf{0} = (0, \ldots, 0)$ certainly satisfies (1), so $\mathbf{0} \in S$.
   (ii) If $\mathbf{x} \in S$ and $\mathbf{y} \in S$, then $\mathbf{a}_i \cdot \mathbf{x} = 0$ and $\mathbf{a}_i \cdot \mathbf{y} = 0$ for each i $(1 \le i \le m)$, so $\mathbf{a}_i \cdot (\mathbf{x} + \mathbf{y}) = 0$ for each i, and hence $\mathbf{x} + \mathbf{y} \in S$.
   (iii) If $\mathbf{x} \in S$ and $c \in \mathbb{R}$, then for each i $(1 \le i \le m)$,

$$\mathbf{a}_i \cdot (c\mathbf{x}) = c(\mathbf{a}_i \cdot \mathbf{x}) = c0 = 0,$$

   so $c\mathbf{x} \in S$.

The subspace S is called the **solution space** of the homogeneous system (1). The techniques of Sections 1.1 and 1.2 allow us to find a spanning set for this space. Recall the procedure for finding the general solution of (1). First we write down the matrix of the system:

$$\begin{pmatrix} a_{11} & a_{12} & \cdots & a_{1n} & 0 \\ a_{21} & a_{22} & \cdots & a_{2n} & 0 \\ & & \cdots & & \\ a_{m1} & a_{m2} & \cdots & a_{mn} & 0 \end{pmatrix}.$$

This matrix is sometimes called the **augmented matrix** in order to distinguish it from the **coefficient matrix** of the system, which omits the last column. By performing a sequence of row operations on this matrix we obtain a matrix in row echelon form

$$\begin{pmatrix} 1 & * & 0 & * & 0 & & 0 & * & 0 \\ & & 1 & * & 0 & & 0 & * & 0 \\ & & & & 1 & & 0 & * & 0 \\ & & & & & \ddots & \vdots & & 0 \\ & & & & & & 1 & * & 0 \end{pmatrix}$$

where the asterisks * denote possibly nonzero entries. The system of equations that has this as its matrix has the same solution set as (1), but now the solutions are immediate: if the corner 1's are in the k_1st, k_2nd, ..., k_rth columns, then we can solve for the unknowns $x_{k_1}, \ldots, x_{k_r}$ in terms of the remaining unknowns, whose values may be assigned arbitrarily. Usually we assign to these remaining unknowns the values $c_1, c_2, \ldots, c_{n-r}$ (there are $n - r$ of them since the total number of unknowns is n). Then we factor out the c_i's and write the general solution in the form

$$\mathbf{x} = c_1\mathbf{v}_1 + c_2\mathbf{v}_2 + \cdots + c_{n-r}\mathbf{v}_{n-r}.$$

We conclude that for any choice of real numbers $c_1, \ldots, c_{n-r}$ we get a solution $\mathbf{x} = c_1\mathbf{v}_1 + \cdots + c_{n-r}\mathbf{v}_{n-r}$ of (1), and that every solution of (1) is of this form for some choice of real numbers $c_1, \ldots, c_{n-r}$. Moreover, the set $\{\mathbf{v}_1, \ldots, \mathbf{v}_{n-r}\}$ is uniquely determined by the procedure. We call $\{\mathbf{v}_1, \ldots, \mathbf{v}_{n-r}\}$ the **canonical spanning set** for the solution space S. The number r of corner 1's, which is equal to the number of nonzero rows in the row echelon matrix, is called the **rank** of the system. Intuitively, r measures the number of "independent" equations in the system.

Example 1. Consider the system

$$\begin{aligned}
x_1 - x_2 + x_3 + x_4 - 2x_5 &= 0 \\
-2x_1 + 2x_2 - x_3 \qquad\quad + x_5 &= 0 \\
x_1 - x_2 + 2x_3 + 3x_4 - 5x_5 &= 0.
\end{aligned}$$

The matrix for this system reduces as follows:

$$\begin{pmatrix} 1 & -1 & 1 & 1 & -2 & 0 \\ -2 & 2 & -1 & 0 & 1 & 0 \\ 1 & -1 & 2 & 3 & -5 & 0 \end{pmatrix} \rightarrow \begin{pmatrix} 1 & -1 & 0 & -1 & 1 & 0 \\ 0 & 0 & 1 & 2 & -3 & 0 \\ 0 & 0 & 0 & 0 & 0 & 0 \end{pmatrix}$$

Thus the system has rank 2 and its solution space will have $3 = 5 - 2$ vectors in its canonical spanning set. The system corresponding to the reduced matrix is

$$\begin{aligned}
x_1 - x_2 \qquad - x_4 + x_5 &= 0 \\
x_3 + 2x_4 - 3x_5 &= 0 \\
0 &= 0
\end{aligned}$$

with general solution

$$\begin{aligned}
\mathbf{x} &= (c_1 + c_2 - c_3, c_1, -2c_2 + 3c_3, c_2, c_3) \\
&= c_1(1, 1, 0, 0, 0) + c_2(1, 0, -2, 1, 0) + c_3(-1, 0, 3, 0, 1)
\end{aligned}$$

obtained by setting $x_2 = c_1$, $x_4 = c_2$, $x_5 = c_3$, and solving for x_1 and x_3. Thus the canonical spanning set is $\{(1, 1, 0, 0, 0), (1, 0, -2, 1, 0), (-1, 0, 3, 0, 1)\}$. ∎

We summarize the results of this discussion in a theorem.

Theorem 1. *The set of solutions of the homogeneous linear system*

$$a_{11}x_1 + a_{12}x_2 + \cdots + a_{1n}x_n = 0$$

$$\cdots$$

$$a_{m1}x_1 + a_{m2}x_2 + \cdots + a_{mn}x_n = 0$$

is a subspace of $\mathbb{R}^n$. There is a spanning set for this subspace consisting of $n - r$ nonzero vectors, where r is the rank of the system. In particular, if $r < n$ then the system has a nonzero solution.

Corollary. *Every homogeneous system of m linear equations in n unknowns with $n > m$ has a nonzero solution.*

Proof The rank r of the system is less than or equal to the number m of equations, so $r \leq m < n$. ∎

Consider now the system

$$a_{11}x_1 + a_{12}x_2 + \cdots + a_{1n}x_n = b_1$$

(2)
$$\cdots$$

$$a_{m1}x_1 + a_{m2}x_2 + \cdots + a_{mn}x_n = b_m.$$

Unless all the b_i's are zero, the zero vector in $\mathbb{R}^n$ is not a solution of (2) and hence the solution set of (2) is not a subspace of $\mathbb{R}^n$. However, we can associate with (2) the homogeneous system

$$a_{11}x_1 + a_{12}x_2 + \cdots + a_{1n}x_n = 0$$

(1)
$$\cdots$$

$$a_{m1}x_1 + a_{m2}x_2 + \cdots + a_{mn}x_n = 0$$

obtained by replacing each b_i by 0. The relation between the solution sets of (1) and (2) is described in the following theorem.

Theorem 2. *Suppose **u** is a solution of (2). Then the solution set of (2) is precisely the set of all vectors in $\mathbb{R}^n$ of the form $\mathbf{x} = \mathbf{u} + \mathbf{v}$, where **v** is a solution of the associated homogeneous system (1) (see Figure 2.6). In particular, the general solu-*

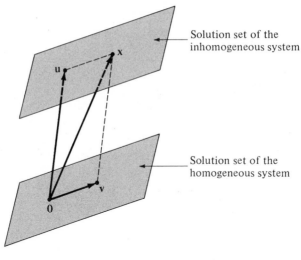

Solution set of the inhomogeneous system

Solution set of the homogeneous system

FIGURE 2.6

tion of (2) *can be written in the form*

$$\mathbf{x} = \mathbf{u} + c_1\mathbf{v}_1 + c_2\mathbf{v}_2 + \cdots + c_{n-r}\mathbf{v}_{n-r}$$

where $\mathbf{v}_1, \ldots, \mathbf{v}_{n-r}$ *is the canonical spanning set for the solution space of* (1).

Proof Suppose $\mathbf{x}$ is any solution of (2). Let $\mathbf{v} = \mathbf{x} - \mathbf{u}$. Then for each $i(1 \le i \le m)$,

$$\mathbf{a}_i \cdot \mathbf{v} = \mathbf{a}_i \cdot (\mathbf{x} - \mathbf{u}) = \mathbf{a}_i \cdot \mathbf{x} - \mathbf{a}_i \cdot \mathbf{u} = b_i - b_i = 0$$

where $\mathbf{a}_i$ is the *i*th row vector in the coefficient matrix for (2). Thus $\mathbf{v}$ is a solution of (1), and $\mathbf{x} = \mathbf{u} + \mathbf{v}$.

Conversely, if $\mathbf{v}$ is any solution of (1), then for each i,

$$\mathbf{a}_i \cdot (\mathbf{u} + \mathbf{v}) = \mathbf{a}_i \cdot \mathbf{u} + \mathbf{a}_i \cdot \mathbf{v} = b_i + 0 = b_i,$$

so $\mathbf{x} = \mathbf{u} + \mathbf{v}$ is a solution of (2).

Thus the solution set of (2) is precisely the set of all vectors $\mathbf{x}$ of the form $\mathbf{x} = \mathbf{u} + \mathbf{v}$ where $\mathbf{v}$ is a solution of (1). But we already know that every solution of (1) is of the form

$$\mathbf{v} = c_1\mathbf{v}_1 + \cdots + c_{n-r}\mathbf{v}_{n-r}$$

where $\{\mathbf{v}_1, \ldots, \mathbf{v}_{n-r}\}$ is the canonical spanning set for the solution space of (1). We conclude, then, that the general solution of (2) is

$$\mathbf{x} = \mathbf{u} + c_1\mathbf{v}_1 + \cdots + c_{n-r}\mathbf{v}_{n-r}. \quad \blacksquare$$

Remark 1. Note that Theorem 2 applies only in the case when the system (2) is consistent. Otherwise, of course, the solution set of (2) is the empty set; that is, there are no solutions.

Remark 2. The solution procedure described in Sections 1.1 and 1.2 leads directly to the general solution of (2) in the form $\mathbf{x} = \mathbf{u} + c_1\mathbf{v}_1 + \cdots + c_{n-r}\mathbf{v}_{n-r}$ where $\mathbf{u}$ is a particular solution of (2) and $\mathbf{v}_1, \ldots, \mathbf{v}_{n-r}$ is the canonical spanning set for the solution space of (1).

Example 2. (Compare this example with Example 1 above.) Consider the system

$$\begin{aligned}
x_1 - x_2 + x_3 + x_4 - 2x_5 &= -1 \\
-2x_1 + 2x_2 - x_3 + x_5 &= 2 \\
x_1 - x_2 + 2x_3 + 3x_4 - 5x_5 &= -1.
\end{aligned}$$

The matrix for this system reduces as follows:

$$\begin{pmatrix} 1 & -1 & 1 & 1 & -2 & -1 \\ -2 & 2 & -1 & 0 & 1 & 2 \\ 1 & -1 & 2 & 3 & -5 & -1 \end{pmatrix} \rightarrow \begin{pmatrix} 1 & -1 & 0 & -1 & 1 & -1 \\ 0 & 0 & 1 & 2 & -3 & 0 \\ 0 & 0 & 0 & 0 & 0 & 0 \end{pmatrix}.$$

The system corresponding to the reduced matrix is

$$\begin{aligned}
x_1 - x_2 - x_4 + x_5 &= -1 \\
x_3 + 2x_4 - 3x_5 &= 0 \\
0 &= 0
\end{aligned}$$

with general solution (obtained by setting $x_2 = c_1$, $x_4 = c_2$, and $x_5 = c_3$)

$$\mathbf{x} = (-1 + c_1 + c_2 - c_3, c_1, -2c_2 + 3c_3, c_2, c_3)$$
$$= (-1, 0, 0, 0, 0) + c_1(1, 1, 0, 0, 0) + c_2(1, 0, -2, 1, 0) + c_3(-1, 0, 3, 0, 1). \quad \blacksquare$$

EXERCISES

1. Find the rank of each of the following homogeneous systems and exhibit the canonical spanning set for each solution space.

   (a) $\quad x_1 + x_2 + x_3 = 0$
   $$2x_1 - x_2 - 7x_3 = 0$$

(b) $x_1 + 6x_2 + 2x_3 - 5x_4 = 0$
       $-x_1 - 6x_2 - x_3 - 3x_4 = 0$
       $2x_1 + 12x_2 + 5x_3 - 18x_4 = 0$

(c) $x_1 + x_2 - x_3 = 0$
       $x_1 \quad + 2x_3 = 0$
       $-x_1 + x_2 + 3x_3 = 0$

(d) $x_1 + 2x_2 \quad - 9x_4 - x_5 = 0$
       $-x_1 - x_2 - x_3 + x_4 - 5x_5 = 0$
       $x_1 + 2x_2 + x_3 - 7x_4 + 4x_5 = 0$

2. Exhibit the general solution of each of the following systems in the form $\mathbf{x} = \mathbf{u} + c_1\mathbf{v}_1 + \cdots + c_{n-r}\mathbf{v}_{n-r}$, where $\mathbf{v}_1, \ldots, \mathbf{v}_{n-r}$ is the canonical spanning set for the solution space of the associated homogeneous system.

(a) $x_1 + 6x_2 + 2x_3 - 5x_4 = -2$
       $-x_1 - 6x_2 - x_3 - 3x_4 = 7$
       $2x_1 + 12x_2 + 5x_3 - 18x_4 = 1$

(b) $x_1 + 2x_2 \quad - 9x_4 - x_5 = 3$
       $-x_1 - x_2 - x_3 + x_4 - 5x_5 = 0$
       $x_1 + 2x_2 + x_3 - 7x_4 + 4x_5 = -8$

3. The equation of a plane through the origin in $\mathbb{R}^3$ may be viewed as a "system" of one equation in three unknowns. Use the method described in this section to find a spanning set for each of the following planes:

(a) $x_1 + x_2 + x_3 = 0$ (c) $2x_1 - 3x_2 + 4x_3 = 0$
(b) $x_2 + x_3 = 0$ (d) $x_3 = 0$

4. Find a spanning set for the hyperplane
$$x_1 + 3x_2 - x_3 - 2x_4 + 4x_5 = 0$$
in $\mathbb{R}^5$.

5. Show that a system of linear equations is consistent if and only if the rank of its coefficient matrix equals the rank of its augmented matrix. (By **rank of a matrix** we mean the number of nonzero rows (= the number of corner 1's) in its row echelon matrix.)

Linear Differential Equations

THREE

3.1 CONSTANT COEFFICIENT LINEAR DIFFERENTIAL EQUATIONS

In Chapter 1 we studied systems of linear *algebraic* equations. We learned how to solve these systems and, in Chapter 2, we learned about the structure of the solution sets. In this chapter, we shall study linear *differential* equations. We shall develop methods for solving some of these equations and study the structure of the solution sets. As we will see, solution sets of linear differential equations have many features in common with solution sets of linear algebraic systems.

We begin by studying *linear differential equations with constant coefficients;* that is, equations of the form

$$a_n y^{(n)} + a_{n-1} y^{(n-1)} + \cdots + a_1 y' + a_0 y = f$$

where $a_0, a_1, \ldots, a_n \in \mathbb{R}$, y is an unknown function, $y^{(k)}$ is its kth derivative, and f is some specified function. If the function f is the zero function, the equation is *homogeneous.* The *order* of the equation is the order of the highest derivative that appears in it. Thus, if $a_n \neq 0$, the equation above has order n.

The following are examples of linear differential equations with constant coefficients:

(1) $$y' + y = x$$

(2) $$y'' - y' - 6y = 0$$

(3) $$8y''' + y' = \sin x.$$

The orders of these equations are, respectively, 1, 2, and 3. Equation (2) is homogeneous.

We shall assume for the time being that the function f is defined and continuous on $\mathbb{R}$. A *solution* of the equation

$$a_n y^{(n)} + \cdots + a_1 y' + a_0 y = f$$

is then a function $y \in \mathcal{F}(\mathbb{R})$ that is n times differentiable and that satisfies the equation. Thus, for example, $y = x - 1$ is a solution of Equation (1), since

$$(x - 1)' + (x - 1) = 1 + (x - 1) = x,$$

73

and $y = e^{3x}$ is a solution of Equation (2), since

$$(e^{3x})'' - (e^{3x})' - 6e^{3x} = 9e^{3x} - 3e^{3x} - 6e^{3x} = 0.$$

The following theorem describes the structure of the solution sets. Compare this theorem with Theorems 1 and 2 of Section 2.4. We shall denote by $\mathcal{C}(\mathbb{R})$ the vector space of all continuous functions with domain $\mathbb{R}$. Note that $\mathcal{C}(\mathbb{R})$ is a subspace of $\mathcal{F}(\mathbb{R})$ (why?).

Theorem.   *The set of solutions of the homogeneous linear differential equation with constant coefficients*

(A) $$a_n y^{(n)} + \cdots + a_1 y' + a_0 y = 0$$

is a subspace of $\mathcal{F}(\mathbb{R})$. Moreover, if $f \in \mathcal{C}(\mathbb{R})$ and if y_P is any solution of

(B) $$a_n y^{(n)} + \cdots + a_1 y' + a_0 y = f,$$

then the solution set of (B) *is precisely the set of all functions in $\mathcal{F}(\mathbb{R})$ of the form $y = y_P + y_H$ where y_H is a solution of the homogeneous equation* (A). *In particular, if $\{y_1, \ldots, y_k\}$ is a spanning set for the solution space of* (A), *then the solution set of* (B) *consists of all functions of the form*

$$y = y_P + c_1 y_1 + \cdots + c_k y_k.$$

Before indicating the proof of the theorem we shall give some examples and make several remarks.

Example 1.   Consider the equation

$$y''' = e^x.$$

This equation is easily solved by three integrations: $y'' = e^x + c_1$, $y' = e^x + c_1 x + c_2$, and, finally,

$$y = e^x + c_1 \frac{x^2}{2} + c_2 x + c_3.$$

Note that $y_P = e^x$ is a particular solution of $y''' = e^x$ and that

$$y_H = c_1 \frac{x^2}{2} + c_2 x + c_3 \cdot 1$$

is the general solution of the homogeneous equation $y''' = 0$. The set $\left\{\dfrac{x^2}{2}, x, 1\right\}$ is a spanning set for the solution space of $y''' = 0$. Note that $\{x^2, x, 1\}$ is another spanning set for the same solution space. (Why?) ∎

Remark. The word "general" is used in linear differential equations exactly as it is used in linear algebraic equations: $y = y_P + c_1y_1 + \cdots + c_ky_k$ is the **general solution** of a given linear differential equation if every solution of the equation is of this form for some choice of $c_1, \ldots, c_k \in \mathbb{R}$.

Example 2. Consider the equation

$$y' + y = e^{-x}.$$

This equation becomes simpler if we multiply both sides by e^x to get

$$e^xy' + e^xy = 1,$$

for then the left hand side is equal to the derivative of e^xy so the equation becomes $(e^xy)' = 1$. Integrating both sides yields $e^xy = x + c$, or

$$y = xe^{-x} + ce^{-x}.$$

Note that xe^{-x} is a particular solution of the given equation and that $\{e^{-x}\}$ is a spanning set for the solution space of the homogeneous equation $y' + y = 0$. ■

Example 3. Consider the equation

$$y'' + y' = 2.$$

One solution of this equation can be discovered by inspection: all we need is to find a function y_P with $y_P' = 2$, for then $y_P'' = 0$ so $y_P'' + y_P' = 2$ as required. The function $y_P = 2x$ will do.

Having found one solution of $y'' + y' = 2$, we can find all solutions by solving the homogeneous equation $y'' + y' = 0$ and using the theorem. But this homogeneous equation can be solved by a simple substitution. Let $u = y'$. Then the equation becomes $u' + u = 0$ with solution (from Example 2) $u = c_1e^{-x}$. Since $y' = u = c_1e^{-x}$, we find that

$$y = -c_1e^{-x} + c_2$$

is the general solution of the homogeneous equation $y'' + y' = 0$. Thus, the set $\{e^{-x}, 1\}$ is a spanning set for the solution space of $y'' + y' = 0$, and

$$y = 2x + c_1e^{-x} + c_2$$

is the general solution of the equation $y'' + y' = 2$. ■

Remarks. (i) Note that the general solution of the homogeneous equation $y'' + y' = 0$ in Example 3 can be written either as $y = -c_1e^{-x} + c_2$ or as $y = c_1e^{-x} + c_2$. This is because c_1 represents an arbitrary real number, and this number can be called $-c_1$ rather than c_1 if we wish.

(ii) Note that the theorem describes a strategy for finding the general solution of a linear differential equation with constant coefficients

$$a_n y^{(n)} + \cdots + a_1 y' + a_0 y = f.$$

Namely, (1) find one *particular solution* y_P of the equation, and (2) find the general solution y_H of the *associated homogeneous equation*

$$a_n y^{(n)} + \cdots + a_1 y' + a_0 y = 0.$$

The general solution of the given equation will then be

$$y = y_P + y_H.$$

(iii) As indicated in the theorem, if $\{y_1, \ldots, y_k\}$ is a spanning set for the solution space of the homogeneous equation

$$a_n y^{(n)} + \cdots + a_1 y' + a_0 y = 0,$$

then the general solution y_H of this homogeneous equation will be

$$y_H = c_1 y_1 + \cdots + c_k y_k.$$

Now, let us see why the theorem is true.

Proof of the Theorem. For simplicity, let us assume that the order n of the equation is 2. First, we want to verify that the solution set of

(A) $$ay'' + by' + cy = 0 \quad (a, b, c \in \mathbb{R})$$

is a subspace of $\mathcal{F}(\mathbb{R})$. Note first that the zero function $y = 0$ certainly satisfies (A). Second, if y_1 and y_2 are solutions of (A), then

$$ay_1'' + by_1' + cy_1 = 0$$

and

$$ay_2'' + by_2' + cy_2 = 0$$

so, adding these equations, we find

$$a(y_1 + y_2)'' + b(y_1 + y_2)' + c(y_1 + y_2) = 0;$$

that is, $y_1 + y_2$ is a solution. Finally, if y is a solution of (A) and r is any real number, then

$$a(ry)'' + b(ry)' + c(ry) = r(ay'' + by' + cy) = r \cdot 0 = 0$$

so ry is also a solution. Thus, the solution set of (A) is indeed a subspace.

Next, we want to show that if y_P is any solution of

(B) $$ay'' + by' + cy = f \quad (a, b, c \in \mathbb{R})$$

then $y_P + y_H$ is also a solution of (B) whenever y_H is a solution of (A). But

$$a(y_P + y_H)'' + b(y_P + y_H)' + c(y_P + y_H)$$
$$= (ay_P'' + by_P' + cy_P) + (ay_H'' + by_H' + cy_H)$$
$$= f + 0 = f,$$

so this assertion is true.

Finally, we must show that *every* solution of (B) is of the form $y_P + y_H$. So let y be any solution of (B), and substitute the function $y - y_P$ into the left side of (A) to get

$$a(y - y_P)'' + b(y - y_P)' + c(y - y_P)$$
$$= (ay'' + by' + cy) - (ay_P'' + by_P' + cy_P)$$
$$= f - f = 0.$$

This shows that $y_H = y - y_P$ is a solution of (A), and $y = y_P + y_H$, as asserted. The proof for equations of arbitrary order is similar. ∎

We close this section by developing an algorithm for solving any first order linear differential equation with constant coefficients,

$$ay' + by = f, \quad a \neq 0.$$

First, rewrite the equation in the form

$$y' + ky = \frac{1}{a}f, \quad \text{where } k = b/a.$$

Then notice that if we multiply the left side of this equation by e^{kx} we get $e^{kx}y' + ke^{kx}y = (e^{kx}y)'$. The solution procedure now becomes clear:

(i) After dividing by the coefficient a, multiply both sides of the equation by the **integrating factor** e^{kx}, where $k = b/a$; the left hand side then becomes $(e^{kx}y)'$.

(ii) Integrate both sides and solve for y.

Example 4. Solve $y' + 2y = e^x$. Here $k = 2$ so the integrating factor is e^{2x}. Multiply both sides of the equation by e^{2x} to get

$$(e^{2x}y)' = e^{2x}e^x = e^{3x}.$$

Integrate:

$$e^{2x}y = \tfrac{1}{3}e^{3x} + c.$$

Solve for y:

$$y = \tfrac{1}{3}e^x + ce^{-2x}.$$

Note that $\tfrac{1}{3}e^x$ is a particular solution of the equation $y' + 2y = e^x$ and that $\{e^{-2x}\}$ is a spanning set for the solution space of the associated homogeneous equation $y' + 2y = 0$. ■

Example 5. Solve $2y' - y = 2e^{\frac{1}{2}x}$. First we must divide by the leading coefficient 2 to get

$$y' - \tfrac{1}{2}y = e^{\frac{1}{2}x}.$$

Multiplying by the integrating factor $e^{-\frac{1}{2}x}$ yields

$$(e^{-\frac{1}{2}x}y)' = 1$$

so

$$e^{-\frac{1}{2}x}y = x + c$$

or

$$y = xe^{\frac{1}{2}x} + ce^{\frac{1}{2}x}. \quad ■$$

EXERCISES

Solve the following differential equations. Find a particular solution by inspection and then use the theorem of this section to describe the general solution.

1. $y' + y = 2$
2. $y' + y = 2x + x^2$
3. $y'' + y' = 5$
4. $y'' + y' = -\sin x + \cos x$
5. $y''' = e^{-x}$

Use an appropriate integrating factor to solve each of the following first order differential equations.

6. $y' - y = e^x$
7. $y' + 3y = 4e^x$
8. $y' + 2y = x$
9. $y' - 2y = e^{-x}$
10. $y' + y = \sin x$

11. $y' - y = e^{kx}$ [You will need to consider the cases $k = 1$ and $k \neq 1$ separately.]

12. $y' + \pi y = \cos \pi x$

13. $2y' + y = x^2$

14. $y' - 3y = c_1 e^{2x}$ (c_1 an arbitrary real number)

15. $y' - 3y = c_1 e^{3x}$ (c_1 an arbitrary real number)

16. The uninhibited growth of a bacteria population is described by the differential equation

$$\frac{dN}{dt} = kN \quad (k \in \mathbb{R})$$

where N is the number of bacteria in the population at time t.

(a) Find the general solution of this differential equation.

(b) Suppose there are 4×10^7 bacteria in the population at time t_0, for some $t_0 \in \mathbb{R}$, and 5×10^7 at time $t_0 + 1$. Use this information to find a numerical value for k.

17. Suppose the bacteria population in Exercise 16 is depleted by the removal of bacteria from the population at the constant rate of r bacteria per unit of time. The population growth is then described by the differential equation

$$\frac{dN}{dt} = kN - r \quad (k, r \in \mathbb{R}).$$

(a) Find the general solution of this differential equation.

(b) Make rough sketches of the graph of N in each of the four cases $r = 0$, $0 < r < kN_0$, $r = kN_0$, and $r > kN_0$, where N_0 is the number of bacteria in the initial population.

18. The temperature T in a house that is cooling due to loss of power is described by the differential equation

$$\frac{dT}{dt} = k(K - T)$$

where $K \in \mathbb{R}$ is the temperature outside the house and k is a constant that depends on how well the house is insulated.

(a) Find the general solution of this differential equation.

(b) If the outside temperature is $-5°C$ and if the house cools from $20°C$ to $15°C$ in five hours, how long will it take for the house temperature to drop from $15°C$ to $0°C$?

3.2 SECOND ORDER HOMOGENEOUS EQUATIONS

The theorem in the first section of this chapter gives us a two-step strategy for solving any linear differential equation with constant coefficients. One step in that process is to solve the associated homogeneous equation. In this section we will learn how to solve second order homogeneous linear differential equations with constant coefficients; that is, equations of the form

(1) $$ay'' + by' + cy = 0,$$

where $a, b, c \in \mathbb{R}$ and $a \neq 0$. If we denote differentiation by the letter D, so that $Dy = y'$ and $D^2y = D(Dy) = y''$, this equation can be rewritten in the form

$$aD^2y + bDy + cy = 0,$$

or

(2) $$(aD^2 + bD + c)y = 0.$$

The expression $aD^2 + bD + c$ is called the general **second order linear differential operator;** it is the rule that assigns to each twice differentiable function f the function $af'' + bf' + cf$.

The advantage of rewriting Equation (1) in the form (2) is that the operator $aD^2 + bD + c$ looks like a polynomial. Certainly there is a polynomial associated with the operator $aD^2 + bD + c$; namely, the polynomial $az^2 + bz + c$, where we have used z as the variable rather than x to avoid confusion. The polynomial $az^2 + bz + c$ is called the **auxiliary polynomial** associated with the differential equation $ay'' + by' + cy = 0$. We shall see that the solution of the differential equation is completely determined by the zeros of this auxiliary polynomial.

The important observation is that constant coefficient linear differential operators multiply just like polynomials. Consider the two first order constant coefficient operators $aD + b$ and $cD + d$. Their product $(aD + b)(cD + d)$ is the second order operator defined by

$$
\begin{aligned}
[(aD + b)(cD + d)]f &= (aD + b)[(cD + d)f] \\
&= (aD + b)(cf' + df) \\
&= af'' + (ad + bc)f' + bdf \\
&= [aD^2 + (ad + bc)D + bd]f;
\end{aligned}
$$

that is,

$$(aD + b)(cD + d) = aD^2 + (ad + bc)D + bd.$$

Notice that this *is* polynomial multiplication, since

$$(az + b)(cz + d) = az^2 + (ad + bc)z + bd.$$

The method for solving second order differential equations is to factor the auxiliary polynomial, if possible, into linear factors. This will correspond to factoring the second order differential operator into a product of two first order differential operators and will reduce the problem of solving the second order equation to a problem of solving two first order equations.

Suppose, then, that the auxiliary polynomial $az^2 + bz + c$ has real zeros λ_1 and λ_2. Then

$$az^2 + bz + c = a(z - \lambda_1)(z - \lambda_2)$$

and hence

$$aD^2 + bD + c = a(D - \lambda_1)(D - \lambda_2)$$

so our equation becomes

$$a(D - \lambda_1)(D - \lambda_2)y = 0$$

or

$$(D - \lambda_1)(D - \lambda_2)y = 0,$$

since $a \neq 0$.

Let $u = (D - \lambda_2)y$. Then $(D - \lambda_1)u = 0$, or $u' - \lambda_1 u = 0$. The general solution of this equation is $u = c_1 e^{\lambda_1 x}$. Since $u = (D - \lambda_2)y$, we now need to solve $(D - \lambda_2)y = c_1 e^{\lambda_1 x}$ or $y' - \lambda_2 y = c_1 e^{\lambda_1 x}$. An integrating factor for this equation is $e^{-\lambda_2 x}$ so, after multiplying by $e^{-\lambda_2 x}$, the equation becomes

$$(e^{-\lambda_2 x}y)' = c_1 e^{(\lambda_1 - \lambda_2)x}.$$

Integrate this to get

$$e^{-\lambda_2 x}y = \begin{cases} \dfrac{c_1}{\lambda_1 - \lambda_2}e^{(\lambda_1 - \lambda_2)x} + c_2, & \text{if } \lambda_1 \neq \lambda_2 \\[2mm] c_1 x + c_2, & \text{if } \lambda_1 = \lambda_2. \end{cases}$$

Now solve for y. Thus, the general solution of $(D - \lambda_1)(D - \lambda_2)y = 0$ is

$$y = \frac{c_1}{\lambda_1 - \lambda_2}e^{\lambda_1 x} + c_2 e^{\lambda_2 x} \quad \text{if } \lambda_1 \neq \lambda_2$$

or

$$y = c_1 x e^{\lambda_2 x} + c_2 e^{\lambda_2 x} \quad \text{if } \lambda_1 = \lambda_2.$$

If c_1 is an arbitrary constant, so is $\dfrac{c_1}{\lambda_1 - \lambda_2}$ $(\lambda_1 \neq \lambda_2)$, so we can summarize our result in the following way.

The set of solutions of the homogeneous second order differential equation $(D - \lambda_1)(D - \lambda_2)y = 0$ is the vector space spanned by

(i) $\{e^{\lambda_1 x}, e^{\lambda_2 x}\}$ if $\lambda_1 \neq \lambda_2$,
(ii) $\{e^{\lambda x}, xe^{\lambda x}\}$ if $\lambda_1 = \lambda_2 = \lambda$.

Example 1. Solve $(D^2 - D - 6)y = 0$. Since $D^2 - D - 6 = (D - 3)(D + 2)$, the solution space is spanned by $\{e^{3x}, e^{-2x}\}$. That is, the general solution is $y = c_1 e^{3x} + c_2 e^{-2x}$ where c_1 and c_2 are arbitrary real numbers. ∎

Example 2.   Solve $y'' - y = 0$. The operator form of the equation is $(D^2 - 1)y = 0$. Since $(D^2 - 1) = (D - 1)(D + 1)$, a spanning set for the solution space is $\{e^x, e^{-x}\}$. The general solution is $y = c_1 e^x + c_2 e^{-x}$. ∎

Example 3.   Solve $y'' - 4y' + 4y = 0$. The operator form is $(D^2 - 4D + 4)y = 0$, and $D^2 - 4D + 4 = (D - 2)^2$, so we are in case (ii). The solution space is spanned by $\{e^{2x}, xe^{2x}\}$. The general solution is $y = c_1 e^{2x} + c_2 x e^{2x}$. ∎

Notice that the numbers λ_1 and λ_2 that appear in our recipe for solving $ay'' + by' + cy = 0$ are just the zeros of the auxiliary polynomial $az^2 + bz + c = a(z - \lambda_1)(z - \lambda_2)$. Sometimes we must use the quadratic formula to find these zeros.

Example 4.   Solve $y'' + y' - y = 0$. By the quadratic formula, the zeros of the auxiliary polynomial $z^2 + z - 1$ are

$$\frac{-1 \pm \sqrt{1 + 4}}{2} = -\tfrac{1}{2} \pm \tfrac{1}{2}\sqrt{5}$$

Thus, the solution space of $y'' + y' - y = 0$ is spanned by

$$\{e^{(-\frac{1}{2}+\frac{1}{2}\sqrt{5})x}, e^{(-\frac{1}{2}-\frac{1}{2}\sqrt{5})x}\}.$$

The general solution is

$$y = c_1 e^{(-\frac{1}{2}+\frac{1}{2}\sqrt{5})x} + c_2 e^{(-\frac{1}{2}-\frac{1}{2}\sqrt{5})x}.$$ ∎

We now know how to solve $ay'' + by' + cy = 0$ whenever the auxiliary polynomial $az^2 + bz + c$ has real zeros. But what if the zeros are not real numbers? In this case, it is helpful to complete the square in the auxiliary polynomial. Since

$$az^2 + bz + c = a\left[z^2 + \frac{b}{a}z + \left(\frac{b}{2a}\right)^2 - \left(\frac{b}{2a}\right)^2 + \frac{c}{a}\right]$$

$$= a\left[\left(z + \frac{b}{2a}\right)^2 - \frac{b^2 - 4ac}{4a^2}\right]$$

we see that, when $b^2 - 4ac < 0$,

$$az^2 + bz + c = a[(z - \alpha)^2 + \beta^2],$$

and hence

$$aD^2 + bD + c = a[(D - \alpha)^2 + \beta^2],$$

where $\alpha = -b/2a$ and $\beta = \sqrt{-(b^2 - 4ac)}/2a$. Thus our differential equation becomes

$$a[(D - \alpha)^2 + \beta^2]y = 0$$

or

$$[(D - \alpha)^2 + \beta^2]y = 0.$$

Notice that $\alpha \pm \beta i$ are the zeros of the auxiliary polynomial $az^2 + bz + c = a[(z - \alpha)^2 + \beta^2]$.

The case $\alpha = 0$ is not difficult. The equation becomes

$$y'' + \beta^2 y = 0$$

where $\beta \neq 0$. It is easy to check that $\cos \beta x$ and $\sin \beta x$ are solutions and therefore so is $c_1 \cos \beta x + c_2 \sin \beta x$ for arbitrary real numbers c_1 and c_2. To see that these are the *only* solutions, notice that if y is any solution, then

$$y'' = -\beta^2 y.$$

If we multiply both sides of this equation by $2y'$, we get

$$2y'y'' = -\beta^2(2y'y)$$

or

$$\frac{d}{dx}((y')^2) = -\beta^2 \frac{d}{dx} y^2$$

so that

$$\frac{d}{dx}((y')^2 + \beta^2 y^2) = 0.$$

But this means that, for any solution y of $y'' + \beta^2 y = 0$, the function $(y')^2 + \beta^2 y^2$ is constant. (Its derivative is zero!) If we let $c_1 = y(0)$ and $c_2 = y'(0)/\beta$, then the function

$$u = y - (c_1 \cos \beta x + c_2 \sin \beta x)$$

not only satisfies $u'' + \beta^2 u = 0$ (why?), but it has the additional property that $u(0) = 0$ and $u'(0) = 0$. Since u satisfies the original equation, it has the property that $(u')^2 + \beta^2 u^2$ is constant, but since $u(0) = u'(0) = 0$, that constant must be zero. This means that u must be the constant function zero (since $(u')^2 + \beta^2 u^2$ can equal zero only when $u' = u = 0$), which in turn implies that

$$y - (c_1 \cos \beta x + c_2 \sin \beta x) = 0 \quad \text{for all } x,$$

hence, that

$$y = c_1 \cos \beta x + c_2 \sin \beta x.$$

Thus, *every solution of* $y'' + \beta^2 y = 0$ *is a linear combination of* $\cos \beta x$ *and* $\sin \beta x$.

Example 5.   Solve $(D^2 + 1)y = 0$. The auxiliary polynomial is $z^2 + 1$, whose zeros are $\pm i$. Thus, $\alpha = 0$ and $\beta = 1$. The solution space is spanned by $\{\cos x, \sin x\}$. The general solution is $y = c_1 \cos x + c_2 \sin x$.   ■

Finally, we deal with the remaining case, that of solving $[(D - \alpha)^2 + \beta^2]y = 0$ when $\alpha \neq 0$ and $\beta \neq 0$. We do this by making use of the identity

$$D^2(e^{-\alpha x}y) = e^{-\alpha x}(D - \alpha)^2 y$$

(see Exercise 23). If we multiply both sides of the equation

$$[(D - \alpha)^2 + \beta^2]y = 0$$

by $e^{-\alpha x}$ and use this identity, our equation becomes

$$D^2(e^{-\alpha x}y) + \beta^2(e^{-\alpha x}y) = 0.$$

Now, if we let $u = e^{-\alpha x}y$, the equation is just $u'' + \beta^2 u = 0$. The general solution u of this equation is $u = c_1 \cos \beta x + c_2 \sin \beta x$. Hence,

$$e^{-\alpha x}y = c_1 \cos \beta x + c_2 \sin \beta x$$

and

$$y = c_1 e^{\alpha x} \cos \beta x + c_2 e^{\alpha x} \sin \beta x.$$

Thus, *if the zeros of the auxiliary polynomial are complex numbers* $\alpha \pm \beta i (\beta \neq 0)$ *then* $\{e^{\alpha x} \cos \beta x, \ e^{\alpha x} \sin \beta x\}$ *is a spanning set for the solution space of* $ay'' + by' + cy = 0$.

Example 6.   Solve $y'' + y' + y = 0$. The auxiliary polynomial is $z^2 + z + 1$ and its zeros are

$$\frac{-1 \pm \sqrt{1 - 4}}{2} = -\frac{1}{2} \pm \frac{\sqrt{3}}{2}i.$$

Thus, the solution space is spanned by $\left\{ e^{-\frac{1}{2}x} \cos \dfrac{\sqrt{3}}{2}x, \ e^{-\frac{1}{2}x} \sin \dfrac{\sqrt{3}}{2}x \right\}$. The general solution is

$$y = c_1 e^{-\frac{1}{2}x} \cos \frac{\sqrt{3}}{2}x + c_2 e^{-\frac{1}{2}x} \sin \frac{\sqrt{3}}{2}x.   ■$$

Here is a summary of the results of this section.

To solve a homogeneous second order linear differential equation with constant coefficients, $ay'' + by' + cy = 0$, proceed as follows:

(1) Find the zeros of the auxiliary polynomial $az^2 + bz + c$, and call these zeros λ_1 and λ_2.

(2) If λ_1 and λ_2 are real and unequal, the solution space is spanned by

$$\{e^{\lambda_1 x}, e^{\lambda_2 x}\}.$$

(3) If λ_1 and λ_2 are real and equal, $\lambda_1 = \lambda_2 = \lambda$, then the solution space is spanned by

$$\{e^{\lambda x}, xe^{\lambda x}\}.$$

(4) If λ_1 and λ_2 are the complex numbers $\alpha \pm \beta i (\beta \neq 0)$, then the solution space is spanned by

$$\{e^{\alpha x} \cos \beta x, e^{\alpha x} \sin \beta x\}.$$

Differential equations problems, especially those that arise from applications, often ask for a particular solution satisfying specified *initial conditions*. That is to say, the values of y and y' at a specific value of x are prescribed. To solve such problems, first find the general solution and then impose the initial conditions to determine c_1 and c_2. We conclude this section with an example of this type.

Example 7.   Find the solution of $y'' + 4y = 0$ that satisfies $y(0) = 2$ and $y'(0) = -1$. First, we notice that the zeros of the auxiliary polynomial $z^2 + 4$ are $\pm 2i$; hence the general solution is

$$y = c_1 \cos 2x + c_2 \sin 2x.$$

We now impose the initial conditions. Since

$$y' = -2c_1 \sin 2x + 2c_2 \cos 2x,$$

these conditions become

$$2 = c_1 \quad \text{and} \quad -1 = 2c_2.$$

Hence, the particular solution we seek is

$$y = 2 \cos 2x - \tfrac{1}{2} \sin 2x. \quad \blacksquare$$

EXERCISES

Find the general solution of each of the following differential equations.

1. $y'' - 6y' + 9y = 0$

2. $y'' - 9y = 0$

3. $(D^2 + D - 6)y = 0$

4. $(D^2 + 2D - 3)y = 0$

5. $y'' + 9y = 0$

6. $y'' + 2y' + 3y = 0$

7. $(D^2 + 2D - 1)y = 0$

8. $3y'' + 5y' - 2y = 0$

9. $7y'' + 20y' - 3y = 0$

10. $(D^2 + 2D + 2)y = 0$

11. $(D^2 + 7)y = 0$

12. $(D^2 - 2\sqrt{2}D + 2)y = 0$

13. $y'' + 4y' = 0$

14. $y'' - 2y' - 4y = 0$

15. $8y'' - 42y' + 10y = 0$

16. $2y'' + y' + y = 0$

Find particular solutions to the following differential equations that satisfy the stated initial conditions.

17. $y'' + 9y = 0$, $y(0) = 1$, $y'(0) = -2$

18. $y'' - 2y' - 8y = 0$, $y(0) = 0$, $y'(0) = 1$

19. $y'' - 2y' + 2y = 0$, $y(0) = 5$, $y'(0) = 4$

20. $y'' - 16y = 0$, $y(0) = -2$, $y'(0) = 10$

21. (a) Show that if a is any real number, then the first order differential equation $a_1 y' + a_0 y = 0$ has exactly one solution satisfying the initial condition $y(0) = a$.

(b) Show that if a and b are real numbers, then the second order differential equation $a_2 y'' + a_1 y' + a_0 y = 0$ has exactly one solution satisfying the initial conditions $y(0) = a$, $y'(0) = b$.

22. Show that $(D - \lambda_1)(D - \lambda_2) = (D - \lambda_2)(D - \lambda_1)$ where λ_1 and λ_2 are real numbers.

23. Show that if $y \in \mathcal{F}(\mathbb{R})$ is any twice differentiable function of x and $\alpha \in \mathbb{R}$, then
$$D^2(e^{-\alpha x}y) = e^{-\alpha x}(D - \alpha)^2 y.$$

24. Show that the general solution of the differential equation $y'' + \omega^2 y = 0$ ($\omega \in \mathbb{R}$) can be written in the form
$$y = A \cos(\omega t + \varphi)$$
where A and φ are arbitrary constants. (The constant A is the *amplitude* of the solution, ω is its *frequency,* and φ is its *phase.*

25. According to Hooke's Law, if an object is attached to a wall by a spring (see Figure 3.1), then the spring exerts on the object a restoring force proportional

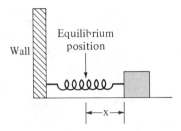

FIGURE 3.1

to the displacement of the object from equilibrium: if x denotes displacement from equilibrium, then Force $= -kx$ where $k > 0$ is the **spring constant.** According to Newton's Law, the acceleration of an object is proportional to the force acting on it: $x'' = \frac{1}{m} \cdot$ Force (or Force $= mx''$) where $m \in \mathbb{R}$ is the **mass** of the object and prime denotes differentiation with respect to time.

(a) Assume that no other forces act on the object (e.g., no friction, no air resistance) so that the equation of motion is $mx'' = -kx$. Find the position $x(t)$ of the body at time t if $x(0) = 0$ and $x'(0) = 1$.

(b) Now assume that the object is immersed in a viscous medium (e.g., oil), which exerts on the object a resisting force proportional to velocity: Force $= -vx'$ where $v \in \mathbb{R}$ is the **coefficient of viscosity.** The total force is then Force $= -vx' - kx$ so the equation of motion becomes

$$mx'' = -vx' - kx,$$

or

$$mx'' + vx' + kx = 0.$$

Find x if $x(0) = 0$, $x'(0) = 1$, and sketch its graph.

3.3 HIGHER ORDER HOMOGENEOUS EQUATIONS

In the previous section, we learned how to solve homogeneous second order linear differential equations with constant coefficients. Let us turn now to the problem of solving higher order equations of the same type. Our discussion focuses on equations of order three, where it is not hard to see what to do. When this case has been understood, a procedure for solving nth order homogeneous linear differential equations with constant coefficients will be relatively clear. At the end of the section, we shall describe this procedure.

Suppose we wish to solve the differential equation

$$a_3 y''' + a_2 y'' + a_1 y' + a_0 y = 0,$$

where $a_0, a_1, a_2, a_3 \in \mathbb{R}$, $a_3 \neq 0$. In operator notation, this equation is written as

(1) $$(a_3 D^3 + a_2 D^2 + a_1 D + a_0)y = 0.$$

The third order linear differential operator $a_3 D^3 + a_2 D^2 + a_1 D + a_0$ always factors into a product of a first order operator and a second order operator, because linear differential operators with constant coefficients multiply just like polynomials, and every degree three polynomial with real coefficients has at least one real zero. Suppose that λ_1 is a real zero of the auxiliary polynomial

$$p(z) = a_3 z^3 + a_2 z^2 + a_1 z + a_0$$

of (1). Then this polynomial factors as

$$p(z) = (b_2 z^2 + b_1 z + b_0)(z - \lambda_1)$$

and equation (1) becomes

$$(b_2 D^2 + b_1 D + b_0)(D - \lambda_1)y = 0$$

where b_2, b_1 and b_0 are real numbers. Let $u = (D - \lambda_1)y$. Then the equation (1) becomes

$$(b_2 D^2 + b_1 D + b_0)u = 0.$$

In the previous section, we learned that the solution space of this equation is spanned by two functions, which we call y_1 and y_2. Then

$$u = c_1 y_1 + c_2 y_2, \text{ where } c_1 \text{ and } c_2 \text{ are real numbers;}$$

hence we need only solve the first order equation

$$(D - \lambda_1)y = c_1 y_1 + c_2 y_2.$$

Multiplying this equation by the integrating factor $e^{-\lambda_1 x}$, we get

$$(e^{-\lambda_1 x} y)' = c_1 e^{-\lambda_1 x} y_1 + c_2 e^{-\lambda_1 x} y_2$$

and then y can be found by integrating both sides. Thus,

$$y = e^{\lambda_1 x} \int (c_1 e^{-\lambda_1 x} y_1 + c_2 e^{-\lambda_1 x} y_2)\, dx.$$

The precise form that this integral takes depends on what the functions y_1 and y_2 are, and they, in turn, are determined by the zeros of the polynomial $b_2 z^2 + b_1 z + b_0$. Suppose these zeros are λ_2 and λ_3. We must consider four cases.

Case 1. If λ_1, λ_2, and λ_3 are real and distinct, then $y_1 = e^{\lambda_2 x}$ and $y_2 = e^{\lambda_3 x}$, hence

$$y = e^{\lambda_1 x} \int (c_1 e^{(\lambda_2 - \lambda_1)x} + c_2 e^{(\lambda_3 - \lambda_1)x}) \, dx$$

$$= \frac{c_1}{\lambda_2 - \lambda_1} e^{\lambda_2 x} + \frac{c_2}{\lambda_3 - \lambda_1} e^{\lambda_3 x} + c_3 e^{\lambda_1 x}.$$

Case 2. If the zeros λ_1, λ_2, and λ_3 are real but two are equal, say $\lambda_1 = \lambda_2$, then $y_1 = e^{\lambda_1 x}$, $y_2 = e^{\lambda_3 x}$, and

$$y = e^{\lambda_1 x} \int (c_1 + c_2 e^{(\lambda_3 - \lambda_1)x}) \, dx$$

$$= c_1 x e^{\lambda_1 x} + \frac{c_2}{\lambda_3 - \lambda_1} e^{\lambda_3 x} + c_3 e^{\lambda_1 x}.$$

Case 3. If $\lambda_1 = \lambda_2 = \lambda_3$, then $y_1 = e^{\lambda_1 x}$ and $y_2 = x e^{\lambda_1 x}$, hence

$$y = e^{\lambda_1 x} \int (c_1 + c_2 x) \, dx = c_1 x e^{\lambda_1 x} + \tfrac{1}{2} c_2 x^2 e^{\lambda_1 x} + c_3 e^{\lambda_1 x}.$$

Case 4. If λ_2 and λ_3 are the complex numbers $\alpha \pm \beta i$, $\beta \neq 0$, then $y_1 = e^{\alpha x} \cos \beta x$ and $y_2 = e^{\alpha x} \sin \beta x$, hence

$$y = e^{\lambda_1 x} \int (c_1 e^{(\alpha - \lambda_1)x} \cos \beta x + c_2 e^{(\alpha - \lambda_1)x} \sin \beta x) \, dx.$$

Integration by parts yields

$$\int e^{(\alpha - \lambda_1)x} \cos \beta x \, dx = \frac{\alpha - \lambda_1}{(\alpha - \lambda_1)^2 + \beta^2} e^{(\alpha - \lambda_1)x} \cos \beta x +$$

$$\frac{\beta}{(\alpha - \lambda_1)^2 + \beta^2} e^{(\alpha - \lambda_1)x} \sin \beta x$$

and

$$\int e^{(\alpha - \lambda_1)x} \sin \beta x \, dx = \frac{\alpha - \lambda_1}{(\alpha - \lambda_1)^2 + \beta^2} e^{(\alpha - \lambda_1)x} \sin \beta x -$$

$$\frac{\beta}{(\alpha - \lambda_1)^2 + \beta^2} e^{(\alpha - \lambda_1)x} \cos \beta x,$$

hence y is a linear combination of

$$e^{\alpha x} \cos \beta x, \quad e^{\alpha x} \sin \beta x, \quad \text{and} \quad e^{\lambda_1 x}.$$

These results may be summarized as follows.

To solve the homogeneous third order linear differential equation $a_3 y''' + a_2 y'' + a_1 y' + a_0 y = 0$, where a_3, a_2, a_1, a_0 are constants, first find the zeros $\lambda_1, \lambda_2, \lambda_3$ of the auxiliary polynomial $a_3 z^3 + a_2 z^2 + a_1 z + a_0$. Then

(i)   if $\lambda_1, \lambda_2, \lambda_3$ are real and distinct, $\{e^{\lambda_1 x}, e^{\lambda_2 x}, e^{\lambda_3 x}\}$ spans the solution space;

(ii) if $\lambda_1 = \lambda_2 \neq \lambda_3$, then $\{e^{\lambda_1 x}, xe^{\lambda_1 x}, e^{\lambda_3 x}\}$ spans the solution space;

(iii) if $\lambda_1 = \lambda_2 = \lambda_3$, then $\{e^{\lambda_1 x}, xe^{\lambda_1 x}, x^2 e^{\lambda_1 x}\}$ spans the solution space;

and

(iv) if λ_1 is real and λ_2 and λ_3 are the complex numbers $\alpha \pm \beta i$, $\beta \neq 0$, then

$$\{e^{\lambda_1 x}, e^{\alpha x} \cos \beta x, e^{\alpha x} \sin \beta x\}$$

spans the solution space.

Example 1. Solve $y''' - y = 0$. The auxiliary polynomial is $z^3 - 1$, one of whose zeros is clearly the number 1. Divide $z^3 - 1$ by $z - 1$ to get the factorization $z^3 - 1 = (z - 1)(z^2 + z + 1)$. The quadratic factor has zeros $-\frac{1}{2} \pm \frac{\sqrt{3}}{2}i$; hence we are in case (iv), and $\left\{ e^x, e^{-\frac{1}{2}x} \cos \frac{\sqrt{3}}{2}x, e^{-\frac{1}{2}x} \sin \frac{\sqrt{3}}{2}x \right\}$ is a spanning set for the solution space. The general solution is

$$y = c_1 e^x + c_2 e^{-\frac{1}{2}x} \cos \frac{\sqrt{3}}{2}x + c_3 e^{-\frac{1}{2}x} \sin \frac{\sqrt{3}}{2}x. \quad \blacksquare$$

Example 2. Solve $y''' - y'' - y' + y = 0$. The auxiliary polynomial is $z^3 - z^2 - z + 1$, one of whose zeros is 1. Divide by $z - 1$ to get the factorization $(z - 1)(z^2 - 1) = (z - 1)(z - 1)(z + 1)$. We are in case (ii) with $\lambda_1 = \lambda_2 = 1$, $\lambda_3 = -1$, so

$$\{e^x, xe^x, e^{-x}\}$$

is a spanning set for the solution space, and the general solution is

$$y = c_1 e^x + c_2 xe^x + c_3 e^{-x}. \quad \blacksquare$$

The factorization of third degree polynomials is not always so easy. However, if a polynomial with integer coefficients, $a_n z^n + a_{n-1} z^{n-1} + \cdots + a_1 z + a_0$, has a *rational* zero, then it must be among the set of quotients $\pm r/s$, where r is an integer factor of a_0 and s is an integer factor of a_n.

Example 3. Solve $y''' - 3y'' + 3y' - y = 0$. The auxiliary polynomial is $z^3 - 3z^2 + 3z - 1$. If there is a rational zero, then it must be either 1 or -1, by the observation preceding this example. Try both, find that 1 is a zero, and hence find the factorization $(z - 1)(z^2 - 2z + 1) = (z - 1)^3$. We are in case (iii), so the general solution is

$$y = c_1 e^x + c_2 x e^x + c_3 x^2 e^x. \quad \blacksquare$$

Example 4. Solve $(D^3 - 4D^2 + D + 6)y = 0$. The auxiliary polynomial is $z^3 - 4z^2 + z + 6$. Its rational zeros must be among the numbers ± 1, ± 2, ± 3, ± 6. Try them. You will find that -1 is a zero, hence you will be led to the factorization $(z + 1)(z^2 - 5z + 6) = (z + 1)(z - 2)(z - 3)$. Thus, the general solution is

$$y = c_1 e^{-x} + c_2 e^{2x} + c_3 e^{3x}. \quad \blacksquare$$

We will not give a detailed proof of the algorithm for solving higher order equations. The pattern is clear. Here is what you do:

To solve a homogeneous nth order linear differential equation with constant coefficients,

(2) $\qquad a_n y^{(n)} + a_{n-1} y^{(n-1)} + \cdots + a_1 y' + a_0 y = 0,$

first find the zeros of the auxiliary polynomial $a_n z^n + a_{n-1} z^{n-1} + \cdots + a_1 z + a_0$. Some of these zeros may be real numbers, and some may be complex numbers, but the complex ones will occur in pairs of the form $\alpha \pm \beta i$, where α and β are real and $\beta \neq 0$. Construct a spanning set for the solution space of equation (2) as follows:

(i) If λ is a real zero of the auxiliary polynomial, and $z - \lambda$ occurs exactly once as a factor, include $e^{\lambda x}$ in the spanning set.

(ii) If λ is a real zero and $z - \lambda$ occurs exactly k times as a factor of the auxiliary polynomial, include the k functions $e^{\lambda x}$, $x e^{\lambda x}, \ldots, x^{k-1} e^{\lambda x}$ in the spanning set.

(iii) If $\alpha \pm \beta i$ are complex zeros of the auxiliary polynomial and $(z - \alpha)^2 + \beta^2$ occurs exactly once as a factor, include the two functions

$$e^{\alpha x} \cos \beta x \quad \text{and} \quad e^{\alpha x} \sin \beta x$$

in the spanning set.

(iv)   If the pair of complex zeros $\alpha \pm \beta i$ occurs exactly k times (that is, if $(z - \alpha)^2 + \beta^2$ occurs exactly k times as a factor of the auxiliary polynomial), then include the $2k$ functions

$$e^{\alpha x} \cos \beta x, \ e^{\alpha x} \sin \beta x, \ xe^{\alpha x} \cos \beta x, \ xe^{\alpha x} \sin \beta x, \ \ldots,$$

$$x^{k-1}e^{\alpha x} \cos \beta x, \ x^{k-1}e^{\alpha x} \sin \beta x$$

in the spanning set.

The general solution of (2) is then an arbitrary linear combination of the functions in the spanning set you have constructed.

Example 5.   Solve $(D - 2)^2(D + 3)^3(D^2 - 2D + 2)^2 y = 0$. This is a ninth order equation whose operator is already factored (fortunately!). The zeros of the auxiliary polynomial are:

2 (occurring twice), -3 (occurring three times)

and the pair $1 \pm i$ (occurring twice). A spanning set for the solution space is

$$\{e^{2x}, xe^{2x}, e^{-3x}, xe^{-3x}, x^2e^{-3x},$$

$$e^x \cos x, e^x \sin x, xe^x \cos x, xe^x \sin x\}.$$

The general solution is

$$y = c_1 e^{2x} + c_2 x e^{2x} + c_3 e^{-3x} + c_4 x e^{-3x} + c_5 x^2 e^{-3x} + c_6 e^x \cos x +$$

$$c_7 e^x \sin x + c_8 x e^x \cos x + c_9 x e^x \sin x. \ \blacksquare$$

Example 6.   Solve $y^{(4)} - 16y = 0$. The auxiliary polynomial is $z^4 - 16 = (z^2 - 4)(z^2 + 4)$, hence its zeros are ± 2 and $\pm 2i$. A spanning set for the solution space is

$$\{e^{2x}, e^{-2x}, \cos 2x, \sin 2x\}.$$

The general solution is

$$y = c_1 e^{2x} + c_2 e^{-2x} + c_3 \cos 2x + c_4 \sin 2x. \ \blacksquare$$

Example 7.   Solve $y^{(5)} + 4y^{(4)} + 6y''' + 4y'' + y' = 0$. The auxiliary polynomial is $z^5 + 4z^4 + 6z^3 + 4z^2 + z$ so clearly z is a factor. The remaining factor, $z^4 + 4z^3 + 6z^2 + 4z + 1$, equals $(z + 1)^4$ (remember the binomial formula?), so the zeros are 0 and -1 (occurring four times). Since $e^{0x} = 1$, the spanning set for the solution space given by the algorithm is

$$\{1, e^{-x}, xe^{-x}, x^2e^{-x}, x^3e^{-x}\}.$$

The general solution is

$$y = c_1 + c_2 e^{-x} + c_3 x e^{-x} + c_4 x^2 e^{-x} + c_5 x^3 e^{-x}. \quad \blacksquare$$

Sometimes, in solving linear differential equations, we will encounter an auxiliary polynomial with a factor of the form $z^n - a$, where a is some complex number, and we will need to find all the zeros of this polynomial. In other words, we must find all nth roots of a. Let us recall the method for doing this.

If two complex numbers z_1 and z_2 are written in polar form, $z_1 = r_1(\cos \theta_1 + i \sin \theta_1)$ and $z_2 = r_2(\cos \theta_2 + i \sin \theta_2)$, then their product is

$$z_1 z_2 = r_1 r_2 (\cos \theta_1 + i \sin \theta_1)(\cos \theta_2 + i \sin \theta_2)$$
$$= r_1 r_2 [(\cos \theta_1 \cos \theta_2 - \sin \theta_1 \sin \theta_2) +$$
$$i(\sin \theta_1 \cos \theta_2 + \cos \theta_1 \sin \theta_2)]$$
$$= r_1 r_2 [\cos(\theta_1 + \theta_2) + i \sin(\theta_1 + \theta_2)].$$

Hence the multiplication of complex numbers can be described geometrically as follows (see Figure 3.2):

(1)   the length of the product $z_1 z_2$ is the product of the lengths of z_1 and z_2, and

(2)   the angle of $z_1 z_2$ is the sum of the angles of z_1 and z_2.

It follows that if $z = r(\cos \theta + i \sin \theta)$, then $z^2 = r^2(\cos 2\theta + i \sin 2\theta)$, $z^3 = r^3(\cos 3\theta + i \sin 3\theta)$, and so on. Therefore

$$z^n = r^n(\cos n\theta + i \sin n\theta), \text{ for all positive integers } n.$$

This formula, often referred to as DeMoivre's Theorem, is the basis for finding the zeros of $z^n - a$. We proceed as follows:

Write a in polar form: $a = r(\cos \theta + i \sin \theta)$. We seek n complex numbers

$$z_j = r_j(\cos \theta_j + i \sin \theta_j), \quad j = 0, 1, 2, \ldots, n - 1,$$

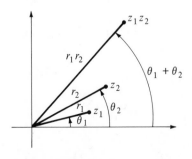

FIGURE 3.2

with the property that $z_j^n = a$ for all j. But

$$z_j^n = r_j^n(\cos n\theta_j + i \sin n\theta_j),$$

by DeMoivre's Theorem; hence $z_j^n = a$ if and only if

$$r_j^n = r \quad \text{and} \quad (\cos n\theta_j, \sin n\theta_j) = (\cos \theta, \sin \theta),$$

for each $j = 0, \ldots, n - 1$. The n solutions of these equations are given by

$$r_j = \sqrt[n]{r}, \text{ and}$$

$$\theta_j = \frac{\theta}{n} + j\frac{2\pi}{n}, \quad j = 0, 1, \ldots, n - 1$$

(see Figure 3.3).

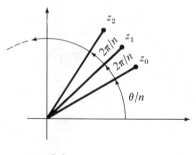

FIGURE 3.3

Example 8.   Solve $(D^6 + 64)y = 0$. The auxiliary polynomial $z^6 + 64$ has as zeros the 6th roots of -64. Since

$$-64 = 64(\cos \pi + i \sin \pi)$$

(see Figure 3.4), we see that the 6th roots of -64 are

(1) $\sqrt[6]{64}\left(\cos \frac{\pi}{6} + i \sin \frac{\pi}{6}\right) = 2\left(\frac{\sqrt{3}}{2} + \frac{1}{2}i\right) = \sqrt{3} + i$

(2) $\sqrt[6]{64}\left(\cos\left(\frac{\pi}{6} + \frac{2\pi}{6}\right) + i \sin\left(\frac{\pi}{6} + \frac{2\pi}{6}\right)\right) = 2(0 + 1i) = 2i$

(3) $\sqrt[6]{64}\left(\cos\left(\frac{\pi}{6} + \frac{4\pi}{6}\right) + i \sin\left(\frac{\pi}{6} + \frac{4\pi}{6}\right)\right) = 2\left(-\frac{\sqrt{3}}{2} + \frac{1}{2}i\right) = -\sqrt{3} + i$

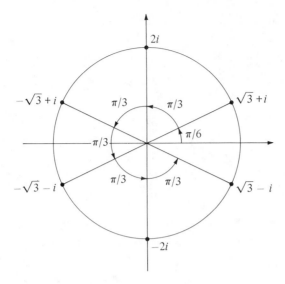

FIGURE 3.4

(4) $\sqrt[6]{64}\left(\cos\left(\dfrac{\pi}{6}+\dfrac{6\pi}{6}\right)+i\sin\left(\dfrac{\pi}{6}+\dfrac{6\pi}{6}\right)\right)=2\left(-\dfrac{\sqrt{3}}{2}-\dfrac{1}{2}i\right)=-\sqrt{3}-i$

(5) $\sqrt[6]{64}\left(\cos\left(\dfrac{\pi}{6}+\dfrac{8\pi}{6}\right)+i\sin\left(\dfrac{\pi}{6}+\dfrac{8\pi}{6}\right)\right)=2(0-i)=-2i$

(6) $\sqrt[6]{64}\left(\cos\left(\dfrac{\pi}{6}+\dfrac{10\pi}{6}\right)+i\sin\left(\dfrac{\pi}{6}+\dfrac{10\pi}{6}\right)\right)=2\left(\dfrac{\sqrt{3}}{2}-\dfrac{1}{2}i\right)=\sqrt{3}-i.$

Notice that these six zeros of z^6+64 occur, as expected, in conjugate pairs $\pm 2i$, $\sqrt{3}\pm i$, $-\sqrt{3}\pm i$. Hence the set

$$\{\cos 2x,\ \sin 2x,\ e^{\sqrt{3}x}\cos x,\ e^{\sqrt{3}x}\sin x,\ e^{-\sqrt{3}x}\cos x,\ e^{-\sqrt{3}x}\sin x\}$$

spans the solution space of the equation $(D^6+64)y=0$. The general solution is

$$y=c_1\cos 2x+c_2\sin 2x+c_3 e^{\sqrt{3}x}\cos x+c_4 e^{\sqrt{3}x}\sin x+$$
$$c_5 e^{-\sqrt{3}x}\cos x+c_6 e^{-\sqrt{3}x}\sin x. \quad \blacksquare$$

EXERCISES

Find the general solution of each of the following differential equations.

1. $(D+5)(D-2)(D-3)y=0$
2. $(D-1)^2(D+3)y=0$
3. $y'''-4y''+4y'=0$

4. $(D - 4)^3 y = 0$

5. $D^4 y = 0$

6. $(D^2 + 9)^2 y = 0$

7. $(D^4 - 1)y = 0$

8. $(D - 1)^2(D + 2)^2(D^2 + 2D + 2)^2 y = 0$

9. $(D^4 + 1)y = 0$

10. $y^{(4)} - 16y = 0$

11. $y^{(6)} - y = 0$

For each of the following differential equations, find the solution that satisfies the indicated initial conditions.

12. $y''' - 4y'' + 4y' = 0$, $y(0) = y'(0) = 1$, $y''(0) = 2$

13. $y^{(4)} - y = 0$, $y(0) = 1$, $y'(0) = -1$, $y''(0) = 2$, $y'''(0) = 0$

3.4 THE METHOD OF UNDETERMINED COEFFICIENTS

In Section 3.1, we learned that the general solution of the constant coefficient linear differential equation

$$a_n y^{(n)} + \cdots + a_1 y' + a_0 y = f$$

is of the form $y = y_P + y_H$, where y_P is a particular solution of the equation and y_H is the general solution of the associated homogeneous equation

$$a_n y^{(n)} + \cdots + a_1 y' + a_0 y = 0.$$

In Sections 3.2 and 3.3, we learned how to find y_H. Now we shall develop a method for finding a particular solution y_P. This method will work only for certain functions f, but these functions are the ones that occur most frequently in applications. (In Chapter 9, we will describe another method for finding y_P, a method that works for a larger class of functions f.)

Before describing the method in general, let's look at an example.

Example 1. Consider the equation $y'' + y = e^x$. In operator notation, this equation becomes $(D^2 + 1)y = e^x$. The success of the method depends on finding a differential operator that sends the function on the right to zero. But we know that $(D - 1)e^x = 0$. So, if we apply the operator $D - 1$ to both sides of the equation

$$(D^2 + 1)y = e^x$$

we get

$$(D - 1)(D^2 + 1)y = 0.$$

Hence every solution of the equation $(D^2 + 1)y = e^x$ also satisfies the homogeneous equation $(D - 1)(D^2 + 1)y = 0$. So, to find a particular solution y_P of our
original equation $y'' + y = e^x$ we need only look among the functions of the form

$$Ae^x + B \cos x + C \sin x,$$

where $A, B, C \in \mathbb{R}$, since these are the only functions that satisfy $(D - 1)(D^2 + 1)y = 0$. But $B \cos x + C \sin x$ is a solution of the homogeneous equation $y'' + y = 0$, so if $Ae^x + B \cos x + C \sin x$ satisfies $y'' + y = e^x$, then

$$(Ae^x + B \cos x + C \sin x) - (B \cos x + C \sin x) = Ae^x$$

does also. Thus there must be a particular solution of $y'' + y = e^x$ of the form

$$y_P = Ae^x.$$

To find A, simply substitute y_P into the equation $y'' + y = e^x$:

$$y_P'' + y_P = Ae^x + Ae^x = 2Ae^x$$

so $y_P'' + y_P = e^x$ if and only if $2Ae^x = e^x$, or $2A = 1$, $A = \frac{1}{2}$. We conclude that

$$y_P = \tfrac{1}{2}e^x$$

is a particular solution of the equation $y'' + y = e^x$. ∎

The method illustrated in Example 1 will work whenever the function f
satisfies some constant coefficient homogeneous linear differential equation. Suppose L is a constant coefficient linear differential operator of order n and that we
want to find a particular solution of the differential equation $Ly = f$, where $f \in \mathcal{C}(\mathbb{R})$. Suppose further that $L_1 f = 0$ where L_1 is some constant coefficient linear
differential operator, say of order k. Then, applying the operator L_1 to both sides
of the differential equation

(1) $$Ly = f$$

yields

(2) $$L_1 Ly = 0.$$

Hence every solution of (1) is also a solution of the differential equation (2), which
is of order $n + k$. But we learned in Section 3 how to find all solutions of (2):
simply find the zeros of the auxiliary polynomial associated with the operator $L_1 L$
and use these zeros to construct a spanning set $\{y_1, \ldots, y_{n+k}\}$ for the solution
space of $L_1 Ly = 0$. Every solution of $L_1 Ly = 0$, and hence every solution of
$Ly = f$, must be a linear combination of these functions $y_1, \ldots, y_{n+k}$. So we can
find a particular solution of $Ly = f$ by substituting the function

$$y_P = A_1 y_1 + \cdots + A_{n+k} y_{n+k} \quad (A_1, \ldots, A_{n+k} \in \mathbb{R})$$

into the equation $Ly = f$ and determining which values of $A_1, \ldots, A_{n+k}$ enable y_P to satisfy this equation. But some of the functions $y_1, \ldots, y_{n+k}$, say, $y_1, \ldots, y_n$, come from zeros of the auxiliary polynomial associated with L. These functions satisfy the homogeneous equation $Ly = 0$, so if we substitute

$$y_P = A_1 y_1 + \cdots + A_n y_n + A_{n+1} y_{n+1} + \cdots + A_{n+k} y_{n+k}$$

into the equation $Ly = f$, and use the fact that

$$L(A_1 y_1 + \cdots + A_n y_n) = 0,$$

we get

$$L(A_{n+1} y_{n+1} + \cdots + A_{n+k} y_{n+k}) = f.$$

So, no matter how the coefficients $A_1, \ldots, A_n$ are chosen, $A_1 y_1 + \cdots + A_{n+k} y_{n+k}$ will satisfy $Ly = f$ if and only if $A_{n+1} y_{n+1} + \cdots + A_{n+k} y_{n+k}$ does. We conclude that there must be a particular solution of $Ly = f$ of the form

$$y_P = A_{n+1} y_{n+1} + \cdots + A_{n+k} y_{n+k}.$$

We summarize this discussion in the following algorithm.

Method of Undetermined Coefficients

To find a particular solution y_P of the differential equation $Ly = f$, where L is a constant coefficient linear differential operator and f is a solution of some constant coefficient homogeneous linear differential equation:

(i)   Find a constant coefficient linear differential operator L_1 such that $L_1 f = 0$.

(ii) Construct a spanning set $\{y_1, \ldots, y_n\}$ for the solution space of the homogeneous equation $Ly = 0$.

(iii) Enlarge this set to obtain a spanning set $\{y_1, \ldots, y_{n+k}\}$ for the solution space of the homogeneous equation $L_1 L y = 0$.

(iv) Set $y_P = A_{n+1} y_{n+1} + \cdots + A_{n+k} y_{n+k}$, where the coefficients $A_{n+1}, \ldots, A_{n+k}$ are real numbers to be determined. Find $A_{n+1}, \ldots, A_{n+k}$ by substituting y_P into the equation $Ly = f$.

Remark. If the auxiliary polynomials of $Ly = 0$ and $L_1 y = 0$ have no zeros in common, then the functions $y_{n+1}, \ldots, y_{n+k}$ all satisfy the equation $L_1 y = 0$, and $\{y_{n+1}, \ldots, y_{n+k}\}$ is simply any spanning set for the solution space of the homogeneous equation $L_1 y = 0$.

Example 2. Solve $y'' - 2y' + y = \sin x$. Here, $L = D^2 - 2D + 1 = (D - 1)^2$ so the solution space of $Ly = 0$ is spanned by

$$\{e^x, xe^x\}.$$

Since $(D^2 + 1)\sin x = 0$, we may take $L_1 = D^2 + 1$. The solution space of $L_1 Ly = 0$, that is, of $(D^2 + 1)(D - 1)^2 y = 0$, is spanned by

$$\{e^x, xe^x, \cos x, \sin x\}.$$

Hence there must be a particular solution of $y'' - 2y' + y = \sin x$ of the form

$$y_P = A \cos x + B \sin x.$$

Substituting y_P into the equation $y'' - 2y' + y = \sin x$, we find

$$(-A \cos x - B \sin x) - 2(-A \sin x + B \cos x) + (A \cos x + B \sin x) = \sin x$$

or

$$2A \sin x - 2B \cos x = \sin x.$$

This equation is satisfied when $2A = 1$ and $-2B = 0$, that is, when $A = \frac{1}{2}$ and $B = 0$, so

$$y_P = \tfrac{1}{2} \cos x$$

is a particular solution of the equation $y'' - 2y' + y = \sin x$. The general solution is

$$y = \tfrac{1}{2} \cos x + c_1 e^x + c_2 x e^x. \quad \blacksquare$$

Example 3. Solve $y'' - 2y' + y = e^x$. As in Example 2, $L = (D - 1)^2$ and the solution space of $Ly = 0$ is spanned by

$$\{e^x, xe^x\}.$$

But now $L_1 = D - 1$, since $(D - 1)e^x = 0$. The solution space of $L_1 Ly = (D - 1)^3 y = 0$ is spanned by

$$\{e^x, xe^x, x^2 e^x\}.$$

Hence there must be a particular solution of the form

$$y_P = Ax^2 e^x.$$

Substituting y_P into $y'' - 2y' + y = e^x$ yields

$$A(2e^x + 4xe^x + x^2e^x) - 2A(2xe^x + x^2e^x) + Ax^2e^x = e^x$$

or

$$2Ae^x = e^x.$$

so $2A = 1$, $A = \frac{1}{2}$. Thus,

$$y_P = \tfrac{1}{2}x^2e^x$$

is a particular solution of $y'' - 2y' + y = e^x$, and

$$y = \tfrac{1}{2}x^2e^x + c_1e^x + c_2xe^x$$

is the general solution. ∎

Example 4. Solve $y'' - 2y' + y = e^x + \sin x$. Here, a simple observation will enable us to use our previous work to write down the solution immediately. Note that if y_1 satisfies $y'' - 2y' + y = e^x$ and y_2 satisfies $y'' - 2y' + y = \sin x$, then $y_1 + y_2$ satisfies $y'' - 2y' + y = e^x + \sin x$. But from Example 3 we know that

$$y_1 = \tfrac{1}{2}x^2e^x$$

satisfies $y'' - 2y' + y = e^x$, and from Example 2 we know that

$$y_2 = \tfrac{1}{2}\cos x$$

satisfies $y'' - 2y' + y = \sin x$. Hence

$$y_P = y_1 + y_2 = \tfrac{1}{2}x^2e^x + \tfrac{1}{2}\cos x$$

is a particular solution of $y'' - 2y' + y = e^x + \sin x$. The general solution is

$$y = \tfrac{1}{2}x^2e^x + \tfrac{1}{2}\cos x + c_1e^x + c_2xe^x. \quad ∎$$

The technique used in Example 4 is sometimes called the **principle of superposition:** if y_1 is a solution of $Ly = f$ and y_2 is a solution of $Ly = g$, then $y_1 + y_2$ is a solution of $Ly = f + g$. This principle can be expressed more generally: *If the functions $y_1, \ldots, y_k$ satisfy the equations $Ly_i = f_i$, $1 \le i \le k$, and if $b_1, \ldots, b_k \in \mathbb{R}$, then the function $y_P = b_1y_1 + \cdots + b_ky_k$ is a solution of the equation $Ly = b_1f_1 + \cdots + b_kf_k$.*

Example 5. Solve $y'' + y = 3e^{2x} + 12e^{-x} - 17x$. The solution space of $y'' + y = (D^2 + 1)y = 0$ is spanned by

$$\{\cos x, \sin x\}.$$

Now $(D - 2)e^{2x} = 0$, and the solution space of $(D - 2)(D^2 + 1)y = 0$ is spanned by

$$\{\cos x, \sin x, e^{2x}\}$$

so we can find a particular solution of $y'' + y = e^{2x}$ of the form $y_1 = Ae^{2x}$. Substituting, we find that $5Ae^{2x} = e^{2x}$ so $A = \frac{1}{5}$ and

$$y_1 = \tfrac{1}{5}e^{2x}$$

satisfies the equation $y'' + y = e^{2x}$. Similarly, we find that

$$y_2 = \tfrac{1}{2}e^{-x}$$

satisfies $y'' + y = e^{-x}$. Finally, we see by inspection that

$$y_3 = x$$

satisfies $y'' + y = x$. We conclude that

$$y_P = 3(\tfrac{1}{5}e^{2x}) + 12(\tfrac{1}{2}e^{-x}) - 17x = \tfrac{3}{5}e^{2x} + 6e^{-x} - 17x$$

is a particular solution of $y'' + y = 3e^{2x} + 12e^{-x} - 17x$, and that the general solution is

$$y = \tfrac{3}{5}e^{2x} + 6e^{-x} - 17x + c_1 \cos x + c_2 \sin x. \quad \blacksquare$$

In order to use the method of undetermined coefficients to find a particular solution of the equation $Ly = f$, we must be able to recognize which functions f are solutions of constant coefficient linear differential equations $L_1 y = 0$, and for each such f we must be able to write down an L_1 such that $L_1 f = 0$. But from Section 3.3 we know that if f satisfies $L_1 f = 0$ for some constant coefficient linear differential operator, then f must be a linear combination of functions appearing in the first column of Table 3.1.

TABLE 3.1

FUNCTION f	OPERATOR L_1 SUCH THAT $L_1 f = 0$
x^s for some integer s	D^{s+1}
e^{kx} for some $k \in \mathbb{R}$	$D - k$
$x^s e^{kx}$ for some integer s and some $k \in \mathbb{R}$	$(D - k)^{s+1}$
$\cos \beta x$ for some $\beta \in \mathbb{R}$	$D^2 + \beta^2$
$\sin \beta x$ for some $\beta \in \mathbb{R}$	$D^2 + \beta^2$
$e^{\alpha x} \cos \beta x$ for some $\alpha, \beta \in \mathbb{R}$	$(D - \alpha)^2 + \beta^2$
$e^{\alpha x} \sin \beta x$ for some $\alpha, \beta \in \mathbb{R}$	$(D - \alpha)^2 + \beta^2$
$x^s e^{\alpha x} \cos \beta x$ for some integer s and some $\alpha, \beta \in \mathbb{R}$	$[(D - \alpha)^2 + \beta^2]^{s+1}$
$x^s e^{\alpha x} \sin \beta x$ for some integer s and some $\alpha, \beta \in \mathbb{R}$	$[(D - \alpha)^2 + \beta^2]^{s+1}$

In the second column of the table we have listed differential operators that send the functions in the first column to zero. This information, and the principle of superposition, should enable you to solve any reasonable equation $Ly = f$, when $L_1 f = 0$ for some L_1.

EXERCISES

Find the general solution of each of the following differential equations.

1. $y'' - y = \sin x$
2. $y'' - y = xe^x$
3. $y'' + 2y' + y = e^x$
4. $y'' + 2y' + y = e^{-x}$
5. $y'' + y = \sin x$
6. $y'' + y' + y = x^3$
7. $y'' - y = e^x \cos x$
8. $y'' - y = e^x \cos x + e^{-x}$
9. $y'' + y' = x + e^{-x}$
10. $y'' + y' + y = e^{-x} \cos x + e^{-x} \sin x$

For each of the following differential equations, find the solution that satisfies the indicated initial conditions.

11. $y'' + y = e^{3x}$, $y(0) = 1$, $y'(0) = 0$
12. $y'' - y = e^{-2x}$, $y(0) = 0$, $y'(0) = 1$
13. $y'' - 5y' + 6y = e^{2x}$, $y(0) = y'(0) = 1$.
14. $y''' - y = e^x$, $y(0) = y'(0) = y''(0) = 0$.
15. $y''' + 3y'' + 3y' + y = e^{2x}$, $y(0) = 1$, $y'(0) = 2$, $y''(0) = -1$.
16. Suppose an object is attached to a wall by a spring as in Exercise 25 of Section 3.2, but now suppose there acts an additional driving force $F = \cos \omega t$. Assuming no viscosity, so that the equation of motion becomes

$$mx'' + kx = \cos \omega t,$$

find x so that $x(0) = x'(0) = 0$. Sketch the graph of x when $\omega = \frac{1}{2}\sqrt{k/m}$ and when $\omega = \sqrt{k/m}$.

Linear Independence

4.1 INDEPENDENCE

In the first three chapters we described techniques for solving systems of linear algebraic equations and techniques for solving linear differential equations with constant coefficients. We found that if the equations are homogeneous, then the solution sets are vector spaces, and our techniques led directly to spanning sets for these solution spaces. In this section, we shall explore the question of whether or not our spanning sets are as small as possible.

If we have one spanning set for a vector space V, it is easy to produce other spanning sets for V. For example, we can always enlarge the given spanning set by adjoining to it additional elements of V. On the other hand, it is sometimes possible to reduce the size of a spanning set by deleting elements.

Example 1. The set $S = \{1, x, x^2, 1 + x^2\}$ spans the vector space $\mathcal{P}^3$. However, each of the subsets $\{1, x, x^2\}$, $\{1, x, 1 + x^2\}$, and $\{x, x^2, 1 + x^2\}$ also spans $\mathcal{P}^3$, so S is not an efficient spanning set for $\mathcal{P}^3$. It contains more elements than are needed to do the job. None of the three-element subsets listed can be reduced in size, however. For example, $\{1, x\}$ does not span $\mathcal{P}^3$, nor does $\{1, 1 + x^2\}$. ■

We are led, therefore, to consider the question: How can we tell if a spanning set is as small as possible?

Suppose S is a spanning set for a vector space V and suppose $\mathbf{v} \in S$. If the set $S - \{\mathbf{v}\}$ obtained by deleting $\mathbf{v}$ from S still spans V then $\mathbf{v}$, being a vector in V, must be a linear combination of the vectors in $S - \{\mathbf{v}\}$. Conversely, if $\mathbf{v} \in S$ is a linear combination of the vectors in $S - \{\mathbf{v}\}$, then $S - \{\mathbf{v}\}$ spans V (see Exercise 14 in Section 2.3). We conclude that *the set obtained by deleting a vector* $\mathbf{v}$ *from a spanning set S is still a spanning set (for the same vector space) if and only if that vector* $\mathbf{v}$ *is a linear combination of the other vectors in S.* This criterion can be reformulated more conveniently as follows.

Theorem 1. *Let V be a vector space and let $\mathbf{v}_1, \ldots, \mathbf{v}_k \in V (k > 1)$. Then one of the vectors in the set $\{\mathbf{v}_1, \ldots, \mathbf{v}_k\}$ is a linear combination of the others if and only if the equation*

$$c_1\mathbf{v}_1 + \cdots + c_k\mathbf{v}_k = 0$$

has a nonzero solution $(c_1, \ldots, c_k) \neq (0, \ldots, 0)$. *Moreover, a particular vector* v_j $(1 \leq j \leq k)$ *is a linear combination of the others if and only if this equation has a solution* $(c_1, \ldots, c_k)$ *with* $c_j \neq 0$.

Proof. If v_j is a linear combination of $\{v_1, \ldots, v_{j-1}, v_{j+1}, \ldots, v_k\}$ then

$$v_j = a_1 v_1 + \cdots + a_{j-1} v_{j-1} + a_{j+1} v_{j+1} + \cdots + a_k v_k$$

for some $a_1, \ldots, a_{j-1}, a_{j+1}, \ldots, a_k \in \mathbb{R}$. Adding $(-1)v_j$ to both sides of this equation yields

$$a_1 v_1 + \cdots + a_{j-1} v_{j-1} + (-1)v_j + a_{j+1} v_{j+1} + \cdots + a_k v_k = 0.$$

Thus

$$(c_1, \ldots, c_k) = (a_1, \ldots, a_{j-1}, -1, a_{j+1}, \ldots, a_k)$$

is a nonzero solution of the equation

$$c_1 v_1 + \cdots + c_k v_k = 0$$

and, in fact, $c_j = -1 \neq 0$.

Conversely, if $c_1 v_1 + \cdots + c_k v_k = 0$ and $c_j \neq 0$ for some j, then

$$-c_j v_j = c_1 v_1 + \cdots + c_{j-1} v_{j-1} + c_{j+1} v_{j+1} + \cdots + c_k v_k$$

so

$$v_j = -(c_1/c_j)v_1 + \cdots + (-c_{j-1}/c_j)v_{j-1} + (-c_{j+1}/c_j)v_{j+1}$$
$$+ \cdots + (-c_k/c_j)v_k;$$

that is, v_j is a linear combination of $\{v_1, \ldots, v_{j-1}, v_{j+1}, \ldots, v_k\}$, as asserted. ∎

A finite set $\{v_1, \ldots, v_k\}$ of vectors is said to be **linearly independent** or, simply, **independent,** if the only solution to the equation

$$c_1 v_1 + \cdots + c_k v_k = 0$$

is the zero solution $(c_1, \ldots, c_k) = (0, \ldots, 0)$. If this equation has a nonzero solution, the set is linearly **dependent.**

Theorem 1 says that, if $k > 1$, *the set* $\{v_1, \ldots, v_k\}$ *is linearly dependent if and only if one of the vectors in the set is a linear combination of the others.* In particular, *a pair of vectors is linearly dependent if and only if one is a scalar multiple of the other.*

If we represent vectors in $\mathbb{R}^2$ or in $\mathbb{R}^3$ as arrows with tails at the origin, we see that a set of two vectors is linearly dependent if and only if both point along the same line. A set of three vectors in $\mathbb{R}^3$ is linearly dependent if and only if all three (viewed as arrows with tails at **0**) lie in the same plane. Note that a set consisting of a single vector is linearly dependent if and only if that vector is the zero vector.

From Theorem 1 and the discussion preceding it we see that *a spanning set for a vector space is linearly independent if and only if it is minimal* (that is, if and only if it contains no proper subset that also spans the space.)

Example 2.   To determine if the set

$$\{(1, 2, 3), (4, 5, 6), (7, 8, 9)\}$$

in $\mathbb{R}^3$ is linearly independent, we must solve the vector equation

$$c_1(1, 2, 3) + c_2(4, 5, 6) + c_3(7, 8, 9) = \mathbf{0}.$$

This equation is the same as the equation

$$(c_1 + 4c_2 + 7c_3, 2c_1 + 5c_2 + 8c_3, 3c_1 + 6c_2 + 9c_3) = (0, 0, 0),$$

which is satisfied if and only if

$$c_1 + 4c_2 + 7c_3 = 0$$
$$2c_1 + 5c_2 + 8c_3 = 0$$
$$3c_1 + 6c_2 + 9c_3 = 0.$$

The solution set of this system is the line $(c_1, c_2, c_3) = t(1, -2, 1)$. In particular, $(1, -2, 1)$ is a nonzero solution, so

$$1(1, 2, 3) + (-2)(4, 5, 6) + 1(7, 8, 9) = \mathbf{0}.$$

Thus the set $\{(1, 2, 3), (4, 5, 6), (7, 8, 9)\}$ is linearly dependent.   ∎

Example 3.   To determine if the set $\{1, x - 1, (x - 1)^2\}$ in $\mathcal{F}(\mathbb{R})$ is linearly independent, we consider the equation

$$c_1 \cdot 1 + c_2(x - 1) + c_3(x - 1)^2 = 0.$$

This equation is equivalent to the equation

$$(c_1 - c_2 + c_3) + (c_2 - 2c_3)x + c_3 x^2 = 0,$$

which is satisfied for all $x \in \mathbb{R}$ if and only if

$$c_1 - c_2 + c_3 = 0$$
$$c_2 - 2c_3 = 0$$
$$c_3 = 0.$$

The only solution to this system is the zero solution $(c_1, c_2, c_3) = (0, 0, 0)$ so the set $\{1, x - 1, (x - 1)^2\}$ is linearly independent. ■

Testing finite subsets of $\mathbb{R}^n$ for linear independence always leads to a homogeneous system of linear equations to solve. Consider the set

$$\{(a_{11}, \ldots, a_{1n}), (a_{21}, \ldots, a_{2n}), \ldots, (a_{m1}, \ldots, a_{mn})\}$$

of m vectors in $\mathbb{R}^n$. This set is linearly dependent if and only if there exists a nonzero solution $(c_1, c_2, \ldots, c_n)$ of the vector equation

$$c_1(a_{11}, \ldots, a_{1n}) + c_2(a_{21}, \ldots, a_{2n}) + \cdots + c_m(a_{m1}, \ldots, a_{mn}) = \mathbf{0}.$$

This vector equation is equivalent to the linear system

$$a_{11}c_1 + a_{21}c_2 + \cdots + a_{m1}c_m = 0$$
$$\cdots$$
$$a_{1n}c_1 + a_{2n}c_2 + \cdots + a_{mn}c_m = 0$$

with coefficient matrix

$$A = \begin{pmatrix} a_{11} & \cdots & a_{m1} \\ & \cdots & \\ a_{1n} & \cdots & a_{mn} \end{pmatrix}.$$

This homogeneous linear system has the zero solution as its only solution if and only if the row echelon matrix

of A has a corner 1 in every column. Thus we have proved the following theorem.

Theorem 2. *A set of m vectors in $\mathbb{R}^n$ is linearly independent if and only if the matrix whose columns are these vectors has m corner 1's in its row echelon matrix.*

Corollary. *Each set of m vectors in $\mathbb{R}^n$, where $m > n$, is linearly dependent.*

Proof. The number of corner 1's in an echelon matrix cannot be larger than the number of rows. ■

The number of corner 1's in the echelon matrix of a matrix A is called the **rank** of A. We can rephrase Theorem 2 in terms of rank as follows: *a set of m vectors in $\mathbb{R}^n$ is linearly independent if and only if the matrix whose columns are these vectors has rank m.*

Example 4. The subset $\{(-1, 2, 0, -3), (-2, 1, 1, 2), (2, 4, -1, 3)\}$ of $\mathbb{R}^4$ is linearly independent because the row echelon matrix of

$$A = \begin{pmatrix} -1 & -2 & 2 \\ 2 & 1 & 4 \\ 0 & 1 & -1 \\ -3 & 2 & 3 \end{pmatrix} \quad \text{is} \quad B = \begin{pmatrix} 1 & 0 & 0 \\ 0 & 1 & 0 \\ 0 & 0 & 1 \\ 0 & 0 & 0 \end{pmatrix}$$

so the rank is 3. ■

Remark. In using this method to test finite subsets of $\mathbb{R}^n$ for linear dependence, it suffices to reduce the matrix to staircase form (it is not necessary to make the nonzero corner entries into 1's or to introduce stacks of zeros above the corner entries). This is true because the rank of a matrix is equal to the number of nonzero rows in its row echelon matrix, and this is the same as the number of nonzero rows in any row equivalent staircase matrix. Thus in Example 2 we can conclude that the matrix A has rank 3 by observing, for example, that A is row equivalent to the staircase matrix

$$\begin{pmatrix} 1 & 2 & -2 \\ 0 & 1 & -1 \\ 0 & 0 & 5 \\ 0 & 0 & 0 \end{pmatrix}.$$

We now state two linear independence tests for finite subsets of function spaces.

Theorem 3. *Let I be an interval in $\mathbb{R}$ and let $f_1, \ldots, f_n \in \mathcal{F}(I)$. If the matrix*

$$\begin{pmatrix} f_1(x_1) & f_2(x_1) & \cdots & f_n(x_1) \\ f_1(x_2) & f_2(x_2) & \cdots & f_n(x_2) \\ & & \cdots & \\ f_1(x_n) & f_2(x_n) & \cdots & f_n(x_n) \end{pmatrix}$$

has rank n for some choice of $x_1, \ldots, x_n \in I$, then $\{f_1, \ldots, f_n\}$ is linearly independent.

Theorem 4. *Let I be an open interval in $\mathbb{R}$ and let $f_1, \ldots, f_n \in \mathcal{F}(I)$. Suppose each f_i is $(n-1)$-times differentiable on I. If the matrix*

$$W_x(f_1, \ldots, f_n) = \begin{pmatrix} f_1(x) & f_2(x) & \cdots & f_n(x) \\ f_1'(x) & f_2'(x) & \cdots & f_n'(x) \\ & & \cdots & \\ f_1^{(n-1)}(x) & f_2^{(n-1)}(x) & \cdots & f_n^{(n-1)}(x) \end{pmatrix}$$

has rank n for some $x \in I$, then $\{f_1, \ldots, f_n\}$ is linearly independent.

Remark. Note that neither of these theorems can be used to establish the linear *dependence* of a set of functions (see Exercise 3).

The matrix $W_x(f_1, \ldots, f_n)$ in Theorem 4 is called the **Wronskian matrix** of $\{f_1, \ldots, f_n\}$ at $x \in I$.
Before proving these theorems we shall consider two examples.

Example 5. Consider the subset $\{1, x, x^2\}$ of the vector space $\mathcal{P}$ of all polynomials. Evaluating $f_1(x) = 1$, $f_2(x) = x$, and $f_3(x) = x^2$ successively at the points $x_1 = -1$, $x_2 = 0$, and $x_3 = 1$, we obtain the matrix

$$\begin{pmatrix} f_1(x_1) & f_2(x_1) & f_3(x_1) \\ f_1(x_2) & f_2(x_2) & f_3(x_2) \\ f_1(x_3) & f_2(x_3) & f_3(x_3) \end{pmatrix} = \begin{pmatrix} 1 & -1 & 1 \\ 1 & 0 & 0 \\ 1 & 1 & 1 \end{pmatrix}.$$

It is easy to check that this matrix has rank 3 so, by Theorem 3, $\{1, x, x^2\}$ is linearly independent. ∎

Example 6. Let a, b, and c be any three distinct real numbers, let I be any open interval containing 0, and consider the set $\{e^{ax}, e^{bx}, e^{cx}\}$ in $\mathcal{F}(I)$. The Wronskian matrix is

$$W_x(e^{ax}, e^{bx}, e^{cx}) = \begin{pmatrix} e^{ax} & e^{bx} & e^{cx} \\ ae^{ax} & be^{bx} & ce^{cx} \\ a^2e^{ax} & b^2e^{bx} & c^2e^{cx} \end{pmatrix}.$$

When $x = 0$ we get

$$W_0(e^{ax}, e^{bx}, e^{cx}) = \begin{pmatrix} 1 & 1 & 1 \\ a & b & c \\ a^2 & b^2 & c^2 \end{pmatrix}.$$

This matrix has rank 3 since it reduces to a staircase form with three nonzero rows:

$$\begin{pmatrix} 1 & 1 & 1 \\ a & b & c \\ a^2 & b^2 & c^2 \end{pmatrix} \rightarrow \begin{pmatrix} 1 & 1 & 1 \\ 0 & b-a & c-a \\ 0 & b^2-a^2 & c_2-a^2 \end{pmatrix}$$

$$\rightarrow \begin{pmatrix} 1 & 1 & 1 \\ 0 & b-a & c-a \\ 0 & 0 & (c-a)(c-b) \end{pmatrix}.$$

(The last row operation used here was: multiply the second row by $b + a$ and subtract from the third row.) Thus, by Theorem 4, $\{e^{ax}, e^{bx}, e^{cx}\}$ is linearly independent. ∎

Proof of Theorem 3.   We must show that the only solution of the equation

$$c_1f_1 + c_2f_2 + \cdots + c_nf_n = 0$$

is the zero solution $(c_1, \ldots, c_n) = (0, \ldots, 0)$. But this equation implies that

$$c_1f_1(x_1) + c_2f_2(x_1) + \cdots + c_nf_n(x_1) = 0$$
$$c_1f_1(x_2) + c_2f_2(x_2) + \cdots + c_nf_n(x_2) = 0$$
$$\cdots$$
$$c_1f_1(x_n) + c_2f_2(x_n) + \cdots + c_nf_n(x_n) = 0.$$

This is a homogeneous system of n linear equations in the n unknowns $c_1, \ldots, c_n$.

The coefficient matrix for the system is

$$\begin{pmatrix} f_1(x_1) & f_2(x_1) & \cdots & f_n(x_1) \\ f_1(x_2) & f_2(x_2) & \cdots & f_n(x_n) \\ & & \cdots & \\ f_1(x_n) & f_2(x_n) & \cdots & f_n(x_n) \end{pmatrix}$$

which, by hypothesis, has rank n. Thus the only solution to this homogeneous system, and hence to the equation

$$c_1 f_1 + c_2 f_2 + \cdots + c_n f_n = 0,$$

is the zero solution $(c_1, \ldots, c_n) = (0, \ldots, 0)$. This says that $\{f_1, \ldots, f_n\}$ is linearly independent. ■

Proof of Theorem 4.   Differentiating the equation

$$c_1 f_1 + c_2 f_2 + \cdots + c_n f_n = 0$$

$n - 1$ times and evaluating at x leads to the homogeneous system of linear equations

$$\begin{aligned} c_1 f_1(x) & + c_2 f_2(x) & + \cdots + c_n f_n(x) & = 0 \\ c_1 f_1'(x) & + c_2 f_2'(x) & + \cdots + c_n f_n'(x) & = 0 \end{aligned}$$

$$\cdots$$

$$c_1 f_1^{(n-1)}(x) + c_2 f_2^{(n-1)}(x) + \cdots + c_n f_n^{(n-1)}(x) = 0$$

whose coefficient matrix is the Wronskian matrix $W_x(f_1, \ldots, f_n)$. Since, by hypothesis, this coefficient matrix has rank n, the only solution to the system, and hence to the equation

$$c_1 f_1 + c_2 f_2 + \cdots + c_n f_n = 0,$$

is the zero solution $(c_1, \ldots, c_n) = (0, \ldots, 0)$. Thus $\{f_1, \ldots, f_n\}$ is linearly independent. ■

EXERCISES

1. Test the following subsets of $\mathbb{R}^3$ for linear independence.
    (a) $\{(1, 5, -4), (1, 6, -4)\}$
    (b) $\{(1, 5, -4), (1, 6, -4), (1, 7, -4)\}$
    (c) $\{(1, 0, 0), (0, 1, 0), (0, 0, 0)\}$

(d) $\{(1, 1, 1), (1, 2, 3), (1, 4, 9)\}$

(e) $\{(1, 2, 3), (1, 4, 9), (1, 8, 27)\}$

(f) $\{(1, 1, 1), (1, 2, 3), (1, 4, 9), (1, 8, 27)\}$

(g) $\{(2, -3, 2), (3, -1, 1), (0, -7, 4)\}$

(h) $\{(-1, 1, 2), (1, 2, 3), (5, 1, 0)\}$

(i) $\{(1, 1, 0), (0, 1, 1), (3, -5, 12)\}$

2. Test the following subsets of $\mathcal{F}(\mathbb{R})$ for linear independence.

(a) $\{e^x, e^{-x}\}$

(b) $\{1, e^x, e^{2x}, e^{3x}\}$

(c) $\{\sin^2 x, \cos^2 x, 1\}$

(d) $\{e^x, xe^x, x^2e^x\}$

(e) $\{1, x - 1, (x - 1)^2, (x - 1)^3\}$

(f) $\{x - 1, x + 1, x^2 + x + 1, x^2 - 2x + 1\}$

(g) $\{1, x, x^2, \ldots, x^n\}$

(h) $\{\sin x, \sin 2x, \sin 3x\}$

(i) $\{1, \cos x, \sin x, \cos 2x, \sin 2x\}$

3. (a) Show that if $f_1(x) = x$ and $f_2(x) = x^2$, then both

$$\begin{pmatrix} f_1(0) & f_2(0) \\ f_1(1) & f_2(1) \end{pmatrix} \quad \text{and} \quad \begin{pmatrix} f_1(0) & f_2(0) \\ f_1'(0) & f_2'(0) \end{pmatrix}$$

have rank <2 even though $\{f_1, f_2\}$ is linearly independent in $\mathcal{F}(\mathbb{R})$.

(b) Show that if $f_1(x) = x^3$ and $f_2(x) = |x^3|$ then $W_x(f_1, f_2)$ has rank <2 for *all* $x \in \mathbb{R}$ even though $\{f_1, f_2\}$ is linearly independent in $\mathcal{F}(\mathbb{R})$.

4. Show that $\{(a_1, a_2), (b_1, b_2)\}$ is linearly independent in $\mathbb{R}^2$ if and only if $a_1 b_2 - a_2 b_1 \neq 0$.

5. Show that if $\mathbf{a}, \mathbf{b} \in \mathbb{R}^3$ then $\{\mathbf{a}, \mathbf{b}\}$ is linearly dependent if and only if $\mathbf{a} \times \mathbf{b} = \mathbf{0}$.

6. Show that if $\mathbf{a}, \mathbf{b}, \mathbf{c} \in \mathbb{R}^3$ then $\{\mathbf{a}, \mathbf{b}, \mathbf{c}\}$ is linearly dependent if and only if $(\mathbf{a} \times \mathbf{b}) \cdot \mathbf{c} = 0$.

7. Show that the nonzero row vectors of any row echelon matrix form a linearly independent set.

8. Suppose $\{\mathbf{v}_1, \ldots, \mathbf{v}_k\}$ is a linearly independent set in the vector space V. Show that every nonempty subset of $\{\mathbf{v}_1, \ldots, \mathbf{v}_k\}$ is also linearly independent.

9. Show that if $\{\mathbf{v}_1, \ldots, \mathbf{v}_n\}$ is a linearly independent set in the vector space V and if $\mathbf{v}_{n+1} \in V$, then $\{\mathbf{v}_1, \ldots, \mathbf{v}_{n+1}\}$ is linearly independent if and only if $\mathbf{v}_{n+1} \notin \mathcal{L}(\mathbf{v}_1, \ldots, \mathbf{v}_n)$.

4.2 BASES

A *basis* for a vector space V is a spanning set that is linearly independent. In other words, a basis is a *minimal* spanning set.

Example 1. $\{1, x, x^2\}$ is a basis for $\mathcal{P}^3$ since it is certainly a spanning set and, by Example 5 of Section 4.1, it is also an independent set. ■

Example 2. $\{(1, 0, 0), (0, 1, 0), (0, 0, 1)\}$ is a basis for $\mathbb{R}^3$. (Why?) This basis is called the **standard basis** for $\mathbb{R}^3$. More generally, the **standard basis** for $\mathbb{R}^n$ is $\{e_1, \ldots, e_n\}$ where

$$e_1 = (1, 0, \ldots, 0), \; e_2 = (0, 1, 0, \ldots, 0), \ldots, e_n = (0, 0, \ldots, 0, 1). \; \blacksquare$$

Example 3. $\{1, x\}$ is *not* a basis for $\mathcal{P}^3$ because, although it is independent, it does not span $\mathcal{P}^3$ $(x^2 \notin \mathcal{L}(1, x))$. $\blacksquare$

Example 4. $\{e^x, e^{-x}, \cosh x\}$ is *not* a basis for the solution space of the differential equation $y'' - y = 0$ because, although it does span the space, it is not linearly independent $(\cosh x = \frac{1}{2}e^x + \frac{1}{2}e^{-x})$. $\blacksquare$

Example 5. Let A be a matrix and let B be its associated row echelon matrix. Then the nonzero rows of B form a basis for the row space of A. Indeed, the row space of B is the same as the row space of A and clearly the nonzero rows of B span the row space of B. Moreover, the nonzero rows of B form a linearly independent set because each of these vectors has its first nonzero entry (a corner 1) in a column where all other row vectors have a zero entry. This is enough to guarantee independence. $\blacksquare$

Example 5 leads to a useful method for finding a basis for a subspace V of $\mathbb{R}^n$ when a spanning set for V is known. Given $v_1, \ldots, v_m \in \mathbb{R}^n$ such that $V = \mathcal{L}(v_1, \ldots, v_m)$, a basis for V can be found by forming the matrix A whose rows are $v_1, \ldots, v_m$ (so that V is the row space of A) and reducing A to its row echelon form B; the nonzero rows of B then form a basis for V.

Example 6. To find a basis for the subspace $V = \mathcal{L}((1, -1, 3, 2), (-1, 3, -2, 2), (2, 1, 2, -1), (-1, 0, 2, 7))$ of $\mathbb{R}^4$, we row reduce

$$A = \begin{pmatrix} 1 & -1 & 3 & 2 \\ -1 & 3 & -2 & 2 \\ 2 & 1 & 2 & -1 \\ -1 & 0 & 2 & 7 \end{pmatrix} \quad \text{to} \quad B = \begin{pmatrix} 1 & 0 & 0 & -3 \\ 0 & 1 & 0 & 1 \\ 0 & 0 & 1 & 2 \\ 0 & 0 & 0 & 0 \end{pmatrix}.$$

Thus $\{(1, 0, 0, -3), (0, 1, 0, 1), (0, 0, 1, 2)\}$ is a basis for V. $\blacksquare$

Sometimes, given a spanning set $\{v_1, \ldots, v_m\}$ for a vector space V, it is required to reduce the given spanning set to a basis. In other words, one seeks a basis for V that is a subset of the given spanning set. This can be accomplished by deleting from the given spanning set any vector that is a linear combination of the others to obtain a smaller spanning set, and then repeating this procedure until a linearly independent set is obtained.

Example 7.   To reduce the spanning set $\{(1, -1, 3, 2), \ (-1, 3, -2, 2),$ $(2, 1, 2, -1), (-1, 0, 2, 7)\}$ for the vector space V of Example 6 to a basis, we first check to see if this set is linearly independent (and hence already a basis). Solving the vector equation

$$c_1(1, -1, 3, 2) + c_2(-1, 3, -2, 2) + c_3(2, 1, 2, -1) + c_4(-1, 0, 2, 7) = \mathbf{0}$$

yields the nonzero solution $(c_1, c_2, c_3, c_4) = (-2, -1, 1, 1)$. Hence the given set is linearly dependent and, since each $c_j \neq 0$, each of the vectors can be expressed as a linear combination of the others. It follows that each of the four sets

$$\{(1, -1, 3, 2), (-1, 3, -2, 2), (2, 1, 2, -1)\},$$
$$\{(1, -1, 3, 2), (-1, 3, -2, 2), (-1, 0, 2, 7)\},$$
$$\{(1, -1, 3, 2), (2, 1, 2, -1), (-1, 0, 2, 7)\},$$

and

$$\{(-1, 3, -2, 2), (2, 1, 2, -1), (-1, 0, 2, 7)\}$$

spans V. Moreover, it is straightforward to check that each of these sets is linearly independent, so each is a basis for V.   ■

In Chapters 2 and 3 we found canonical spanning sets for solution spaces of homogeneous linear systems and for solution spaces of homogeneous linear differential equations. It turns out that these canonical spanning sets are, in fact, bases for the solution spaces.

Theorem 1.   *The canonical spanning set for the solution space of a homogeneous system of linear equations is a basis for that space.*

Proof.   Since the canonical spanning set does span the solution space, we need only check linear independence. But the solution procedure described in Chapter 1 reduces the given system to an equivalent homogeneous linear system whose coefficient matrix B is in row echelon form:

$$B = \begin{pmatrix} 1 & 0 & 0 & 0 \\ & 1 & 0 & 0 \\ & & 1 & 0 \\ & & & \ddots \\ & & & & 1 \end{pmatrix}$$

The general solution of the system is then obtained by solving this equivalent system, expressing each of the unknowns corresponding to a corner 1 in terms of the unknowns not corresponding to corner 1's. These latter unknowns are then specified arbitrarily (as $c_1, \ldots, c_{n-r}$) and the canonical spanning set $\{v_1, \ldots, v_{n-r}\}$ is obtained by expressing this general solution in the form

$$x = c_1 v_1 + \cdots + c_{n-r} v_{n-r}.$$

The vectors $v_1, \ldots, v_{n-r}$ have the property that each has a 1 in some spot where all the others have a zero (these spots correspond to the columns of B that *do not* contain corner 1's). This is sufficient to guarantee linear independence. ■

Remark. Since the canonical spanning set for the solution space of a homogeneous linear system is a basis, we shall henceforth call this spanning set the *canonical basis* for the solution space.

Example 8. Consider the system studied in Example 1 of Section 2.4. We found that the row reduced coefficient matrix for this system is

$$B = \begin{pmatrix} 1 & -1 & 0 & -1 & 1 \\ 0 & 0 & 1 & 2 & -3 \\ 0 & 0 & 0 & 0 & 0 \end{pmatrix}$$

and that the canonical spanning set consists of the three vectors

$$v_1 = (\quad 1, 1, \quad 0, 0, 0)$$
$$v_2 = (\quad 1, 0, -2, 1, 0)$$
$$v_3 = (-1, 0, \quad 3, 0, 1).$$

Notice that the second entries in v_1, v_2, v_3 are 1, 0, 0; the fourth entries in v_1, v_2, v_3 are 0, 1, 0; and the fifth entries in v_1, v_2, v_3 are 0, 0, 1. These entries correspond to the second, fourth, and fifth columns of B, which are the columns without corner ones. The set

$$\{(1, 1, 0, 0, 0), (1, 0, -2, 1, 0), (-1, 0, 3, 0, 1)\}$$

is the canonical basis for the solution space of the given system. ■

Example 9. Find a basis for the hyperplane

$$2x_1 - 3x_2 + 4x_3 - x_4 = 0$$

in $\mathbb{R}^4$. To do this, we view the equation as a system of one equation in four unknowns. The matrix

$$(2 \quad -3 \quad 4 \quad -1 \quad 0) \quad \text{reduces to} \quad (1 \quad -\tfrac{3}{2} \quad 2 \quad -\tfrac{1}{2} \quad 0).$$

The equivalent equation

$$x_1 - \tfrac{3}{2}x_2 + 2x_3 - \tfrac{1}{2}x_4 = 0$$

has general solution (obtained by setting $x_2 = c_1$, $x_3 = c_2$, $x_4 = c_3$)

$$\mathbf{x} = (\tfrac{3}{2}c_1 - 2c_2 + \tfrac{1}{2}c_3, c_1, c_2, c_3)$$
$$= c_1(\tfrac{3}{2}, 1, 0, 0) + c_2(-2, 0, 1, 0) + c_3(\tfrac{1}{2}, 0, 0, 1).$$

Hence the canonical basis for the hyperplane is

$$\{(\tfrac{3}{2}, 1, 0, 0), (-2, 0, 1, 0), (\tfrac{1}{2}, 0, 0, 1)\}. \quad \blacksquare$$

Theorem 2. *The canonical spanning set for the solution space of a homogeneous linear differential equation with constant coefficients is a basis for that space.*

Proof. (For equations of order ≤ 3.) We need only check that the canonical spanning set is linearly independent. For equations of order 1, independence is clear since the canonical spanning set consists of one nonzero function. For equations of order 2, independence is also clear since the canonical spanning set always consists of two functions, neither of which is a multiple of the other. We shall now prove the independence of the canonical spanning set for equations of order 3, using Theorem 4 of the previous section.

Case 1. The zeros $\lambda_1, \lambda_2, \lambda_3$ of the auxiliary polynomial are real and distinct. The canonical spanning set is $\{e^{\lambda_1 x}, e^{\lambda_2 x}, e^{\lambda_3 x}\}$. Independence of this set has already been checked in Example 6 of the previous section.

Case 2. The zeros of the auxiliary polynomial are real numbers $\lambda_1, \lambda_1, \lambda_2$ with $\lambda_2 \neq \lambda_1$. The canonical spanning set is $\{e^{\lambda_1 x}, xe^{\lambda_1 x}, e^{\lambda_2 x}\}$ with Wronskian matrix

$$W_x = \begin{pmatrix} e^{\lambda_1 x} & xe^{\lambda_1 x} & e^{\lambda_2 x} \\ \lambda_1 e^{\lambda_1 x} & (1 + \lambda_1 x)e^{\lambda_1 x} & \lambda_2 e^{\lambda_2 x} \\ \lambda_1^2 e^{\lambda_1 x} & (2\lambda_1 + \lambda_1^2 x)e^{\lambda_1 x} & \lambda_2^2 e^{\lambda_2 x} \end{pmatrix}.$$

Setting $x = 0$ and reducing, we get

$$W_0 = \begin{pmatrix} 1 & 0 & 1 \\ \lambda_1 & 1 & \lambda_2 \\ \lambda_1^2 & 2\lambda_1 & \lambda_2^2 \end{pmatrix} \rightarrow \begin{pmatrix} 1 & 0 & 1 \\ 0 & 1 & \lambda_2 - \lambda_1 \\ 0 & 2\lambda_1 & \lambda_2^2 - \lambda_1^2 \end{pmatrix} \rightarrow \begin{pmatrix} 1 & 0 & 1 \\ 0 & 1 & \lambda_2 - \lambda_1 \\ 0 & 0 & (\lambda_2 - \lambda_1)^2 \end{pmatrix}.$$

Since $\lambda_2 \neq \lambda_1$, this matrix has rank 3 so $\{e^{\lambda_1 x}, xe^{\lambda_1 x}, e^{\lambda_2 x}\}$ is linearly independent.

Case 3. The only zero of the auxiliary polynomial is the real number λ (repeated three times). In this case the canonical spanning set is $\{e^{\lambda x}, xe^{\lambda x}, x^2e^{\lambda x}\}$, and the Wronskian matrix W_x is

$$W_x = \begin{pmatrix} e^{\lambda x} & xe^{\lambda x} & x^2e^{\lambda x} \\ \lambda e^{\lambda x} & (1 + \lambda x)e^{\lambda x} & (2x + \lambda x^2)e^{\lambda x} \\ \lambda^2 e^{\lambda x} & (2\lambda + \lambda^2 x)e^{\lambda x} & (2 + 4\lambda x + \lambda^2 x^2)e^{\lambda x} \end{pmatrix} \quad \text{and} \quad W_0 = \begin{pmatrix} 1 & 0 & 0 \\ \lambda & 1 & 0 \\ \lambda^2 & 2\lambda & 2 \end{pmatrix}.$$

Since W_0 clearly has rank 3, $\{e^{\lambda x}, xe^{\lambda x}, x^2e^{\lambda x}\}$ is linearly independent.

Case 4. The auxiliary polynomial has one real zero λ and a pair $\alpha \pm \beta i$ of complex zeros. The canonical spanning set is $\{e^{\alpha x} \cos \beta x, e^{\alpha x} \sin \beta x, e^{\lambda x}\}$. The Wronskian matrix at $x = 0$ is

$$\begin{pmatrix} 1 & 0 & 1 \\ \alpha & \beta & \lambda \\ \alpha^2 - \beta^2 & 2\alpha\beta & \lambda^2 \end{pmatrix} \quad \text{which reduces to} \quad \begin{pmatrix} 1 & 0 & 1 \\ 0 & \beta & \lambda - \alpha \\ 0 & 0 & (\lambda - \alpha)^2 + \beta^2 \end{pmatrix}.$$

Since $\beta \neq 0$, this matrix has rank 3, so $\{e^{\alpha x} \cos \beta x, e^{\alpha x} \sin \beta x, e^{\lambda x}\}$ is linearly independent.

Thus we have shown that, when the order of the equation is ≤ 3, the canonical spanning set is a basis. The proof when the order is greater than three is similar but, of course, more tedious. You will be asked to check some of the higher order cases in the exercises. ∎

Remark. Since the canonical spanning set for the solution space of a homogeneous linear differential equation with constant coefficients is a basis, we shall henceforth call this spanning set the *canonical basis* for the solution space.

EXERCISES

1. Which of the following sets are bases for $\mathbb{R}^2$?
   (a) $\{(1, 2)\}$
   (b) $\{(1, 2), (1, 3)\}$
   (c) $\{(1, 2), (1, 3), (1, 4)\}$
   (d) $\{(1, 2), (2, 4)\}$
   (e) $\{(1, 2), (2, 4), (-3, -6)\}$

2. Which of the following are bases for $\mathbb{R}^3$?
   (a) $\{(-1, 0, 0), (0, -1, 0), (0, 0, -1)\}$
   (b) $\{(1, 0, 0), (1, 1, 0), (1, 1, 1)\}$
   (c) $\{(0, 1, -1), (1, 0, -1), (1, -1, 0)\}$
   (d) $\{(1, 2, 3), (1, 4, 9)\}$
   (e) $\{(1, 2, 3), (1, 4, 9), (1, 8, 27)\}$

(f) $\{(1, 2, 3), (1, 4, 9), (1, 8, 27), (1, 16, 81)\}$

(g) $\{(1, a, a^2), (1, b, b^2), (1, c, c^2)\}$ where $a, b, c \in \mathbb{R}$ are distinct

3. Which of the following are bases for the indicated vector spaces?

(a) $\{1, x + 1, (x + 1)^2\}$ for $\mathcal{P}^3$

(b) $\{0, x, x^2\}$ for $\mathcal{P}^3$

(c) $\{\cosh x, \sinh x\}$ for the solution space of $y'' - y = 0$

(d) $\{e^x, e^{-x}, \sinh x\}$ for the solution space of $y'' - y = 0$

(e) $\{1, \cos x, \cos 2x, \cos 3x\}$ for $\mathcal{L}(1, \cos x, \cos 2x, \cos 3x)$

4. Find a basis for each of the following vector spaces.

(a) the plane $x_1 + x_2 + x_3 = 0$ in $\mathbb{R}^3$

(b) the hyperplane $x_1 - 2x_2 + 3x_3 + x_4 = 0$ in $\mathbb{R}^4$

(c) the hyperplane $3x_1 - 2x_2 - x_4 = 0$ in $\mathbb{R}^4$

(d) the solution space of the linear system

$$x_1 - 5x_2 - 7x_3 = 0$$
$$2x_1 + 2x_2 - 3x_3 = 0$$

(e) the solution space of the linear system

$$2x_1 - x_2 - x_3 + 2x_4 = 0$$
$$3x_1 + 2x_2 - x_3 - x_4 = 0$$

5. Find a basis for each of the following vector spaces.

(a) $\mathcal{L}((-1, 3, 4), (2, 1, 3), (-4, 5, 5))$

(b) $\mathcal{L}((1, -3, 2), (-1, 3, 2), (2, 3, 4), (7, -3, 14))$

(c) $\mathcal{L}(1, \sin 2x, \cos 2x, \sin^2 x, \cos^2 x)$

(d) $\mathcal{L}(x^2 + x + 1, x^2 - x + 1, x^2 + 1)$

6. Find a basis for the solution space of each of the following homogeneous linear differential equations.

(a) $y' - 16y = 0$ (d) $y''' - y'' + y' - y = 0$

(b) $y'' - 16y = 0$ (e) $4y'' + 4y' + y = 0$

(c) $y^{(4)} - 16y = 0$

7. Show that the canonical spanning set for the solution space of the fourth order linear differential equation

$$a_4 y^{(4)} + a_3 y''' + a_2 y'' + a_1 y' + a_0 y = 0$$

is linearly independent

(a) when the four zeros of the auxiliary polynomial are all equal

(b) when the four zeros of the auxiliary polynomial are real and distinct

4.3 COORDINATES

Given a spanning set $\{v_1, \ldots, v_n\}$ for a vector space V, it is possible to express each vector $v \in V$ as a linear combination

$$v = c_1 v_1 + c_2 v_2 + \cdots + c_n v_n$$

of the vectors in the spanning set. The coefficients $(c_1, \ldots, c_n)$ of this linear combination are, in general, not uniquely determined. The vector equation

$$c_1 \mathbf{v}_1 + \cdots + c_n \mathbf{v}_n = \mathbf{v}$$

will usually have many solutions $(c_1, \ldots, c_n)$. But if the spanning set is a basis, there will be only one solution. This is a consequence of the following theorem.

Theorem 1.   *Let* $\{\mathbf{v}_1, \ldots, \mathbf{v}_n\}$ *be a linearly independent set in a vector space* V. *Suppose*

$$a_1 \mathbf{v}_1 + \cdots + a_n \mathbf{v}_n = b_1 \mathbf{v}_1 + \cdots + b_n \mathbf{v}_n$$

$(a_i, b_i \in \mathbb{R})$. *Then* $a_1 = b_1$, $a_2 = b_2, \ldots, a_n = b_n$.

Proof.   If

$$a_1 \mathbf{v}_1 + \cdots + a_n \mathbf{v}_n = b_1 \mathbf{v}_1 + \cdots + b_n \mathbf{v}_n$$

then

$$(a_1 - b_1)\mathbf{v}_1 + \cdots + (a_n - b_n)\mathbf{v}_n = \mathbf{0}.$$

But, since $\{\mathbf{v}_1, \ldots, \mathbf{v}_n\}$ is linearly independent, this implies that

$$a_1 - b_1 = 0, \ldots, a_n - b_n = 0$$

or

$$a_1 = b_1, \ldots, a_n = b_n. \quad \blacksquare$$

It follows from Theorem 1 that, given a basis $\{\mathbf{v}_1, \ldots, \mathbf{v}_n\}$ for a vector space V, each vector $\mathbf{v} \in V$ determines, and is determined by, a unique n-tuple $(c_1, \ldots, c_n) \in \mathbb{R}^n$ through the equation

$$\mathbf{v} = c_1 \mathbf{v}_1 + \cdots + c_n \mathbf{v}_n.$$

The numbers $c_1, \ldots, c_n$ are called the **coordinates** of $\mathbf{v}$ relative to the given basis, and the n-tuple, $(c_1, \ldots, c_n)$ is called the **coordinate n-tuple** of $\mathbf{v}$ relative to the basis. Note, however, that if the order of the basis vectors is changed, then the order of the coordinates changes correspondingly so the coordinate n-tuple depends on the order in which the basis vectors appear. For this reason, we must consider **ordered bases** $(\mathbf{v}_1, \ldots, \mathbf{v}_n)$ for V: two *ordered* bases $(\mathbf{v}_1, \ldots, \mathbf{v}_n)$ and $(\mathbf{w}_1, \ldots, \mathbf{w}_n)$ are equal if and only if $\mathbf{v}_1 = \mathbf{w}_1, \ldots, \mathbf{v}_n = \mathbf{w}_n$. If $\mathbf{B} = (\mathbf{v}_1, \ldots, \mathbf{v}_n)$ is an ordered basis for V and $\mathbf{v} \in V$, then the numbers $c_1, \ldots, c_n$ such that

$$\mathbf{v} = c_1 \mathbf{v}_1 + \cdots + c_n \mathbf{v}_n$$

are called the **B-coordinates** of $\mathbf{v}$, and the n-tuple $(c_1, \ldots, c_n)$ is called the **B-coordinate n-tuple** of $\mathbf{v}$.

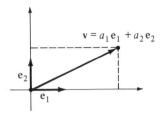

FIGURE 4.1

Example 1. Let $V = \mathbb{R}^n$ and let $\mathbf{v} = (a_1, a_2, \ldots, a_n) \in \mathbb{R}^n$. Then

$$\mathbf{v} = a_1(1, 0, \ldots, 0) + a_2(0, 1, 0, \ldots, 0) + \cdots + a_n(0, 0, \ldots, 0, 1)$$
$$= a_1\mathbf{e}_1 + a_2\mathbf{e}_2 + \cdots + a_n\mathbf{e}_n.$$

Thus the coordinates of $\mathbf{v} \in \mathbb{R}^n$ relative to the standard ordered basis $(\mathbf{e}_1, \ldots, \mathbf{e}_n)$ for $\mathbb{R}^n$ are the usual *Cartesian coordinates* of $\mathbf{v}$ (see Figure 4.1.) ■

Example 2. Let $V = \mathbb{R}^2$ and let $\mathbf{B}_1 = ((1, 0), (0, 1))$, $\mathbf{B}_2 = ((0, 1), (1, 0))$, and $\mathbf{B}_3 = ((0, 1), (1, 1))$. Then $\mathbf{B}_1$, $\mathbf{B}_2$, and $\mathbf{B}_3$ are distinct ordered bases for $\mathbb{R}^2$. If we express the vector $(2, 3) \in \mathbb{R}^2$ in terms of these ordered bases we obtain

(i) $(2, 3) = 2(1, 0) + 3(0, 1)$
(ii) $(2, 3) = 3(0, 1) + 2(1, 0)$
(iii) $(2, 3) = 1(0, 1) + 2(1, 1)$

(see Figure 4.2). Hence $(2, 3)$ has $\mathbf{B}_1$-coordinate pair $(2, 3)$, $\mathbf{B}_2$-coordinate pair $(3, 2)$, and $\mathbf{B}_3$-coordinate pair $(1, 2)$. ■

Example 3. Let V be the plane $x_1 - 3x_2 - 2x_3 = 0$ through $\mathbf{0}$ in $\mathbb{R}^3$. The canonical spanning set for the solution space of the equation $x_1 - 3x_2 - 2x_3 = 0$ is

$$\{\mathbf{v}_1 = (3, 1, 0), \mathbf{v}_2 = (2, 0, 1)\}$$

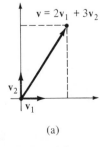

(a)

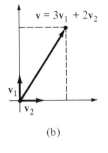

(b)

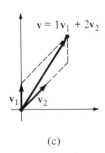

(c)

FIGURE 4.2

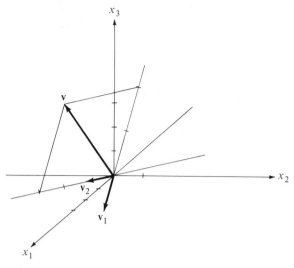

FIGURE 4.3

so $\mathbf{B} = ((3, 1, 0), (2, 0, 1))$ is an ordered basis for V. Using $\mathbf{B}$ we obtain a one-to-one correspondence between V and $\mathbb{R}^2$, assigning to each $\mathbf{v} \in V$ its $\mathbf{B}$-coordinate pair. Thus, for example, the $\mathbf{B}$-coordinate pair of the vector $\mathbf{v} = (0, -2, 3) \in V$ is $(-2, 3)$, since $\mathbf{v} = -2\mathbf{v}_1 + 3\mathbf{v}_2$. The $\mathbf{B}$-coordinate pair of any $(a, b, c) \in V (a = 3b + 2c)$ is just (b, c).

Notice that this one-to-one correspondence between points in the plane V and pairs of real numbers is obtained by a process exactly analagous to the process of setting up a Cartesian coordinate system (see Figure 4.3). We first choose a line in V through $\mathbf{0}$ (the line $\mathbf{x} = t\mathbf{v}_1$) and a point $(\mathbf{v}_1)$ on that line to which we will assign the coordinates $(1, 0)$. Then we choose another line in V (the line $\mathbf{x} = t\mathbf{v}_2$) through $\mathbf{0}$ and a point $(\mathbf{v}_2)$ on that line to which we will assign the coordinates $(0, 1)$. We do not require, however, that the second line be perpendicular to the first line, or that $\|\mathbf{v}_2\| = \|\mathbf{v}_1\|$. The coordinates of any point are then obtained by constructing an appropriate parallelogram (a rectangle in standard Cartesian coordinates.) ■

Example 4. Let $V = \mathcal{P}^3$ and let $\mathbf{B}_1 = (1, x, x^2)$ and $\mathbf{B}_2 = (1, x - 1, (x - 1)^2)$. Then $\mathbf{B}_1$ and $\mathbf{B}_2$ are ordered bases for $\mathcal{P}^3$. The $\mathbf{B}_1$-coordinate triple of

$$p(x) = a_0 + a_1 x + a_2 x^2$$

is clearly (a_0, a_1, a_2), whereas the $\mathbf{B}_2$-coordinate triple of $p(x)$ is $(a_0 + a_1 + a_2, a_1 + 2a_2, a_2)$, since

$$a_0 + a_1 x + a_2 x^2 = (a_0 + a_1 + a_2) + (a_1 + 2a_2)(x - 1) + a_2(x - 1)^2. \quad ■$$

Remark.   One way to find the expansion, in Example 4, for $a_0 + a_1x + a_2x^2$ in powers of $x - 1$ is to write

$$a_0 + a_1x + a_2x^2 = c_0 + c_1(x - 1) + c_2(x - 1)^2,$$

expand the right hand side, equate coefficients of like powers of x, and solve for a_0, a_1, a_2. A better way is simply to find the Taylor series expansion of the function $p(x) = a_0 + a_1x + a_2x^2$ about the point $x_0 = 1$. The coefficients are

$$c_0 = p(1) = a_0 + a_1 + a_2, \ c_1 = p'(1) = a_1 + 2a_2, \text{ and } c_2 = \tfrac{1}{2}p''(1) = a_2.$$

An important property of coordinates is that addition of vectors in V corresponds to addition of their coordinate n-tuples in $\mathbb{R}^n$ and that scalar multiplication in V corresponds to scalar multiplication of coordinate n-tuples in $\mathbb{R}^n$. That is, if $\mathbf{B}$ is an ordered basis for V, if $\mathbf{v}$ and $\mathbf{w} \in V$ have $\mathbf{B}$-coordinate n-tuples $(a_1, \ldots, a_n)$ and $(b_1, \ldots, b_n)$ respectively, and if $c \in \mathbb{R}$, then

$$\mathbf{v} + \mathbf{w} \text{ has } \mathbf{B}\text{-coordinate } n\text{-tuple } (a_1 + b_1, \ldots, a_n + b_n)$$

and

$$c\mathbf{v} \text{ has } \mathbf{B}\text{-coordinate } n\text{-tuple } (ca_1, \ldots, ca_n)$$

(see Exercise 6). Thus the assignment of $\mathbf{B}$-coordinates to each vector in V defines a one-to-one correspondence between V and $\mathbb{R}^n$ that respects vector addition and scalar multiplication. Such a correspondence allows vector computations in V to be converted into numerical computations in $\mathbb{R}^n$.

Example 5.   The **standard ordered basis** for $\mathcal{P}^n$ is the ordered basis $\mathbf{B} = (1, x, x^2, \ldots, x^{n-1})$. The $\mathbf{B}$-coordinate n-tuple of $p(x) = a_0 + a_1x + \cdots + a_{n-1}x^{n-1}$ is $(a_0, a_1, \ldots, a_{n-1})$. The observation that the correspondence

$$p(x) \leftrightarrow (a_0, a_1, \ldots, a_{n-1})$$

respects addition and scalar multiplication amounts to the familiar fact that polynomials can be added to one another and can be multiplied by real numbers simply by performing the corresponding operations on the coefficients. ∎

EXERCISES

1.   Find the $\mathbf{B}$-coordinate pair for $\mathbf{v} \in V$ relative to the ordered basis $\mathbf{B}$ for V when
   (a) $V = \mathbb{R}^2$, $\mathbf{B} = ((1, 2), (1, 3))$, $\mathbf{v} = (1, 4)$
   (b) V is the plane $x_1 - 3x_2 + 5x_3 = 0$, $\mathbf{B} = ((3, 1, 0), (-5, 0, 1))$,
         $\mathbf{v} = (7, -1, -2)$

    (c) V is the plane $x_1 - 3x_2 + 5x_3 = 0$, $\mathbf{B} = ((0, 5, 3), (-5, 0, 1))$,
       $\mathbf{v} = (7, -1, -2)$

    (d) $V = \mathcal{P}^2$, $\mathbf{B} = (x - 1, x - 2)$, $\mathbf{v} = 1 + x$

2.   What is the $\mathbf{B}$-coordinate of $\mathbf{w} = a\mathbf{v}$ relative to the ordered basis $\mathbf{B} = (\mathbf{v})$ for the line $\mathbf{x} = t\mathbf{v}$? Relative to the ordered basis $\mathbf{B} = (2\mathbf{v})$?

3.   Find the vector $\mathbf{v} \in V$ whose $\mathbf{B}$-coordinate pair is $(3, -2)$ when
    (a) $V = \mathbb{R}^2$, $\mathbf{B} = ((1, 2), (1, 3))$
    (b) V is the plane $x_1 - 3x_2 + 5x_3 = 0$, $\mathbf{B} = ((-2, 1, 1), (3, 1, 0))$
    (c) $V = \mathcal{P}^2$, $\mathbf{B} = (x - 1, x + 1)$
    (d) V is the solution space of $y'' - y = 0$, $\mathbf{B} = (e^x, e^{-x})$

4.   Find the $\mathbf{B}$-coordinates of each of the following vectors relative to the given ordered basis $\mathbf{B}$.
    (a) $(3, -2, 1)$ relative to $\mathbf{B} = ((1, 1, 1), (1, 1, 0)\ (1, 0, 0))$ in $\mathbb{R}^3$
    (b) $x^2 + 2x + 1$ relative to $\mathbf{B} = (x^2, 2x - 1, 3)$ in $\mathcal{P}^3$
    (c) $(-1, 1, 4)$ relative to $\mathbf{B} = ((1, 1, 0), (1, 0\ 1), (0, 1, 1))$ in $\mathbb{R}^3$
    (d) $\sinh x$ relative to $\mathbf{B} = (e^x, e^{-x})$ in $\mathcal{L}(e^x, e^{-x})$
    (e) $\cos^2 \dfrac{x}{2}$ relative to $\mathbf{B} = (1, \sin x, \cos x)$ in the solution space of
$$D(D^2 + 1)y = 0$$

5.   (a)   Show that, for each $a \in \mathbb{R}$, the ordered set
$$\mathbf{B} = (1, x - a, (x - a)^2, \ldots, (x - a)^{n-1})$$
   is an ordered basis for $\mathcal{P}^n$.
    (b)   Find the $\mathbf{B}$-coordinates ($\mathbf{B}$ as in part (a)) of $p(x) \in \mathcal{P}^n$ in terms of the value of p and its derivatives at a.

6.   Show that if $\mathbf{B}$ is an ordered basis for a vector space V, if $\mathbf{v}$ and $\mathbf{w} \in V$ have $\mathbf{B}$-coordinate n-tuples $(a_1, \ldots, a_n)$ and $(b_1, \ldots, b_n)$ respectively, and if $c \in \mathbb{R}$, then
    (a) $\mathbf{v} + \mathbf{w}$ has $\mathbf{B}$-coordinate n-tuple $(a_1 + b_1, \ldots, a_n + b_n)$
    (b) $c\mathbf{v}$ has $\mathbf{B}$-coordinate n-tuple $(ca_1, \ldots, ca_n)$

4.4 DIMENSION

    In Section 4.2 we studied bases for vector spaces, and we saw that they are not unique. Indeed, we saw many instances where several different bases for the same vector space arose rather naturally (e.g., see Examples 6 and 7). You may have noticed, however, that each basis for a given vector space has had the same number of elements. We shall now show that, in general, *if a vector space has a basis with n elements, then every basis for that vector space has n elements.* This fact is a corollary of the following theorem.

Theorem 1. *Suppose $\{\mathbf{v}_1, \ldots, \mathbf{v}_m\}$ is a spanning set for the vector space V and suppose $\{\mathbf{w}_1, \ldots, \mathbf{w}_n\}$ is a linearly independent set in V. Then $m \geq n$.*

Proof. Since $\{v_1, \ldots, v_m\}$ spans V, each vector w_i can be expressed as a linear combination of $v_1, \ldots, v_m$. Thus there exist real numbers $a_{11}, a_{21}, \ldots, a_{mn}$ such that

$$w_1 = a_{11}v_1 + a_{21}v_2 + \cdots + a_{m1}v_m$$

$$w_2 = a_{12}v_1 + a_{22}v_2 + \cdots + a_{m2}v_m$$

(1)

$$\cdots$$

$$w_n = a_{1n}v_1 + a_{2n}v_2 + \cdots + a_{mn}v_m.$$

Since $\{w_1, \ldots, w_n\}$ is linearly independent, the only solution $(c_1, \ldots, c_n)$ of the equation

(2)
$$c_1 w_1 + \cdots + c_n w_n = 0$$

is the zero solution, $(c_1, \ldots, c_n) = (0, \ldots, 0)$. But, multiplying the ith equation in (1) by $c_i (1 \leq i \leq n)$ and adding the resulting equations, we obtain

$$c_1 w_1 + \cdots + c_n w_n = (a_{11} c_1 + \cdots + a_{1n} c_n) v_1$$

$$+ \cdots$$

$$+ (a_{m1} c_1 + \cdots + a_{mn} c_n) v_m.$$

Thus $(c_1, \ldots, c_n)$ is a solution of (2) if it is a solution of the system

$$a_{11} c_1 + \cdots + a_{1n} c_n = 0$$

$$\cdots$$

$$a_{m1} c_1 + \cdots + a_{mn} c_n = 0.$$

By the corollary to Theorem 1 of Section 2.4, this system has a nonzero solution whenever $m < n$. Thus the independence of $\{w_1, \ldots, w_n\}$ implies $m \geq n$. ∎

Corollary. *Suppose $\{v_1, \ldots, v_n\}$ and $\{w_1, \ldots, w_m\}$ are both bases for the same vector space V. Then $n = m$.*

Proof. Since $\{v_1, \ldots, v_n\}$ spans V and $\{w_1, \ldots, w_m\}$ is linearly independent, $m \leq n$. On the other hand, since $\{w_1, \ldots, w_m\}$ spans V and $\{v_1, \ldots, v_n\}$ is linearly independent, $n \leq m$. Thus $m = n$. ∎

Thus we have shown that any two bases for the same vector space V must have the same number of elements. This number, written **dim V,** is called the **dimension** of the vector space V.

Remark. Not every vector space has a (finite) basis. For example, we saw in Section 2.3, that the vector space $\mathcal{P}$ of all polynomials with real coefficients has no finite spanning set and hence it has no finite basis. Vector spaces that do have finite bases are called *finite dimensional,* and vector spaces that do not have finite bases are called *infinite dimensional.* Thus the space $\mathcal{P}$ of polynomials is an infinite dimensional vector space.

Remark. In defining the notions of spanning set, basis, and dimension, we have deliberately ignored the very special vector space $V = \{\mathbf{0}\}$, which consists of the zero vector alone. Certainly the subset $\{\mathbf{0}\}$ spans V, but it is not independent, since, for example $1 \cdot \mathbf{0} = \mathbf{0}$, and $1 \neq 0$. Thus V appears not to have a basis. We remedy this by *defining* the dimension of $\{\mathbf{0}\}$ to be 0. Thus, the only vector space of dimension 0 is a vector space whose only element is the zero vector. Some authors define the empty subset $\varnothing$ of V to be a basis for V. This is consistent with our definition of dim V, since the number of elements in the empty set $\varnothing$ is 0, the dimension of V.

Example 1. The dimension of $\mathbb{R}^n$ is n, since the standard basis $\{\mathbf{e}_1, \ldots, \mathbf{e}_n\}$ for $\mathbb{R}^n$ has n elements. ■

Example 2. The dimension of the space $\mathcal{P}^n$ of all polynomials of degree less than n is also n, since $\{1, x, x^2, \ldots, x^{n-1}\}$ is a basis for $\mathcal{P}^n$ with n elements. ■

Example 3. The dimension of the row space of a matrix A is equal to the rank of A, since the rank of a matrix A is equal to the number of nonzero rows in its row echelon matrix and these nonzero rows form a basis for the row space of A. ■

Example 4. The solution space of a homogeneous system of m linear equations in n unknowns has dimension $n - r$, where r is the rank of the system, because the canonical spanning set for the solution space is a basis and it has $n - r$ elements. In particular, the dimension of a hyperplane $a_1 x_1 + \cdots + a_n x_n = 0$ through $\mathbf{0}$ in $\mathbb{R}^n$ is $n - 1$. ■

Example 5. The solution space of a homogeneous linear differential equation with constant coefficients has dimension n, where n is the order of the equation. This is because the canonical spanning set for the solution space is a basis and it has n elements. ■

The following theorem is an easy but very useful consequence of Theorem 1.

Theorem 2. *Let V be a finite dimensional vector space and suppose $\{\mathbf{v}_1, \ldots, \mathbf{v}_n\}$ is a subset of V.*
     (i) *If $n = \dim V$ and $\{\mathbf{v}_1, \ldots, \mathbf{v}_n\}$ is linearly independent, then $\{\mathbf{v}_1, \ldots, \mathbf{v}_n\}$ is a basis for V.*
     (ii) *If $n = \dim V$ and $\{\mathbf{v}_1, \ldots, \mathbf{v}_n\}$ spans V, then $\{\mathbf{v}_1, \ldots, \mathbf{v}_n\}$ is a basis for V.*

Proof. (i) If $\{v_1, \ldots, v_n\}$ is linearly independent but is not a basis for V, then it must not span V. Hence there must be a vector $v_{n+1} \in V$ that is not in $\mathcal{L}(v_1, \ldots, v_n)$. But then $\{v_1, \ldots, v_{n+1}\}$ would be an independent set in V with more than $n = \dim V$ elements; this is impossible by Theorem 1. Hence $\{v_1, \ldots, v_n\}$ must be a basis for V.

(ii) If $\{v_1, \ldots, v_n\}$ spans V but is not a basis for V, then it can be reduced to a basis for V by deleting elements. But this basis will have fewer than n elements, which is impossible by the Corollary to Theorem 1. Hence $\{v_1, \ldots, v_n\}$ must be a basis for V. ■

Example 6. To show that $\{\cosh x, \sinh x\}$ is a basis for the solution space of the linear differential equation $y'' - y = 0$, it suffices to observe that the dimension of the solution space is 2 and that $\{\cosh x, \sinh x\}$ contains 2 elements and is a linearly independent subset of this space. ■

Theorem 3. *A set of m vectors in $\mathbb{R}^n$ is linearly independent if and only if the matrix whose rows are these vectors has rank m.*

Proof. Let A be the matrix whose rows are the given vectors. If these vectors form a linearly independent set, then they form a basis for the row space of A so, by Example 3, rank $A = m$. Conversely, if A has rank m then the dimension of the row space of A is m (Example 3) so the row vectors of A form a basis for the row space, by Theorem 2(ii), and in particular they form a linearly independent set. ■

Example 7. To show that the set

$$\{(-1, 2, 0, -3), (-2, 1, 1, 2), (2, 4, -1, 3)\}$$

is linearly independent, we need only reduce the matrix

$$A = \begin{pmatrix} -1 & 2 & 0 & -3 \\ -2 & 1 & 1 & 2 \\ 2 & 4 & -1 & 3 \end{pmatrix} \quad \text{to} \quad \begin{pmatrix} 1 & -2 & 0 & 3 \\ 0 & 1 & 0 & 1 \\ 0 & 0 & 1 & 11 \end{pmatrix}$$

and observe that the rank is 3. ■

It it interesting to compare Theorem 3 above with Theorem 2 of Section 4.1, which states that a set of vectors in $\mathbb{R}^n$ is linearly independent if and only if the matrix whose *columns* are the given vectors has rank m. The theorem of this section is easier to apply because it uses a matrix with fewer rows. (If $m > n$ then we know that any set of m vectors in $\mathbb{R}^n$ must be linearly dependent, so in all interesting problems we will have $m \leq n$.)

In Section 4.2 you learned how to construct a basis for a finite dimensional vector space by deleting vectors from a linearly dependent spanning set. The next theorem shows that it is also possible to construct a basis for a finite dimensional vector space by enlarging a linearly independent set.

Theorem 4.   *Let $\{\mathbf{v}_1, \ldots, \mathbf{v}_k\}$ be a linearly independent subset of a finite dimensional vector space V. Then there is a basis for V containing $\{\mathbf{v}_1, \ldots, \mathbf{v}_k\}$ as a subset.*

Proof.   If $\{\mathbf{v}_1, \ldots, \mathbf{v}_k\}$ spans V, then it is itself a basis. If not, we can find $\mathbf{v}_{k+1} \in V$ such that $\mathbf{v}_{k+1} \notin \mathcal{L}(\mathbf{v}_1, \ldots, \mathbf{v}_k)$. The set $\{\mathbf{v}_1, \ldots, \mathbf{v}_{k+1}\}$ will be linearly independent. If it spans V, then we are done. If not, we can find $\mathbf{v}_{k+2} \in V$ such that $\mathbf{v}_{k+2} \notin \mathcal{L}(\mathbf{v}_1, \ldots, \mathbf{v}_{k+1})$ and get a linearly independent set $\{\mathbf{v}_1, \ldots, \mathbf{v}_{k+2}\}$. Continuing in this way, we will finally obtain a linearly independent set $\{\mathbf{v}_1, \ldots, \mathbf{v}_k, \mathbf{v}_{k+1}, \ldots, \mathbf{v}_n\}$ where $n = \dim V$. By Theorem 2, this set will be a basis for V. ∎

Remark.   In practice, the easiest way to enlarge an independent set $\{\mathbf{v}_1, \ldots, \mathbf{v}_k\}$ to a basis is as follows. Start with any basis $\{\mathbf{w}_1, \ldots, \mathbf{w}_n\}$ for V. Find the first vector in the list $\mathbf{w}_1, \ldots, \mathbf{w}_n$ that does not lie in $\mathcal{L}(\mathbf{v}_1, \ldots, \mathbf{v}_k)$; call it $\mathbf{v}_{k+1}$. Then continue on through the list $\mathbf{w}_1, \ldots, \mathbf{w}_n$ to find a vector that does not lie in $\mathcal{L}(\mathbf{v}_1, \ldots, \mathbf{v}_{k+1})$; call it $\mathbf{v}_{k+2}$. Continue in this way until your enlarged set has $n = \dim V$ elements. This set will be the required basis.

Example 8.   Find a basis for $\mathbb{R}^3$ containing the independent set $\{(0, 1, 1), (1, 1, 1)\}$. We shall enlarge this by adjoining elements of the standard basis $\{\mathbf{e}_1, \mathbf{e}_2, \mathbf{e}_3\}$ for $\mathbb{R}^3$. Since

$$\mathbf{e}_1 = (1, 0, 0) = (-1)(0, 1, 1) + 1(1, 1, 1) \in \mathcal{L}((0, 1, 1), (1, 1, 1))$$

we discard $\mathbf{e}_1$. Since the equation

$$\mathbf{e}_2 = (0, 1, 0) = c_1(0, 1, 1) + c_2(1, 1, 1)$$

has no solution, $\mathbf{e}_2 \notin \mathcal{L}((0, 1, 1), (1, 1, 1))$ so we adjoin $\mathbf{e}_2$ to our given independent set to obtain the independent set

$$\{(0, 1, 1), (1, 1, 1), (0, 1, 0)\}.$$

Since this independent set has $3 = \dim \mathbb{R}^3$ elements, it is a basis for $\mathbb{R}^3$. ∎

Theorem 5.   *Let V be a finite dimensional vector space and let W be a subspace of V. Then W is also finite dimensional, and $\dim W \leq \dim V$. Moreover, if $\dim W = \dim V$ then $W = V$.*

Proof. If $W = \{0\}$ then certainly W is finite dimensional and dim $W = 0 \leq$ dim V. If $W \neq \{0\}$, we proceed as in the proof of Theorem 4. First take $\mathbf{w}_1 \in W$ with $\mathbf{w}_1 \neq \mathbf{0}$. If $\mathcal{L}(\mathbf{w}_1) = W$, then $\{\mathbf{w}_1\}$ is a basis for W, so W is finite dimensional. If $\mathcal{L}(\mathbf{w}_1) \neq W$, take $\mathbf{w}_2 \in W$ with $\mathbf{w}_2 \notin \mathcal{L}(\mathbf{w}_1)$. Then $\{\mathbf{w}_1, \mathbf{w}_2\}$ is independent. If $\mathcal{L}(\mathbf{w}_1, \mathbf{w}_2) = W$ then $\{\mathbf{w}_1, \mathbf{w}_2\}$ is a basis for W and so W is finite dimensional. If $\mathcal{L}(\mathbf{w}_1, \mathbf{w}_2) \neq W$, take $\mathbf{w}_3 \in W$ with $\mathbf{w}_3 \notin \mathcal{L}(\mathbf{w}_1, \mathbf{w}_2)$. Continuing in this way, we obtain successively larger independent sets $\{\mathbf{w}_1, \ldots, \mathbf{w}_k\}$ in W. The process will stop when $\mathcal{L}(\mathbf{w}_1, \ldots, \mathbf{w}_k) = W$, and this must happen for some $k \leq n = $ dim V because it is impossible to find more than n vectors in V that form a linearly independent set. Thus W is finite dimensional, $\{\mathbf{w}_1, \ldots, \mathbf{w}_k\}$ is a basis for W, and dim $W = k \leq$ dim V.

Finally, if dim $W = $ dim V, then $k = n$ and $\{\mathbf{w}_1, \ldots, \mathbf{w}_n\}$ is a basis for V, by Theorem 2(i). Hence $V = \mathcal{L}(\mathbf{w}_1, \ldots, \mathbf{w}_n) = W$. ∎

Example 9. The vector space $\mathcal{F}(\mathbb{R})$ of real valued functions with domain $\mathbb{R}$ is infinite dimensional. If it were finite dimensional, then each of its subspaces would also be finite dimensional. But the space $\mathcal{P}$ of all polynomial functions is a subspace of $\mathcal{F}(\mathbb{R})$ and, as we have seen, $\mathcal{P}$ is not finite dimensional. ∎

EXERCISES

1. Use Theorem 3 to test each of the following sets for linear independence.
   (a) $\{(1, 5, -4), (1, 6, -4)\}$
   (b) $\{(1, 5, -4), (1, 6, -4), (1, 7, -4)\}$
   (c) $\{(1, 1, 1), (1, 2, 3), (1, 4, 9)\}$
   (d) $\{(1, 2, 3), (1, 4, 9), (1, 8, 27)\}$
   (e) $\{(2, 1, -3, 4), (3, 0, 2, -1), (1, 2, -8, 9)\}$
   (f) $\{(0, 1, -1, 5), (4, 3, 2, 1), (1, 2, 3, 4)\}$

2. Find a basis for $\mathbb{R}^3$ containing the given independent set.
   (a) $\{(1, 1, 2)\}$
   (b) $\{(1, 0, 1), (1, 2, 1)\}$
   (c) $\{(1, 1, 1), (1, 1, 0), (0, 1, 1)\}$

3. Find a basis for $\mathcal{P}^3$ containing the given independent set.
   (a) $\{1, x - 1\}$
   (b) $\{x^2 - 1, x^2 + 2x + 1\}$
   (c) $\{2, x^2 + 2\}$

4. Find the dimension of each of the following vector spaces.
   (a) the hyperplane $3x_2 - x_3 + x_4 = 0$ in $\mathbb{R}^4$
   (b) the subspace of $\mathfrak{M}_{m \times n}$ consisting of all diagonal matrices
   $$\begin{pmatrix} \lambda_1 & \cdots & 0 \\ \vdots & \lambda_2 & \vdots \\ 0 & \cdots & \lambda_n \end{pmatrix}, \lambda_1, \ldots, \lambda_n \in \mathbb{R}$$

(c) $\mathfrak{M}_{m \times n}$, the vector space of $m \times n$ matrices

(d) The space of solutions of the differential equation $(D^3 - D^2 + D)y = 0$

(e) The subspace of $\mathbb{R}^4$ consisting of all solutions of the linear system

$$
\begin{aligned}
x_1 - x_2 + x_3 \quad\quad &= 0 \\
x_1 \quad\quad + x_3 - x_4 &= 0
\end{aligned}
$$

(f) The subspace of $\mathbb{R}^n$ consisting of all vectors $\mathbf{x}$ such that $\mathbf{a}_1 \cdot \mathbf{x} = 0$, $\mathbf{a}_2 \cdot \mathbf{x} = 0, \ldots, \mathbf{a}_k \cdot \mathbf{x} = 0$, where $\{\mathbf{a}_1, \mathbf{a}_2, \ldots, \mathbf{a}_k\}$ is a linearly independent set in $\mathbb{R}^n$

5. Show that the vector space $\mathfrak{D}$ of differentiable real valued functions with domain $\mathbb{R}$ is infinite dimensional.

6. Suppose that V is a vector space of dimension 5, and that S is a subset of V containing 6 elements.

   (a) Show by example that S can be a spanning set for V

   (b) Show by example that S need not be a spanning set for V

7. Find all bases for the vector space $\mathbb{R}^1$.

8. Let V be an n-dimensional vector space. Show that V has a subspace of each dimension $0, 1, 2, \ldots, n$. Show also that V has only one subspace of dimension n, namely V itself.

9. Let V be an n-dimensional vector space. Show that if V has exactly one subspace of dimension k, then $k = 0$ or n.

10. Let V be the subset of $\mathbb{R}^\infty$ consisting of all convergent sequences:

$$V = \{(a_1, a_2, a_3, \ldots) \,|\, \lim a_n \text{ exists}\}.$$

   (a) Show that V is a subspace of $\mathbb{R}^\infty$.

   (b) Show that V is infinite dimensional.

Determinants

FIVE

5.1 ELEMENTARY PROPERTIES

You may have already studied determinants of 2×2 matrices, and perhaps even 3×3 matrices. In this section we shall study determinants of $n \times n$ matrices. We shall see that determinants are useful for testing sets of vectors for linear independence. In particular, we shall see that the determinant of a matrix is nonzero if and only if its row vectors form a linearly independent set.

We begin by reviewing 2×2 determinants.

The **determinant** of a 2×2 matrix

$$A = \begin{pmatrix} a & b \\ c & d \end{pmatrix}, \text{ where } a, b, c, \text{ and } d \in \mathbb{R},$$

is defined by the formula

$$\det A = ad - bc.$$

Notice that det is a function that assigns to each 2×2 matrix a real number. This function has the following four important properties:

(i) If B is obtained from A by multiplying a row of A by a real number r, then $\det B = r \det A$.

(ii) If B is obtained from A by interchanging the rows, then $\det B = -\det A$.

(iii) If B is obtained from A by adding a scalar multiple of one row to the other, then $\det B = \det A$.

(iv) $\det \begin{pmatrix} 1 & 0 \\ 0 & 1 \end{pmatrix} = 1.$

These four properties are easy to check. We have

$$\det \begin{pmatrix} ra & rb \\ c & d \end{pmatrix} = (ra)d - (rb)c = r(ad - bc) = r \det \begin{pmatrix} a & b \\ c & d \end{pmatrix}$$

and

$$\det \begin{pmatrix} a & b \\ rc & rd \end{pmatrix} = a(rd) - b(rc) = r(ad - bc) = r \det \begin{pmatrix} a & b \\ c & d \end{pmatrix},$$

which proves (i). Similarly,

$$\det \begin{pmatrix} c & d \\ a & b \end{pmatrix} = cb - da = -(ad - bc) = -\det \begin{pmatrix} a & b \\ c & d \end{pmatrix}$$

which proves (ii). To prove (iii), we compute

$$\det \begin{pmatrix} a + rc & b + rd \\ c & d \end{pmatrix} = (a + rc)d - (b + rd)c = ad - bc = \det \begin{pmatrix} a & b \\ c & d \end{pmatrix},$$

which shows that the determinant is unchanged when we add a scalar multiple of the second row to the first. The proof that the determinant is unchanged when we add a scalar multiple of the first row to the second is similar. Finally, property (iv) follows directly from the definition, since $\det \begin{pmatrix} 1 & 0 \\ 0 & 1 \end{pmatrix} = 1 \cdot 1 - 0 \cdot 0 = 1$.

The four properties (i)–(iv) are especially important because they characterize the determinant of 2×2 matrices.

Theorem 1.   *The determinant* $\det: \mathfrak{M}_{2 \times 2} \to \mathbb{R}$ *is the only real valued function on* $\mathfrak{M}_{2 \times 2}$ *having properties* (i), (ii), (iii), *and* (iv).

Proof.   Let $f: \mathfrak{M}_{2 \times 2} \to \mathbb{R}$ be any function with the properties

(i) $f(B) = rf(A)$ whenever B is obtained from A by multiplying a row of A by a real number, r,

(ii) $f(B) = -f(A)$ whenever B is obtained from A by interchanging the rows,

(iii) $f(B) = f(A)$ whenever B is obtained from A by adding a scalar multiple of one row to the other, and

(iv) $f \begin{pmatrix} 1 & 0 \\ 0 & 1 \end{pmatrix} = 1.$

We shall show that $f(A) = \det A$ for all $A \in \mathfrak{M}_{2 \times 2}$.

In the following computations, we indicate above each equality sign which of the properties (i)–(iv) we are using.

Let

$$A = \begin{pmatrix} a & b \\ c & d \end{pmatrix}.$$

If $a \neq 0$, then

$$f(A) \stackrel{(i)}{=} af \begin{pmatrix} 1 & \dfrac{b}{a} \\ c & d \end{pmatrix} \stackrel{(iii)}{=} af \begin{pmatrix} 1 & \dfrac{b}{a} \\ 0 & d - \dfrac{cb}{a} \end{pmatrix} \stackrel{(i)}{=} a\left(d - \dfrac{bc}{a}\right) f \begin{pmatrix} 1 & \dfrac{b}{a} \\ 0 & 1 \end{pmatrix}$$

$$\stackrel{(iii)}{=} (ad - bc) f \begin{pmatrix} 1 & 0 \\ 0 & 1 \end{pmatrix} \stackrel{(iv)}{=} ad - bc = \det A.$$

If $a = 0$, then

$$f(A) = f \begin{pmatrix} 0 & b \\ c & d \end{pmatrix} \stackrel{(i)}{=} bf \begin{pmatrix} 0 & 1 \\ c & d \end{pmatrix} \stackrel{(iii)}{=} bf \begin{pmatrix} 0 & 1 \\ c & 0 \end{pmatrix} \stackrel{(i)}{=} bcf \begin{pmatrix} 0 & 1 \\ 1 & 0 \end{pmatrix}$$

$$\stackrel{(ii)}{=} -bcf \begin{pmatrix} 1 & 0 \\ 0 & 1 \end{pmatrix} \stackrel{(iv)}{=} -bc = \det A.$$

So, in both cases, $f(A) = \det A.$ ∎

The determinant of an $n \times n$ matrix is also a real number. The function det: $\mathfrak{M}_{n \times n} \to \mathbb{R}$ has properties similar to (i), (ii), (iii), and (iv) for det: $\mathfrak{M}_{2 \times 2} \to \mathbb{R}$. Rather than write down a formula now for the determinant of an $n \times n$ matrix, we will define the determinant by those properties.

Theorem 2.   *There is one and only one function* det: $\mathfrak{M}_{n \times n} \to \mathbb{R}$ *with the following four properties:*
  (P_1) *if* $B \in \mathfrak{M}_{n \times n}$ *is obtained from* $A \in \mathfrak{M}_{n \times n}$ *by multiplying a row of A by a real number c, then* $\det B = c \det A$,
  (P_2) *if* $B \in \mathfrak{M}_{n \times n}$ *is obtained from* $A \in \mathfrak{M}_{n \times n}$ *by interchanging two rows, then* $\det B = -\det A$,
  (P_3) *if* $B \in \mathfrak{M}_{n \times n}$ *is obtained from* $A \in \mathfrak{M}_{n \times n}$ *by adding a scalar multiple of one row to another, then* $\det B = \det A$, *and*
  (P_4) $\det I = 1$, *where*

$$I = \begin{pmatrix} 1 & 0 & \cdots & 0 \\ 0 & 1 & \cdots & 0 \\ & & \cdots & \\ 0 & 0 & \cdots & 1 \end{pmatrix}.$$

To help you remember these four properties of det, we shall list them again in symbols. If $\mathbf{a}_1, \mathbf{a}_2, \ldots, \mathbf{a}_n$ are the row vectors of A, so that

$$A = \begin{pmatrix} \mathbf{a}_1 \\ \mathbf{a}_2 \\ \vdots \\ \mathbf{a}_n \end{pmatrix},$$

then

$$(P_1) \quad \det \begin{pmatrix} \mathbf{a}_1 \\ \vdots \\ c\mathbf{a}_i \\ \vdots \\ \mathbf{a}_n \end{pmatrix} = c \det \begin{pmatrix} \mathbf{a}_1 \\ \vdots \\ \mathbf{a}_i \\ \vdots \\ \mathbf{a}_n \end{pmatrix},$$

$$(P_2) \quad \det \begin{pmatrix} \vdots \\ \mathbf{a}_i \\ \vdots \\ \mathbf{a}_j \\ \vdots \end{pmatrix} = -\det \begin{pmatrix} \vdots \\ \mathbf{a}_j \\ \vdots \\ \mathbf{a}_i \\ \vdots \end{pmatrix}, \quad \text{for } i \neq j,$$

$$(P_3) \quad \det \begin{pmatrix} \vdots \\ \mathbf{a}_i + c\mathbf{a}_j \\ \vdots \\ \mathbf{a}_j \\ \vdots \end{pmatrix} = \det \begin{pmatrix} \vdots \\ \mathbf{a}_i \\ \vdots \\ \mathbf{a}_j \\ \vdots \end{pmatrix}, \quad \text{for } i \neq j,$$

and

$$(P_4) \quad \det \begin{pmatrix} \mathbf{e}_1 \\ \vdots \\ \mathbf{e}_n \end{pmatrix} = 1.$$

The function det whose existence and uniqueness are asserted in Theorem 2 is called the **determinant function** on $\mathfrak{M}_{n \times n}$. For each $A \in \mathfrak{M}_{n \times n}$, the real number $\det A$ is called the **determinant** of A.

The proof of Theorem 2 is more difficult, of course, than the proof of Theorem 1. We shall postpone the proof until Section 5.3. In this section we shall derive some additional properties of determinants, develop some computational procedures, and discuss some applications.

First, notice that *if $A \in \mathfrak{M}_{n \times n}$ has two rows that are equal, then* $\det A = 0$. For if B is obtained from A by interchanging the equal rows, then $B = A$ so $\det B = \det A$ but, by P_2, we must also have $\det B = -\det A$. This can happen only if $\det A = \det B = 0$.

Next, notice that *if A has a row of zeros, then* $\det A = 0$. For if we add any other row of A to the row of zeros we will get a matrix B with two equal rows and hence, using P_3,

$$\det A = \det B = 0.$$

Now let us look at some matrices whose determinants are not zero.

Example 1.   An $n \times n$ matrix

$$A = \begin{pmatrix} a_{11} & \cdots & a_{1n} \\ & \cdots & \\ a_{n1} & \cdots & a_{nn} \end{pmatrix}$$

is called **lower triangular** if $a_{ij} = 0$ whenever $i < j$. Thus a lower triangular matrix is a matrix of the form

$$A = \begin{pmatrix} a_{11} & 0 & \cdots & 0 \\ a_{21} & a_{22} & \cdots & 0 \\ & & \cdots & \\ a_{n1} & a_{n2} & \cdots & a_{nn} \end{pmatrix}.$$

We can compute the determinant of a lower triangular matrix by alternately using properties P_1 and P_3 of determinants and then, finally, applying P_4 to get

$$\det \begin{pmatrix} a_{11} & 0 & \cdots & 0 \\ a_{21} & a_{22} & \cdots & 0 \\ & & \cdots & \\ a_{n1} & a_{n2} & \cdots & a_{nn} \end{pmatrix} = a_{11} \det \begin{pmatrix} 1 & 0 & \cdots & 0 \\ a_{21} & a_{22} & \cdots & 0 \\ & & \cdots & \\ a_{n1} & a_{n2} & \cdots & a_{nn} \end{pmatrix}$$

$$= a_{11} \det \begin{pmatrix} 1 & 0 & \cdots & 0 \\ 0 & a_{22} & \cdots & 0 \\ & & \cdots & \\ 0 & a_{n2} & \cdots & a_{nn} \end{pmatrix} = a_{11}a_{22} \det \begin{pmatrix} 1 & 0 & \cdots & 0 \\ 0 & 1 & \cdots & 0 \\ & & \cdots & \\ 0 & a_{n2} & \cdots & a_{nn} \end{pmatrix}$$

$$= a_{11}a_{22} \det \begin{pmatrix} 1 & 0 & \cdots & 0 \\ 0 & 1 & \cdots & 0 \\ & & \cdots & \\ 0 & 0 & \cdots & a_{nn} \end{pmatrix} = \cdots = a_{11}a_{22} \cdots a_{nn} \det \begin{pmatrix} 1 & 0 & \cdots & 0 \\ 0 & 1 & \cdots & 0 \\ & & \cdots & \\ 0 & 0 & \cdots & 1 \end{pmatrix}$$

$$= a_{11}a_{22} \cdots a_{nn}.$$

The entries $a_{11}, a_{22}, \ldots, a_{nn}$ are called the **diagonal entries** of A. Thus we have shown that *the determinant of any lower triangular matrix is equal to the product of its diagonal entries.* ■

Example 2. An $n \times n$ matrix

$$A = \begin{pmatrix} a_{11} & \cdots & a_{1n} \\ & \cdots & \\ a_{n1} & \cdots & a_{nn} \end{pmatrix}$$

is **upper triangular** if $a_{ij} = 0$ whenever $i > j$. Thus an upper triangular matrix is a matrix of the form

$$A = \begin{pmatrix} a_{11} & a_{12} & \cdots & a_{1n} \\ 0 & a_{22} & \cdots & a_{2n} \\ & & \cdots & \\ 0 & 0 & \cdots & a_{nn} \end{pmatrix}.$$

We can compute the determinant of an upper triangular matrix in the same way that we computed the determinant of a lower triangular matrix in Example 1, except that we must work from the bottom row up rather than from the top row down. Thus

$$\det \begin{pmatrix} a_{11} & a_{12} & \cdots & a_{1n} \\ 0 & a_{22} & \cdots & a_{2n} \\ & & \cdots & \\ 0 & 0 & \cdots & a_{nn} \end{pmatrix} = a_{nn} \det \begin{pmatrix} a_{11} & a_{12} & \cdots & a_{1n} \\ 0 & a_{22} & \cdots & a_{2n} \\ & & \cdots & \\ 0 & 0 & \cdots & 1 \end{pmatrix}$$

$$= a_{nn} \det \begin{pmatrix} a_{11} & a_{12} & \cdots & 0 \\ 0 & a_{22} & \cdots & 0 \\ & & \cdots & \\ 0 & 0 & \cdots & 1 \end{pmatrix}$$

$$= \cdots = a_{nn} \cdots a_{22} a_{11} \det \begin{pmatrix} 1 & 0 & \cdots & 0 \\ 0 & 1 & \cdots & 0 \\ & & \cdots & \\ 0 & 0 & \cdots & 1 \end{pmatrix}$$

$$= a_{11} a_{22} \cdots a_{nn}.$$

We see, therefore, that *the determinant of any upper triangular matrix is also equal to the product of its diagonal elements.* ∎

The properties P_1, P_2, and P_3 tell us how determinants behave relative to row operations. Hence we can compute the determinant of any $n \times n$ matrix by applying row operations, keeping track of the effect each row operation has on the determinant, until the matrix is transformed into a matrix whose determinant we know. Since we already know how to apply row operations to transform a matrix into an upper triangular matrix (an echelon matrix is upper triangular!) this process is a familiar one. But note that we need not carry the process all the way to the echelon matrix. We can stop as soon as the matrix is upper triangular.

Example 3

$$\det\begin{pmatrix} 1 & 2 & 3 \\ 4 & 5 & 6 \\ 7 & 8 & 9 \end{pmatrix} = \det\begin{pmatrix} 1 & 2 & 3 \\ 0 & -3 & -6 \\ 0 & -6 & -12 \end{pmatrix} = \det\begin{pmatrix} 1 & 2 & 3 \\ 0 & -3 & -6 \\ 0 & 0 & 0 \end{pmatrix} = (1)(-3)(0) = 0. \quad \blacksquare$$

Example 4

$$\det\begin{pmatrix} 1 & 3 & -2 \\ 2 & 0 & 3 \\ 1 & -1 & 2 \end{pmatrix} = \det\begin{pmatrix} 1 & 3 & -2 \\ 0 & -6 & 7 \\ 0 & -4 & 4 \end{pmatrix} = -\det\begin{pmatrix} 1 & 3 & -2 \\ 0 & -4 & 4 \\ 0 & -6 & 7 \end{pmatrix}$$

$$= -(-4)\det\begin{pmatrix} 1 & 3 & -2 \\ 0 & 1 & -1 \\ 0 & -6 & 7 \end{pmatrix} = 4\det\begin{pmatrix} 1 & 3 & -2 \\ 0 & 1 & -1 \\ 0 & 0 & 1 \end{pmatrix}$$

$$= (4)(1)(1)(1) = 4. \quad \blacksquare$$

In the next section we will discuss another method for evaluating determinants. But now let us look at two theorems that show why determinants are important.

Theorem 3. *Let $A \in \mathfrak{M}_{n \times n}$. If $\det A \neq 0$ then*

(i) *A has rank n,*

(ii) *the row vectors of A form a linearly independent set,*

(iii) *the row vectors of A span $\mathbb{R}^n$,*

(iv) *the column vectors of A form a linearly independent set, and*

(v) *the column vectors of A span $\mathbb{R}^n$.*

Conversely, if any of these five conditions holds, then $\det A \neq 0$.

Proof. (i) Since the determinant is unaffected by the addition of one row of a matrix to another, changes sign when two rows are interchanged, and gets multiplied by c whenever a row is multiplied by a nonzero scalar c, we see that $\det A = 0$ if and only if $\det B = 0$ where B is any matrix obtained from A by row operations. In particular, $\det A = 0$ if and only if $\det B = 0$ where B is the echelon matrix of A. If A has rank n then $B = I$, so $\det B = 1 \neq 0$ and hence $\det A \neq 0$. If A has rank $<n$ then B has a row of zeros, so $\det B = 0$ and hence $\det A = 0$. We can conclude, then, that $\det A \neq 0$ if and only if A has rank n.

(ii) The n row vectors of A form a linearly independent set if and only if A has rank n, by Theorem 3 of Section 4.4. Hence (ii) follows from (i).

(iii)   The subspace of $\mathbb{R}^n$ spanned by the row vectors of A is equal to $\mathbb{R}^n$ if and only if it has dimension n, by Theorem 5 of Section 4.4. But this space has dimension n if and only if the n row vectors spanning it form a linearly independent set (see Theorem 2 of Section 4.4). Hence (iii) follows from (ii).

(iv)   The n column vectors of A form a linearly independent set if and only if A has rank n, by Theorem 2 of Section 4.1. Hence (iv) follows from (i).

(v)   This follows from (iv) in the same way that (iii) follows from (ii). Simply replace "row" by "column" everywhere in the proof of (iii).   ■

Example 5.   The set $\{(1, 2, 3), (0, 1, 2), (0, 0, 1)\}$ is linearly independent because the matrix

$$\begin{pmatrix} 1 & 2 & 3 \\ 0 & 1 & 2 \\ 0 & 0 & 1 \end{pmatrix}$$

with these vectors as row vectors has determinant $1 \neq 0$.   ■

Theorem 4.   *Let I be an open interval in $\mathbb{R}$ and let $f_1, \ldots, f_n \in \mathcal{F}(I)$. Suppose each f_i is $(n - 1)$-times differentiable on I. If*

$$\det \begin{pmatrix} f_1(x) & f_2(x) & \cdots & f_n(x) \\ f_1'(x) & f_2'(x) & \cdots & f_n'(x) \\ & & \cdots & \\ f_1^{(n-1)}(x) & f_2^{(n-1)}(x) & \cdots & f_n^{(n-1)}(x) \end{pmatrix} \neq 0$$

for some $x \in I$, then $\{f_1, \ldots, f_n\}$ is linearly independent.

Proof.   If the determinant is nonzero, then the matrix has rank n, by Theorem 3. But the matrix here is the Wronskian matrix, and we know from Theorem 4 of Section 4.1 that if the Wronskian matrix has rank n for some $x \in I$ then $\{f_1, \ldots, f_n\}$ is linearly independent.   ■

The determinant that appears in Theorem 4 is called the **Wronskian determinant** of $\{f_1, \ldots, f_n\}$ at $x \in I$. It has many interesting properties, some of which you will find in the exercises.

The converse of Theorem 4 is false. There do exist functions $f_1, f_2, \ldots, f_n$ that form a linearly independent set but for which the Wronskian determinant is identically zero (see Exercise 3 of Section 4.1). However, when the functions $f_1, f_2, \ldots, f_n$ are solutions on I of an nth order homogeneous constant coefficient linear differential equation, then linear independence does imply that the Wronskian determinant is nonzero, for every $x \in I$ (see Exercise 11).

EXERCISES

1. Evaluate:

(a) $\det \begin{pmatrix} 1 & 3 \\ 4 & 2 \end{pmatrix}$ (b) $\det \begin{pmatrix} -1 & -1 \\ 4 & -1 \end{pmatrix}$ (c) $\det \begin{pmatrix} 0 & -3 \\ 2 & 1 \end{pmatrix}$

(d) $\det \begin{pmatrix} 0 & 1 \\ 1 & 0 \end{pmatrix}$ (e) $\det \begin{pmatrix} -1 & 3 \\ 3 & -1 \end{pmatrix}$ (f) $\det \begin{pmatrix} 5 & -3 \\ 2 & -1 \end{pmatrix}$

2. Evaluate:

(a) $\det \begin{pmatrix} \cos\theta & -\sin\theta \\ \sin\theta & \cos\theta \end{pmatrix}$ (b) $\det \begin{pmatrix} 1 & 1 \\ a & b \end{pmatrix}$ (c) $\det \begin{pmatrix} 1 & a \\ 0 & 1 \end{pmatrix}$

3. Find the determinant of each of the following matrices:

(a) $\begin{vmatrix} 2 & 0 & 0 \\ 1 & -1 & 5 \\ 2 & 3 & -1 \end{vmatrix}$ (b) $\begin{vmatrix} 2 & 1 & 5 \\ 1 & 0 & 3 \\ -1 & 2 & 0 \end{vmatrix}$ (c) $\begin{vmatrix} 0 & -1 & 1 \\ 1 & 0 & 3 \\ 2 & -1 & 0 \end{vmatrix}$

(d) $\begin{vmatrix} 0 & 1 & 0 \\ 0 & 0 & 1 \\ 1 & 0 & 0 \end{vmatrix}$ (e) $\begin{vmatrix} 7 & -1 & 5 \\ 3 & 4 & -5 \\ 2 & 3 & 0 \end{vmatrix}$ (f) $\begin{vmatrix} 2 & 1 & 3 \\ -1 & 3 & -5 \\ 0 & 2 & -2 \end{vmatrix}$

4. Find the determinant of each of the following matrices:

(a) $\begin{vmatrix} 1 & a & b \\ 0 & 1 & c \\ 0 & 0 & 1 \end{vmatrix}$ (b) $\begin{vmatrix} 1 & 1 & 1 \\ a & b & c \\ a^2 & b^2 & c^2 \end{vmatrix}$

5. Evaluate:

(a) $\det \begin{vmatrix} 1 & -1 & 1 & -1 \\ 1 & 2 & 4 & 8 \\ 1 & -2 & 4 & -8 \\ 1 & 1 & 1 & 1 \end{vmatrix}$ (b) $\det \begin{vmatrix} 1 & 2 & 3 & 4 \\ 0 & 1 & 2 & 3 \\ 0 & 0 & 1 & 2 \\ 0 & 0 & 0 & 1 \end{vmatrix}$

(c) $\det \begin{vmatrix} 0 & 1 & -2 & 3 \\ -1 & 0 & 1 & 2 \\ 2 & -1 & 0 & 1 \\ -3 & -2 & -1 & 0 \end{vmatrix}$ (d) $\det \begin{vmatrix} 3 & 1 & 2 & 0 \\ -2 & -1 & 5 & -2 \\ 1 & -3 & 1 & 1 \\ 4 & 1 & 2 & -3 \end{vmatrix}$

(e) $\det \begin{vmatrix} 1 & 1 & 1 & 1 \\ 1 & 2 & 4 & 8 \\ 1 & 3 & 9 & 27 \\ 1 & 4 & 16 & 64 \end{vmatrix}$

6. Show that if

$$A = \begin{pmatrix} 0 & \cdots & 0 & 1 \\ 0 & \cdots & 1 & 0 \\ & \cdots & & \\ 1 & \cdots & 0 & 0 \end{pmatrix}$$

$(a_{ij} = 1$ if $i + j = n + 1$, and $a_{ij} = 0$ otherwise) then $\det A = (-1)^{n(n-1)/2}$.

7. Let $f : \mathbb{R} \to \mathbb{R}$ be defined by

$$f(x) = \det \begin{pmatrix} 1 & x & x^2 \\ 1 & 1 & 1 \\ 1 & 2 & 4 \end{pmatrix}.$$

Without evaluating the determinant, find two values of x for which $f(x) = 0$.

8. Let $A \in \mathfrak{M}_{n \times n}$. Then the row vectors of A form a linearly independent set if and only if the column vectors of A form a linearly independent set. Why?

9. Let a, b, c, and $d \in \mathfrak{F}(I)$ be differentiable functions defined on an open interval I. Define $f \in \mathfrak{F}(I)$ by

$$f(x) = \det \begin{pmatrix} a(x) & b(x) \\ c(x) & d(x) \end{pmatrix}.$$

Show that

$$f'(x) = \det \begin{pmatrix} a'(x) & b'(x) \\ c(x) & d(x) \end{pmatrix} + \det \begin{pmatrix} a(x) & b(x) \\ c'(x) & d'(x) \end{pmatrix}.$$

10. Let f_1 and f_2 be differentiable functions on an open interval I. Show that if $f_1(x) \neq 0$ for all $x \in I$ and the Wronskian determinant

$$\det \begin{pmatrix} f_1(x) & f_2(x) \\ f_1'(x) & f_2'(x) \end{pmatrix}$$

is identically zero on the interval I, then $\{f_1, f_2\}$ is linearly dependent. [Hint: Consider $(f_2/f_1)'$.] (This exercise proves a partial converse of Theorem 4 of Section 4.1.)

11. Suppose y_1 and y_2 are solutions of the second order constant coefficient linear differential equation $ay'' + by' + cy = 0$. Let $W \in \mathfrak{F}(\mathbb{R})$ denote the Wronskian determinant of $\{y_1, y_2\}$. Thus

$$W(x) = \det \begin{pmatrix} y_1(x) & y_2(x) \\ y_1'(x) & y_2'(x) \end{pmatrix}.$$

(a) Show that $W'(x) = -(b/a)W(x)$. [Hint: Use Exercise 9 and the fact that $y_i'' = -(b/a)y_i' - (c/a)y_i$ for $i = 1, 2$.]

(b) Show that if $W(x) \neq 0$ for some $x \in \mathbb{R}$, then $W(x) \neq 0$ for all $x \in \mathbb{R}$. [Hint: Think about solutions of the first order linear differential equation $y' + (b/a)y = 0$.]

(c) Show that if $W(0) = 0$ then $\{y_1, y_2\}$ is linearly dependent. [Hint: First show that $c_1(y_1(0), y_1'(0)) + c_2(y_2(0), y_2'(0)) = \mathbf{0}$ for some $c_1, c_2 \in \mathbb{R}$. Then set $y = c_1 y_1 + c_2 y_2$ and use Exercise 21(b) of Section 3.2.] (The results of this exercise generalize to similar results for solutions of nth order linear differential equations.)

5.2 EVALUATING DETERMINANTS BY MINORS

In the previous section we discussed the evaluation of determinants using row operations. In this section, we shall describe another way to evaluate determinants. We shall derive a formula that expresses the determinant of any $n \times n$ matrix A as a linear combination of the determinants of certain $(n - 1) \times (n - 1)$ submatrices of A. Since we already know an easy rule for finding the determinants of 2×2 matrices, this formula will enable us to evaluate the determinant of any 3×3 matrix. Once we can do that, we will be able to use our formula to evaluate the determinant of any 4×4 matrix, and so on.

The $(n - 1) \times (n - 1)$ matrices that appear in the formula for det A are the minors of A. For each (i, j), $1 \le i, j \le n$, the (i, j)-**minor** of

$$A = \begin{pmatrix} a_{11} & \cdots & a_{1n} \\ & \cdots & \\ a_{n1} & \cdots & a_{nn} \end{pmatrix}$$

is the matrix A_{ij} obtained by deleting from A the ith row and the jth column:

$$A_{ij} = \begin{pmatrix} a_{11} & \cdots & a_{1j} & \cdots & a_{1n} \\ & & \cdots & & \\ a_{i1} & \cdots & a_{ij} & \cdots & a_{in} \\ & & \cdots & & \\ a_{n1} & \cdots & a_{nj} & \cdots & a_{nn} \end{pmatrix}.$$

Thus the $(2, 2)$-minor of

$$A = \begin{pmatrix} 1 & 3 & 4 \\ 2 & 1 & -1 \\ 0 & 5 & 7 \end{pmatrix}$$

is

$$A_{22} = \begin{pmatrix} 1 & 3 & 4 \\ 2 & 1 & -1 \\ 0 & 5 & 7 \end{pmatrix} = \begin{pmatrix} 1 & 4 \\ 0 & 7 \end{pmatrix}$$

and the $(1, 3)$-minor of A is

$$A_{13} = \begin{pmatrix} 1 & 3 & 4 \\ 2 & 1 & -1 \\ 0 & 5 & 7 \end{pmatrix} = \begin{pmatrix} 2 & 1 \\ 0 & 5 \end{pmatrix}.$$

Theorem 1. (*Expansion of* det A *by minors along with ith row.*) *Let $A \in \mathfrak{M}_{n \times n}$ and let i be any integer between 1 and n. Then*

$$\det A = \sum_{j=1}^{n} (-1)^{i+j} a_{ij} \det A_{ij}$$

where A_{ij} is the (i, j)-minor of A.

Before proving this theorem, we shall work out a few examples.

Example 1.   Let

$$A = \begin{pmatrix} 2 & 3 & -3 \\ 1 & -2 & 1 \\ 4 & -1 & -1 \end{pmatrix}.$$

The formula for expansion by minors along the first row says that

$$\det A = (-1)^{1+1}a_{11}\det A_{11} + (-1)^{1+2}a_{12}\det A_{12} + (-1)^{1+3}a_{13}\det A_{13}.$$

Hence

$$\det A = (-1)^2(2)\det \begin{pmatrix} 2 & 3 & -3 \\ 1 & -2 & 1 \\ 4 & -1 & -1 \end{pmatrix} + (-1)^3(3)\det \begin{pmatrix} 2 & 3 & -3 \\ 1 & -2 & 1 \\ 4 & -1 & -1 \end{pmatrix}$$

$$+ (-1)^4(-3)\det \begin{pmatrix} 2 & 3 & -3 \\ 1 & -2 & 1 \\ 4 & -1 & -1 \end{pmatrix}$$

$$= 2\det \begin{pmatrix} -2 & 1 \\ -1 & -1 \end{pmatrix} - 3\det \begin{pmatrix} 1 & 1 \\ 4 & -1 \end{pmatrix} - 3\det \begin{pmatrix} 1 & -2 \\ 4 & -1 \end{pmatrix}$$

$$= 2[(-2)(-1) - (1)(-1)] - 3[(1)(-1) - (1)(4)] - 3[(1)(-1) - (-2)(4)]$$

$$= 0.$$

Since the determinant of this matrix is zero, we can conclude that its rows form a linearly dependent set. Do you see a dependence relation?   ◼

Example 2.   Let's again evaluate the determinant of the matrix A in Example 1, this time expanding by minors along the second row. We get

$$\det A = (-1)^{2+1}(1)\det \begin{pmatrix} 2 & 3 & -3 \\ 1 & -2 & 1 \\ 4 & -1 & -1 \end{pmatrix} + (-1)^{2+2}(-2)\det \begin{pmatrix} 2 & 3 & -3 \\ 1 & -2 & 1 \\ 4 & -1 & -1 \end{pmatrix}$$

$$+ (-1)^{2+3}(1)\det \begin{pmatrix} 2 & 3 & -3 \\ 1 & -2 & 1 \\ 4 & -1 & -1 \end{pmatrix}$$

$$= -\det \begin{pmatrix} 3 & -3 \\ -1 & -1 \end{pmatrix} - 2 \det \begin{pmatrix} 2 & -3 \\ 4 & -1 \end{pmatrix} - \det \begin{pmatrix} 2 & 3 \\ 4 & -1 \end{pmatrix}$$

$$= 6 - 20 + 14 = 0.$$

Thus, expansion by minors along the second row yields the same answer as expansion by minors along the first row. You may check that expansion by minors along the third row of this matrix also yields zero. ◾

Example 3. Now let us work out a formula that will display the determinant of any 3×3 matrix directly in terms of the entries. Let

$$A = \begin{pmatrix} a_{11} & a_{12} & a_{13} \\ a_{21} & a_{22} & a_{23} \\ a_{31} & a_{32} & a_{33} \end{pmatrix}.$$

Then

$$\det A = a_{11} \det \begin{pmatrix} a_{22} & a_{23} \\ a_{32} & a_{33} \end{pmatrix} - a_{12} \det \begin{pmatrix} a_{21} & a_{23} \\ a_{31} & a_{33} \end{pmatrix}$$

$$+ a_{13} \det \begin{pmatrix} a_{21} & a_{22} \\ a_{31} & a_{32} \end{pmatrix}$$

or

$$\det \begin{pmatrix} a_{11} & a_{12} & a_{13} \\ a_{21} & a_{22} & a_{23} \\ a_{31} & a_{32} & a_{33} \end{pmatrix} = a_{11}a_{22}a_{33} - a_{11}a_{23}a_{32} - a_{12}a_{21}a_{33}$$

$$+ a_{12}a_{23}a_{31} + a_{13}a_{21}a_{32} - a_{13}a_{22}a_{31}.$$

You may have seen this formula before. It is often remembered by writing the matrix A with its first two columns reproduced to give the three by five matrix

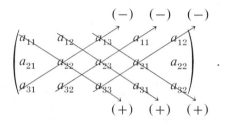

If we superimpose six arrows on the matrix, as shown, then the determinant of A is equal to the sum of the products of the entries hit by the arrows going downward minus the sum of the products of the entries hit by the arrows going upward.

A similar mnemonic device works for the determinant of a 2×2 matrix:

$$\det \begin{pmatrix} a_{11} & a_{12} \\ a_{21} & a_{22} \end{pmatrix} = a_{11}a_{22} - a_{12}a_{21}.$$

$(-)$ ... $(+)$

But be careful. This mnemonic procedure *does not generalize* to determinants of $n \times n$ matrices for $n > 3$. ■

Example 4. Expanding by minors along the fourth row, we find that

$$\det \begin{pmatrix} a & 0 & 0 & b \\ 0 & a & b & 0 \\ 0 & b & a & 0 \\ b & 0 & 0 & a \end{pmatrix} = -b \det \begin{pmatrix} 0 & 0 & b \\ a & b & 0 \\ b & a & 0 \end{pmatrix} + a \det \begin{pmatrix} a & 0 & 0 \\ 0 & a & b \\ 0 & b & a \end{pmatrix}$$

$$= -b^2 \det \begin{pmatrix} a & b \\ b & a \end{pmatrix} + a^2 \det \begin{pmatrix} a & b \\ b & a \end{pmatrix}$$

$$= a^4 - 2a^2b^2 + b^4. \quad ■$$

Expansion by minors is the most efficient way to evaluate the determinant of a 2×2 matrix. It is simply the formula

$$\det \begin{pmatrix} a_{11} & a_{12} \\ a_{21} & a_{22} \end{pmatrix} = a_{11}a_{22} - a_{12}a_{21}.$$

However, for $n \times n$ matrices with $n > 2$, it is usually much more efficient to use row operations to evaluate determinants than to expand by minors.

The remainder of this section will be devoted to proving Theorem 1. The proof is fairly long, although not particularly hard.

In order to prove Theorem 1, we will need to use the following additional property of determinants.

Theorem 2. *The determinant function* det: $\mathfrak{M}_{n \times n} \to \mathbb{R}$ *is additive in each row; that is, for each i, $1 \leq i \leq n$,*

$$\det \begin{pmatrix} \mathbf{a}_1 \\ \vdots \\ \mathbf{a}_{i-1} \\ \mathbf{v} + \mathbf{w} \\ \mathbf{a}_{i+1} \\ \vdots \\ \mathbf{a}_n \end{pmatrix} = \det \begin{pmatrix} \mathbf{a}_1 \\ \vdots \\ \mathbf{a}_{i-1} \\ \mathbf{v} \\ \mathbf{a}_{i+1} \\ \vdots \\ \mathbf{a}_n \end{pmatrix} + \det \begin{pmatrix} \mathbf{a}_1 \\ \vdots \\ \mathbf{a}_{i-1} \\ \mathbf{w} \\ \mathbf{a}_{i+1} \\ \vdots \\ \mathbf{a}_n \end{pmatrix}.$$

Proof. Let $\mathbf{a}_1, \ldots, \mathbf{a}_{i-1}, \mathbf{a}_{i+1}, \ldots, \mathbf{a}_n$ be fixed vectors in $\mathbb{R}^n$. Define a function f: $\mathbb{R}^n \to \mathbb{R}$ by

$$f(\mathbf{v}) = \det \begin{pmatrix} \mathbf{a}_1 \\ \cdots \\ \mathbf{a}_{i-1} \\ \mathbf{v} \\ \mathbf{a}_{i+1} \\ \cdots \\ \mathbf{a}_n \end{pmatrix}.$$

We must show that

$$f(\mathbf{v} + \mathbf{w}) = f(\mathbf{v}) + f(\mathbf{w})$$

for all $\mathbf{v}$ and $\mathbf{w}$ in $\mathbb{R}^n$.

Notice first that if $\{\mathbf{a}_1, \ldots, \mathbf{a}_{i-1}, \mathbf{a}_{i+1}, \ldots, \mathbf{a}_n\}$ is linearly dependent, then $f(\mathbf{v}) = 0$ for all $\mathbf{v} \in \mathbb{R}^n$, since any matrix whose rows form a linearly dependent set has determinant zero. Thus f is the zero function and, in particular,

$$f(\mathbf{v} + \mathbf{w}) = 0 = f(\mathbf{v}) + f(\mathbf{w})$$

for all $\mathbf{v}$ and $\mathbf{w}$ in $\mathbb{R}^n$.

So let us assume that $\{\mathbf{a}_1, \ldots, \mathbf{a}_{i-1}, \mathbf{a}_{i+1}, \ldots, \mathbf{a}_n\}$ is linearly independent. The proof breaks up into two cases, depending on whether or not $\mathbf{w}$ is in $\mathcal{L}(\mathbf{a}_1, \ldots, \mathbf{a}_{i-1}, \mathbf{a}_{i+1}, \ldots, \mathbf{a}_n)$. We shall need to use the following facts about the function f.

(i) $f(\mathbf{0}) = 0$. This is because any matrix with a zero row has determinant zero.

(ii) $f(c\mathbf{v}) = cf(\mathbf{v})$ for all $c \in \mathbb{R}$ and all $\mathbf{v} \in \mathbb{R}^n$. This follows from property P_1 of the determinant.

(iii) $f(\mathbf{v} + \sum_{j \neq i} c_j \mathbf{a}_j) = f(\mathbf{v})$ for all $\mathbf{v} \in \mathbb{R}^n$ and all $c_1, \ldots, c_{i-1}, c_{i+1}, \ldots,$ $c_n \in \mathbb{R}$. This follows from property P_3 of the determinant.

Case I. Suppose $\mathbf{w} \in \mathcal{L}(\mathbf{a}_1, \ldots, \mathbf{a}_{i-1}, \mathbf{a}_{i+1}, \ldots, \mathbf{a}_n)$. Then $\mathbf{w} = \sum_{j \neq i} c_j \mathbf{a}_j$ for some $c_1, \ldots, c_{i-1}, c_{i+1}, \ldots, c_n \in \mathbb{R}$. Hence

$$f(\mathbf{v} + \mathbf{w}) = f\left(\mathbf{v} + \sum_{j \neq i} c_j \mathbf{a}_j\right) \overset{\text{(iii)}}{=\!=} f(\mathbf{v}) \overset{\text{(i)}}{=\!=} f(\mathbf{v}) + f(\mathbf{0})$$

$$\overset{\text{(iii)}}{=\!=} f(\mathbf{v}) + f\left(\mathbf{0} + \sum_{j \neq i} c_j \mathbf{a}_j\right) = f(\mathbf{v}) + f(\mathbf{w}).$$

Case II. Suppose $\mathbf{w} \notin \mathcal{L}(\mathbf{a}_1, \ldots, \mathbf{a}_{i-1}, \mathbf{a}_{i+1}, \ldots, \mathbf{a}_n)$. Then $\{\mathbf{a}_1, \ldots, \mathbf{a}_{i-1}, \mathbf{a}_{i+1}, \ldots, \mathbf{a}_n, \mathbf{w}\}$ is linearly independent and hence is a basis for $\mathbb{R}^n$. Therefore,

$\mathbf{v} = \sum_{j \neq i} c_j \mathbf{a}_j + c\mathbf{w}$ for some $c_1, \ldots, c_{i-1}, c_{i+1}, \ldots, c_n, c \in \mathbb{R}$. It follows that

$$f(\mathbf{v} + \mathbf{w}) = f\left(\sum_{j \neq i} c_j \mathbf{a}_j + c\mathbf{w} + \mathbf{w}\right) \overset{\text{(iii)}}{=} f(c\mathbf{w} + \mathbf{w}) = f((c + 1)\mathbf{w})$$

$$\overset{\text{(i)}}{=} (c + 1)f(\mathbf{w}) = cf(\mathbf{w}) + f(\mathbf{w}) \overset{\text{(i)}}{=} f(c\mathbf{w}) + f(\mathbf{w})$$

$$\overset{\text{(iii)}}{=} f\left(\sum_{j \neq i} c_j \mathbf{a}_j + c\mathbf{w}\right) + f(\mathbf{w}) = f(\mathbf{v}) + f(\mathbf{w}).$$

Hence, in both cases, our assertion is proved. ∎

Proof of Theorem 1. We shall break the proof into several steps, each of which is a special case of the theorem.
    Step 1. For $A \in \mathfrak{M}_{(n-1)\times(n-1)}$,

$$\det \begin{pmatrix} 1 & 0 & \cdots & 0 \\ 0 & & & \\ \vdots & & A & \\ 0 & & & \end{pmatrix} = \det A.$$

Proof. Consider the function $f: \mathfrak{M}_{(n-1)\times(n-1)} \to \mathbb{R}$ defined by

$$f(A) = \det \begin{pmatrix} 1 & 0 & \cdots & 0 \\ 0 & & & \\ \vdots & & A & \\ 0 & & & \end{pmatrix}.$$

Notice that, because of the properties of $\det: \mathfrak{M}_{n \times n} \to \mathbb{R}$,

    (i) if $B \in \mathfrak{M}_{(n-1)\times(n-1)}$ is obtained from $A \in \mathfrak{M}_{(n-1)\times(n-1)}$ by multiplying a row of A by a real number c, then $f(B) = cf(A)$;

    (ii) if B is obtained from A by interchanging two rows, then $f(B) = -f(A)$;

    (iii) if B is obtained from A by adding a scalar multiple of one row to another, then $f(B) = f(A)$; and

$$\text{(iv)}\quad f\begin{pmatrix} 1 & \cdots & 0 \\ & \cdots & \\ 0 & \cdots & 1 \end{pmatrix} = \det \begin{pmatrix} 1 & 0 & \cdots & 0 \\ 0 & 1 & \cdots & 0 \\ & & \cdots & \\ 0 & 0 & \cdots & 1 \end{pmatrix} = 1.$$

But, according to the theorem of the previous section, there is only one real valued function on $\mathfrak{M}_{(n-1)\times(n-1)}$ with properties (i)–(iv), namely, $\det: \mathfrak{M}_{(n-1)\times(n-1)} \to \mathbb{R}$.

We conclude, then, that $f = \det$; that is,

$$f(A) = \det A$$

for each $A \in \mathfrak{M}_{(n-1)\times(n-1)}$. In other words,

$$\det \begin{pmatrix} 1 & 0 & \cdots & 0 \\ 0 & & & \\ \vdots & & A & \\ 0 & & & \end{pmatrix} = \det A.$$

Step 2.

$$\det \begin{pmatrix} 1 & 0 & \cdots & 0 \\ a_{21} & a_{22} & \cdots & a_{2n} \\ & & \cdots & \\ a_{n1} & a_{n2} & \cdots & a_{nn} \end{pmatrix} = \det \begin{pmatrix} 1 & 0 & \cdots & 0 \\ a_{21} & a_{22} & \cdots & a_{2n} \\ & & \cdots & \\ a_{n1} & a_{n2} & \cdots & a_{nn} \end{pmatrix}.$$

Proof. By successively adding appropriate multiples of the first row of the matrix

$$\begin{pmatrix} 1 & 0 & \cdots & 0 \\ a_{21} & a_{22} & \cdots & a_{2n} \\ & & \cdots & \\ a_{n1} & a_{n2} & \cdots & a_{nn} \end{pmatrix}$$

to its other rows, we can row reduce this matrix to

$$\begin{pmatrix} 1 & 0 & \cdots & 0 \\ 0 & a_{22} & \cdots & a_{2n} \\ & & \cdots & \\ 0 & a_{n2} & \cdots & a_{nn} \end{pmatrix}.$$

But we know that matrices related by row operations of this type have the same determinant. Hence, using Step 1, we see that

$$\det \begin{pmatrix} 1 & 0 & \cdots & 0 \\ a_{21} & a_{22} & \cdots & a_{2n} \\ & & \cdots & \\ a_{n1} & a_{n2} & \cdots & a_{nn} \end{pmatrix} = \det \begin{pmatrix} 1 & 0 & \cdots & 0 \\ 0 & a_{22} & \cdots & a_{2n} \\ & & \cdots & \\ 0 & a_{n2} & \cdots & a_{nn} \end{pmatrix}$$

$$= \det \begin{pmatrix} a_{22} & \cdots & a_{2n} \\ & \cdots & \\ a_{n2} & \cdots & a_{nn} \end{pmatrix} = \det \begin{pmatrix} 1 & 0 & \cdots & 0 \\ a_{21} & a_{22} & \cdots & a_{2n} \\ & & \cdots & \\ a_{n1} & a_{n2} & \cdots & a_{nn} \end{pmatrix}.$$

Step 3.

$$\det\begin{pmatrix} 0 & \cdots & 1 & \cdots & 0 \\ a_{21} & \cdots & a_{2j} & \cdots & a_{2n} \\ & & \cdots & & \\ a_{n1} & \cdots & a_{nj} & \cdots & a_{nn} \end{pmatrix} = (-1)^{1+j}\det\begin{pmatrix} 0 & \cdots & 1 & \cdots & 0 \\ a_{21} & \cdots & a_{2j} & \cdots & a_{2n} \\ & & \cdots & & \\ a_{n1} & \cdots & a_{nj} & \cdots & a_{nn} \end{pmatrix}.$$

Proof. Let

$$A = \begin{pmatrix} 0 & \cdots & 1 & \cdots & 0 \\ a_{21} & \cdots & a_{2j} & \cdots & a_{2n} \\ & & \cdots & & \\ a_{n1} & \cdots & a_{nj} & \cdots & a_{nn} \end{pmatrix}.$$

Then, by successively adding appropriate multiples of the first row of this matrix to its other rows and using the fact that these operations do not change the determinant, we see that

$$\det A = \det\begin{pmatrix} 0 & \cdots & 1 & \overset{j\text{th column}}{\cdots} & 0 \\ a_{21} & \cdots & 0 & \cdots & a_{2n} \\ & & \cdots & & \\ a_{n1} & \cdots & 0 & \cdots & a_{nn} \end{pmatrix}.$$

Now, by interchanging the first row in this matrix successively with each of the first $j - 1$ rows directly below it and using the fact that each of these operations changes the sign of the determinant, we find that

$$\det A = (-1)^{j-1}\det\begin{pmatrix} a_{21} & \cdots & 0 & \overset{j\text{th column}}{\cdots} & a_{2n} \\ & & \cdots & & \\ 0 & \cdots & 1 & \cdots & 0 \quad\leftarrow j\text{th row} \\ & & \cdots & & \\ a_{n1} & \cdots & 0 & \cdots & a_{nn} \end{pmatrix}.$$

Finally, an argument identical to that used in proving Step 1 shows that

$$\det A = (-1)^{j-1}\det\begin{pmatrix} a_{21} & \cdots & 0 & \cdots & a_{2n} \\ & & \cdots & & \\ 0 & \cdots & 1 & \cdots & 0 \\ & & \cdots & & \\ a_{n1} & \cdots & 0 & \cdots & a_{nn} \end{pmatrix}.$$

Since $(-1)^{j-1} = (-1)^{j-1}(-1)^2 = (-1)^{1+j}$, we conclude that

$$\det A = (-1)^{1+j} \det \begin{pmatrix} 0 & \cdots & 1 & \cdots & 0 \\ a_{21} & \cdots & a_{2j} & \cdots & a_{2n} \\ & & \cdots & & \\ a_{n1} & \cdots & a_{nj} & \cdots & a_{nn} \end{pmatrix}.$$

Step 4. (Expansion by minors along the first row.)

$$\det A = \det \begin{pmatrix} a_{11} & \cdots & a_{1n} \\ & \cdots & \\ a_{n1} & \cdots & a_{nn} \end{pmatrix} = \sum_{j=1}^{n} (-1)^{1+j} a_{1j} \det \begin{pmatrix} a_{11} & \cdots & a_{1j} & \cdots & a_{1n} \\ a_{21} & \cdots & a_{2j} & \cdots & a_{2n} \\ & & \cdots & & \\ a_{n1} & \cdots & a_{nj} & \cdots & a_{nn} \end{pmatrix}.$$

Proof. Repeated application of Theorem 2 yields

$$\det A = \sum_{j=1}^{n} \det \begin{pmatrix} 0 & \cdots & a_{1j} & \cdots & 0 \\ a_{21} & \cdots & a_{2j} & \cdots & a_{2n} \\ & & \cdots & & \\ a_{n1} & \cdots & a_{nj} & \cdots & a_{nn} \end{pmatrix}$$

$$= \sum_{j=1}^{n} a_{1j} \det \begin{pmatrix} 0 & \cdots & 1 & \cdots & 0 \\ a_{21} & \cdots & a_{2j} & \cdots & a_{2n} \\ & & \cdots & & \\ a_{n1} & \cdots & a_{nj} & \cdots & a_{nn} \end{pmatrix}$$

$$= \sum_{j=1}^{n} (-1)^{1+j} a_{1j} \det \begin{pmatrix} 0 & \cdots & 1 & \cdots & 0 \\ a_{21} & \cdots & a_{2j} & \cdots & a_{2n} \\ & & \cdots & & \\ a_{n1} & \cdots & a_{nj} & \cdots & a_{nn} \end{pmatrix} \quad \text{(by Step 3)}$$

$$= \sum_{j=1}^{n} (-1)^{1+j} a_{1j} \det \begin{pmatrix} a_{11} & \cdots & a_{1j} & \cdots & a_{1n} \\ a_{21} & \cdots & a_{2j} & \cdots & a_{2n} \\ & & \cdots & & \\ a_{n1} & \cdots & a_{nj} & \cdots & a_{nn} \end{pmatrix}$$

$$= \sum_{j=1}^{n} (-1)^{1+j} a_{1j} \det A_{1j}.$$

Step 5. We are now ready to prove the formula for the expansion of det A by minors along the ith row. All we have to do is interchange the ith row of A with each of the rows above it and then expand by minors along the first row as in Step 4. We get

$$\det A = \det \begin{pmatrix} a_{11} & \cdots & a_{1n} \\ & \cdots & \\ a_{n1} & \cdots & a_{nn} \end{pmatrix}$$

$$= (-1)^{i-1} \det \begin{pmatrix} a_{i1} & \cdots & a_{in} \\ a_{11} & \cdots & a_{1n} \\ & \cdots & \\ a_{i1} & \cdots & a_{in} \\ & \cdots & \\ a_{n1} & \cdots & a_{nn} \end{pmatrix}$$

$$= (-1)^{i-1} \sum_{j=1}^{n} (-1)^{1+j} a_{ij} \det \begin{pmatrix} a_{11} & \cdots & a_{1j} & \cdots & a_{1n} \\ a_{i1} & \cdots & a_{ij} & \cdots & a_{in} \\ a_{n1} & \cdots & a_{nj} & \cdots & a_{nn} \end{pmatrix} \quad \text{(by Step 4)}$$

$$= \sum_{j=1}^{n} (-1)^{i+j} a_{ij} \det A_{ij}. \quad \blacksquare$$

EXERCISES

1. Find the determinant of each of the following matrices, using expansion by minors along a row of your choice.

(a) $\begin{pmatrix} 2 & -1 & 3 \\ 4 & 1 & 2 \\ -2 & 3 & 4 \end{pmatrix}$ (b) $\begin{pmatrix} -1 & 1 & 2 \\ 5 & 0 & 3 \\ 1 & 4 & 10 \end{pmatrix}$

(c) $\begin{pmatrix} 3 & 0 & -3 \\ -1 & 2 & 4 \\ 8 & 1 & 5 \end{pmatrix}$ (d) $\begin{pmatrix} 1 & 8 & 1 \\ 2 & 1 & 2 \\ 3 & -4 & -1 \end{pmatrix}$

(e) $\begin{pmatrix} 1 & -1 & 0 & 1 \\ 0 & 1 & 2 & 4 \\ -1 & 3 & 3 & 2 \\ 1 & -1 & 2 & 1 \end{pmatrix}$ (f) $\begin{pmatrix} 3 & 1 & 4 & 1 \\ 5 & 9 & 2 & 7 \\ 2 & 7 & 1 & 8 \\ 2 & 8 & 1 & 8 \end{pmatrix}$

2. Find the determinant of each of the following matrices.

(a) $\begin{pmatrix} a & b & 1 \\ c & 1 & 0 \\ 1 & 0 & 0 \end{pmatrix}$ (b) $\begin{pmatrix} \lambda & 2 & -2 \\ 3 & \lambda-1 & -3 \\ 1 & -1 & \lambda-3 \end{pmatrix}$

(c) $\begin{pmatrix} a_{11} & a_{12} & 0 & 0 \\ a_{21} & a_{22} & 0 & 0 \\ 0 & 0 & b_{11} & b_{12} \\ 0 & 0 & b_{21} & b_{22} \end{pmatrix}$

3. Find the determinants of

(a) $\begin{vmatrix} 0 & a & b \\ -a & 0 & c \\ -b & -c & 0 \end{vmatrix}$ (b) $\begin{vmatrix} 0 & a & b \\ a & 0 & c \\ b & c & 0 \end{vmatrix}$

(c) $\begin{vmatrix} 0 & a & b & c \\ -a & 0 & d & e \\ -b & -d & 0 & f \\ -c & -e & -f & 0 \end{vmatrix}$

4. (a) Show that if $A \in \mathfrak{M}_{n\times n}$ has a column of zeros, then $\det A = 0$.

(b) Show that

$$\det \begin{pmatrix} 1 & a_1 & \cdots & a_n \\ 0 & b_{11} & \cdots & b_{1n} \\ & & \cdots & \\ 0 & b_{n1} & \cdots & b_{nn} \end{pmatrix} = \det \begin{pmatrix} b_{11} & \cdots & b_{1n} \\ & \cdots & \\ b_{n1} & \cdots & b_{nn} \end{pmatrix}.$$

5. (a) Let $\mathbf{a} = (a_1, a_2)$ and $\mathbf{b} = (b_1, b_2)$ be distinct points in $\mathbb{R}^2$. Show that the equation

$$\det \begin{pmatrix} 1 & x_1 & x_2 \\ 1 & a_1 & a_2 \\ 1 & b_1 & b_2 \end{pmatrix} = 0$$

describes the line in $\mathbb{R}^2$ passing through $\mathbf{a}$ and $\mathbf{b}$.

(b) Find a determinant equation, analogous to that in part (a), describing the plane in $\mathbb{R}^3$ passing through the three noncollinear points $\mathbf{a}$, $\mathbf{b}$, and $\mathbf{c}$ in $\mathbb{R}^3$.

(c) Generalizing (a) and (b), find a determinant equation describing the hyperplane in $\mathbb{R}^n$ passing through n given points in $\mathbb{R}^n$.

6. (a) Show that, if $\mathbf{u}$, $\mathbf{v}$, and $\mathbf{w}$ are vectors in $\mathbb{R}^3$, then

$$\det \begin{pmatrix} \mathbf{u} \\ \mathbf{v} \\ \mathbf{w} \end{pmatrix} = (\mathbf{u} \times \mathbf{v}) \cdot \mathbf{w}.$$

(b) Show that, if $\mathbf{u}$, $\mathbf{v}$, and $\mathbf{w}$ are in $\mathbb{R}^3$, then the absolute value of

$$\det \begin{pmatrix} \mathbf{u} \\ \mathbf{v} \\ \mathbf{w} \end{pmatrix}$$

is equal to the volume of the parallelepiped spanned by $\mathbf{u}$, $\mathbf{v}$, and $\mathbf{w}$.

(c) Show that, if $\mathbf{a}$ and $\mathbf{b}$ are in $\mathbb{R}^2$, then the absolute value of $\det \begin{pmatrix} \mathbf{a} \\ \mathbf{b} \end{pmatrix}$ is equal to the area of the parallelogram spanned by $\mathbf{a}$ and $\mathbf{b}$. [Hint: Take $\mathbf{u} = (0, 0, 1)$, $\mathbf{v} = (a_1, a_2, 0)$, $\mathbf{w} = (b_1, b_2, 0)$, and apply part (b).]

7. Let

$$A = \begin{pmatrix} a_{11} & \cdots & a_{1n} \\ & \cdots & \\ a_{n1} & \cdots & a_{nn} \end{pmatrix}$$

be a matrix whose entries a_{ij} are differentiable functions on some open interval I. Show that

$$(\det A)' = \sum_{i=1}^{n} \det \begin{pmatrix} a_{11} & \cdots & a_{1n} \\ & \cdots & \\ a'_{i1} & \cdots & a'_{in} \\ & \cdots & \\ a_{n1} & \cdots & a_{nn} \end{pmatrix}.$$

[Hint: Use induction.]

8. Suppose $y_1, \ldots, y_n \in \mathcal{F}(\mathbb{R})$ are solutions of the nth order homogeneous constant coefficient linear differential equation

$$a_n y^{(n)} + \cdots + a_1 y' + a_0 y = 0.$$

(a) Show that the Wronskian determinant

$$W = \det \begin{pmatrix} y_1 & y_2 & \cdots & y_n \\ y'_1 & y'_2 & \cdots & y'_n \\ & & \cdots & \\ y_1^{(n-1)} & y_2^{(n-1)} & \cdots & y_n^{(n-1)} \end{pmatrix}$$

satisfies the differential equation $W' + (a_{n-1}/a_n) W = 0$. [Hint: Use Exercise 7.]

(b) Show that if $W(x) \neq 0$ for some $x \in \mathbb{R}$, then $W(x) \neq 0$ for all $x \in \mathbb{R}$.

(c) Show that if $W(0) \neq 0$, then $\{y_1, \ldots, y_n\}$ is linearly independent and hence is a basis for the solution space of the equation $a_n y^{(n)} + \cdots + a_1 y' + a_0 y = 0$.

5.3 PERMUTATIONS AND DETERMINANTS

In this section we shall derive another formula for the determinant of an $n \times n$ matrix. This formula expresses the determinant of $A \in \mathfrak{M}_{n \times n}$ directly in terms of the entries in A. With this formula we will be able to prove the theorem of Section 1 on the existence and uniqueness of determinants.

The formula that we shall derive involves permutations. Recall that a **permutation** of the integers $\{1, 2, \ldots, n\}$ is a one-to-one function $\sigma: \{1, 2, \ldots, n\} \to \{1, 2, \ldots, n\}$. A permutation σ is completely determined by its **value vector** $(\sigma(1), \sigma(2), \ldots, \sigma(n))$. There are two permutations of the integers $\{1, 2\}$, with value vectors $(1, 2)$ and $(2, 1)$. There are six permutations of $\{1, 2, 3\}$, with value vectors $(1, 2, 3)$, $(2, 3, 1)$, $(3, 1, 2)$, $(2, 1, 3)$, $(1, 3, 2)$, and $(3, 2, 1)$. We shall often identify a permutation with its value vector and write, for example, $\sigma = (2, 3, 1)$. We shall denote the set of all permutations of $\{1, \ldots, n\}$ by S_n.

Given $\sigma \in S_n$, the matrix

$$M_\sigma = \begin{pmatrix} \mathbf{e}_{\sigma(1)} \\ \cdots \\ \mathbf{e}_{\sigma(n)} \end{pmatrix},$$

where $\{\mathbf{e}_1, \ldots, \mathbf{e}_n\}$ is the standard basis for $\mathbb{R}^n$, is called **the matrix of the permutation** σ. The determinant of M_σ, which we shall see is always ± 1, is called the **sign** of the permutation σ, and we write

$$\text{sign } \sigma = \det M_\sigma.$$

Example 1. The permutations $(1, 2)$ and $(2, 1)$ in S_2 have matrices

$$M_{(1,2)} = \begin{pmatrix} \mathbf{e}_1 \\ \mathbf{e}_2 \end{pmatrix} = \begin{pmatrix} 1 & 0 \\ 0 & 1 \end{pmatrix} \quad \text{and} \quad M_{(2,1)} = \begin{pmatrix} \mathbf{e}_2 \\ \mathbf{e}_1 \end{pmatrix} = \begin{pmatrix} 0 & 1 \\ 1 & 0 \end{pmatrix}.$$

Computing determinants, we see that

$$\text{sign } (1, 2) = \det M_{(1,2)} = +1$$
$$\text{sign } (2, 1) = \det M_{(2,1)} = -1. \quad \blacksquare$$

There is another method for computing the sign of a permutation, one that does not require evaluating a determinant. Given a permutation $\sigma \in S_n$, we say that a pair $(\sigma(i), \sigma(j))$ is an **inversion** in σ if $i < j$ but $\sigma(i) > \sigma(j)$. In other words, a pair (a, b) of integers is an inversion in σ if both a and b appear in the value vector of σ, with a appearing before b but $a > b$. Thus the permutation $\sigma = (2, 3, 1)$ contains two inversions, $(2, 1)$ and $(3, 1)$; the permutation $(3, 2, 1)$ contains three inversions, $(3, 2)$, $(3, 1)$, and $(2, 1)$; and the permutation $(1, 2, 3)$ contains no inversions.

Theorem 1. *Let $\sigma \in S_n$. Then*

$$\text{sign } \sigma = (-1)^k$$

where k is the number of inversions in σ.

Proof. Let $\sigma \in S_n$. Then

$$\text{sign } \sigma = \det M_\sigma = \det \begin{pmatrix} \mathbf{e}_{\sigma(1)} \\ \cdots \\ \mathbf{e}_{\sigma(n)} \end{pmatrix}.$$

The vector $\mathbf{e}_n$ is somewhere among the row vectors of M_σ. By interchanging $\mathbf{e}_n$ successively with each of the rows below it, we see that

$$\text{sign } \sigma = (-1)^{k_n} \det \begin{pmatrix} \mathbf{e}_{\sigma(1)} \\ \cdots \\ \mathbf{e}_n \\ \cdots \\ \mathbf{e}_{\sigma(n)} \\ \mathbf{e}_n \end{pmatrix}$$

where k_n is the number of inversions in σ of the form (n, b). Similarly, by locating $\mathbf{e}_{n-1}$ and moving it down to the $(n-1)$st row, we find that

$$\text{sign } \sigma = (-1)^{k_{n-1}}(-1)^{k_n} \det \begin{pmatrix} \mathbf{e}_{\sigma(1)} \\ \cdots \\ \\ \cdots \\ \\ \cdots \\ \mathbf{e}_{\sigma(n)} \\ \mathbf{e}_{n-1} \\ \mathbf{e}_n \end{pmatrix}$$

where k_{n-1} is the number of inversions in σ of the form $(n-1, b)$ and where the shaded rectangles indicate the spots where $\mathbf{e}_{n-1}$ and $\mathbf{e}_n$ appeared in M_σ. Continuing in this way we find that

$$\text{sign } \sigma = (-1)^{k_2}(-1)^{k_3} \cdots (-1)^{k_n} \det \begin{pmatrix} \mathbf{e}_1 \\ \cdots \\ \mathbf{e}_n \end{pmatrix}$$

$$= (-1)^{k_2 + k_3 + \cdots + k_n} \det I$$

$$= (-1)^{k(\sigma)}$$

where k_i, for $2 \leq i \leq n$, is the number of inversions in σ of the form (i, b) and $k(\sigma) = k_2 + \cdots + k_n$ is the total number of inversions in σ.   ∎

Example 2.   Let us use Theorem 1 to compute the sign of each of the permutations in S_3. We find

$$\text{sign } (1, 2, 3) = (-1)^0 = +1 \qquad \text{sign } (2, 1, 3) = (-1)^1 = -1$$

$$\text{sign } (2, 3, 1) = (-1)^2 = +1 \qquad \text{sign } (1, 3, 2) = (-1)^1 = -1$$

$$\text{sign } (3, 1, 2) = (-1)^2 = +1 \qquad \text{sign } (3, 2, 1) = (-1)^3 = -1. \quad ■$$

We are now ready to derive the formula that expresses the determinant of a matrix directly in terms of its entries.

Theorem 2. *Let*

$$A = \begin{pmatrix} a_{11} & \cdots & a_{1n} \\ & \cdots & \\ a_{n1} & \cdots & a_{nn} \end{pmatrix}.$$

Then

$$\det A = \sum_{\sigma \in S_n} (\text{sign } \sigma) a_{1\sigma(1)} \cdots a_{n\sigma(n)}.$$

Proof. First, use the additivity of the determinant in the first row (Theorem 2 of Section 5.2) and property P_1 of determinants to get

$$\det A = \det \begin{pmatrix} \mathbf{a}_1 \\ \mathbf{a}_2 \\ \cdots \\ \mathbf{a}_n \end{pmatrix} = \det \begin{pmatrix} \sum_{j=1}^{n} a_{1j}\mathbf{e}_j \\ \mathbf{a}_2 \\ \cdots \\ \mathbf{a}_n \end{pmatrix} = \sum_{j=1}^{n} \det \begin{pmatrix} a_{1j}\mathbf{e}_j \\ \mathbf{a}_2 \\ \cdots \\ \mathbf{a}_n \end{pmatrix} = \sum_{j=1}^{n} a_{1j} \det \begin{pmatrix} \mathbf{e}_j \\ \mathbf{a}_2 \\ \cdots \\ \mathbf{a}_n \end{pmatrix}.$$

Now call the summation index in the above formula j_1 instead of j and repeat this process on the second row to get

$$\det A = \sum_{j_1=1}^{n} a_{1j_1} \det \begin{pmatrix} \mathbf{e}_{j_1} \\ \mathbf{a}_2 \\ \mathbf{a}_3 \\ \cdots \\ \mathbf{a}_n \end{pmatrix} = \sum_{j_1=1}^{n} a_{1j_1} \det \begin{pmatrix} \mathbf{e}_{j_1} \\ \sum_{j_2=1}^{n} a_{2j_2}\mathbf{e}_{j_2} \\ \mathbf{a}_3 \\ \cdots \\ \mathbf{a}_n \end{pmatrix}$$

$$= \sum_{j_1, j_2=1}^{n} a_{1j_1} a_{2j_2} \det \begin{pmatrix} \mathbf{e}_{j_1} \\ \mathbf{e}_{j_2} \\ \mathbf{a}_3 \\ \cdots \\ \mathbf{a}_n \end{pmatrix}.$$

Continuing in this way we find that

$$\det A = \sum_{j_1,\dots,j_n=1}^{n} a_{1j_1} \cdots a_{nj_n} \det \begin{pmatrix} \mathbf{e}_{j_1} \\ \cdots \\ \mathbf{e}_{j_n} \end{pmatrix}.$$

But

$$\det \begin{pmatrix} \mathbf{e}_{j_1} \\ \cdots \\ \mathbf{e}_{j_n} \end{pmatrix} = 0$$

unless the integers $j_1, \dots, j_n$ are distinct, since a matrix with two equal rows has determinant zero. Therefore, the only terms in the above sum that are nonzero are those for which $(j_1, \dots, j_n) = (\sigma(1), \dots, \sigma(n))$ for some $\sigma \in S_n$. Hence

$$\det A = \sum_{\sigma \in S_n} a_{1\sigma(1)} \cdots a_{n\sigma(n)} \det \begin{pmatrix} \mathbf{e}_{\sigma(1)} \\ \cdots \\ \mathbf{e}_{\sigma(n)} \end{pmatrix}$$

$$= \sum_{\sigma \in S_n} (\text{sign } \sigma) a_{1\sigma(1)} \cdots a_{n\sigma(n)}. \quad \blacksquare$$

When $n = 2$, the formula of Theorem 2 reduces to the familiar formula

$$\det \begin{pmatrix} a_{11} & a_{12} \\ a_{21} & a_{22} \end{pmatrix} = a_{11}a_{22} - a_{12}a_{21}.$$

When $n = 3$, we get the formula for the determinant of a 3×3 matrix found in Example 3 of the previous section. When $n = 4$, the formula of Theorem 2 expresses the determinant of a 4×4 matrix as a sum of 24 terms. For an $n \times n$ matrix, the sum has $n!$ terms, so you can see that this formula is not very helpful for computing the determinant of a very large matrix. But it does have important theoretical consequences.

Given $A \in \mathfrak{M}_{m \times n}$, the **transpose** of A is the matrix $A^t \in \mathfrak{M}_{n \times m}$ whose (i, j)-entry is the (j, i)-entry of A, for each (i, j) $(1 \le i \le m, 1 \le j \le n)$. When $m = n$, we can use Theorem 2 to prove that A and A^t have the same determinant.

Theorem 3. *Let $A \in \mathfrak{M}_{n \times n}$. Then $\det A^t = \det A$.*

Proof. Let

$$A = \begin{pmatrix} a_{11} & a_{12} & \cdots & a_{1n} \\ a_{21} & a_{22} & \cdots & a_{2n} \\ & & \cdots & \\ a_{n1} & a_{n2} & \cdots & a_{nn} \end{pmatrix}. \text{ Then } A^t = \begin{pmatrix} a_{11} & a_{21} & \cdots & a_{n1} \\ a_{12} & a_{22} & \cdots & a_{n2} \\ & & \cdots & \\ a_{1n} & a_{2n} & \cdots & a_{nn} \end{pmatrix}.$$

By Theorem 2,

$$\det A^t = \sum_{\sigma \in S_n} (\text{sign } \sigma) a_{\sigma(1)1} \cdots a_{\sigma(n)n}.$$

First, let us reorder the factors in each product $a_{\sigma(1)1} \cdots a_{\sigma(n)n}$ so that the first subscripts rather than the second are in increasing order. Notice that when $\sigma(j) = 1$ then $j = \sigma^{-1}(1)$ so $a_{\sigma(j)j} = a_{1\sigma^{-1}(1)}$. Similarly, for each $k (1 \le k \le n)$, when $\sigma(j) = k$ then $j = \sigma^{-1}(k)$ so $a_{\sigma(j)j} = a_{k\sigma^{-1}(k)}$. It follows that

$$\det A^t = \sum_{\sigma \in S_n} (\text{sign } \sigma) a_{1\sigma^{-1}(1)} \cdots a_{n\sigma^{-1}(n)}.$$

Next, notice that, for each $\sigma \in S_n$, sign $\sigma = \text{sign } \sigma^{-1}$. Indeed, the statement that $i < j$, $\sigma(i) > \sigma(j)$ is the same as the statement that $\sigma^{-1}(k) < \sigma^{-1}(l)$, $k > l$ where $k = \sigma(i)$ and $l = \sigma(j)$. Hence σ and σ^{-1} have the same number of inversions and thus have the same sign. It follows that

$$\det A^t = \sum_{\sigma \in S_n} (\text{sign } \sigma^{-1}) a_{1\sigma^{-1}(1)} \cdots a_{n\sigma^{-1}(n)}.$$

Now, as σ runs through S_n so does σ^{-1}. So if we replace σ by $\tau = \sigma^{-1}$ as summation index in the above formula, we get

$$\det A^t = \sum_{\tau \in S_n} (\text{sign } \tau) a_{1\tau(1)} \cdots a_{n\tau(n)} = \det A. \quad \blacksquare$$

Theorem 4. (*Expansion of* det A *by minors along the* jth *column.*) *Let*

$$A = \begin{pmatrix} a_{11} & \cdots & a_{1n} \\ & \cdots & \\ a_{n1} & \cdots & a_{nn} \end{pmatrix}.$$

Then, for each $j (1 \le j \le n)$,

$$\det A = \sum_{i=1}^{n} (-1)^{i+j} a_{ij} \det A_{ij},$$

where A_{ij} *is the* (i, j)-*minor of* A.

Proof. The formula for expansion of det A by minors along the jth column is just the formula for expansion of det A^t along the jth row. Since det $A = \det A^t$, the theorem is proved. ■

Example 3. Expanding by minors along the second column we find that

$$\det \begin{pmatrix} 1 & -1 & 2 \\ 5 & 4 & -1 \\ -3 & 0 & 7 \end{pmatrix} = (-1)^{1+2}(-1)\det \begin{pmatrix} 1 & -1 & 2 \\ 5 & 4 & -1 \\ -3 & 0 & 7 \end{pmatrix}$$

$$+(-1)^{2+2}(4)\det \begin{pmatrix} 1 & -1 & 2 \\ 5 & 4 & -1 \\ -3 & 0 & 7 \end{pmatrix}$$

$$+(-1)^{3+2}(0)\det \begin{pmatrix} 1 & -1 & 2 \\ 5 & 4 & -1 \\ -3 & 0 & 7 \end{pmatrix}$$

$$= \det \begin{pmatrix} 5 & -1 \\ -3 & 7 \end{pmatrix} + 4\det \begin{pmatrix} 1 & 2 \\ -3 & 7 \end{pmatrix}$$

$$= 32 + 4(13) = 84. \quad \blacksquare$$

We conclude this chapter with a proof of Theorem 2 of Section 5.1 on the existence and uniqueness of the determinant function.

Recall the statement of that theorem.

Theorem. *There exists one and only one function* det: $\mathfrak{M}_{n\times n} \to \mathbb{R}$ *with the following four properties:*

(P_1) *if* $B \in \mathfrak{M}_{n\times n}$ *is obtained from* $A \in \mathfrak{M}_{n\times n}$ *by multiplying a row of A by* $c \in \mathbb{R}$ *then* det $B = c$ det A

(P_2) *if* $B \in \mathfrak{M}_{n\times n}$ *is obtained from* $A \in \mathfrak{M}_{n\times n}$ *by interchanging two rows then* det $B = -$det A

(P_3) *if* $B \in \mathfrak{M}_{n\times n}$ *is obtained from* $A \in \mathfrak{M}_{n\times n}$ *by adding a scalar multiple of one row to another then* det $B = $ det A

(P_4) det $I = 1$.

Proof. That there is at most one function det satisfying properties P_1–P_4 is immediate from Theorem 2 of this section. Theorem 2 was proved using only properties P_1–P_4. It gives an explicit formula for what any function satisfying P_1–P_4 must be.

The formula of Theorem 2 also tells us how to prove the existence of the determinant function. Simply define a function det: $\mathfrak{M}_{n\times n} \to \mathbb{R}$ by

$$\det \begin{pmatrix} a_{11} & \cdots & a_{1n} \\ & \cdots & \\ a_{n1} & \cdots & a_{nn} \end{pmatrix} = \sum_{\sigma \in S_n} (\text{sign } \sigma) a_{1\sigma(1)} \cdots a_{n\sigma(n)}$$

and check that this function does indeed have the required properties P_1–P_4.

There is one subtle point that needs to be mentioned here. If we use the above formula to *define* the determinant, then we cannot use the determinant to define the sign of a permutation as we did earlier. That would be circular. So, in this proof, we must use as definition of the sign of a permutation the formula

$$\text{sign } \sigma = (-1)^{k(\sigma)}$$

where $k(\sigma)$ is the number of inversions in σ.

Now let us verify that the function $\det: \mathfrak{M}_{n \times n} \to \mathbb{R}$ defined by the above formula does have property P_1. Suppose B is obtained from A by multiplying the kth row of A by c. Then $b_{ij} = a_{ij}$ whenever $i \neq k$ and $b_{kj} = ca_{kj}$ for all j. Hence

$$\det B = \sum_{\sigma \in S_n} (\text{sign } \sigma) b_{1\sigma(1)} \cdots b_{k\sigma(k)} \cdots b_{n\sigma(n)}$$

$$= \sum_{\sigma \in S_n} (\text{sign } \sigma) a_{1\sigma(1)} \cdots (ca_{k\sigma(k)}) \cdots a_{n\sigma(n)}$$

$$= c \sum_{\sigma \in S_n} (\text{sign } \sigma) a_{1\sigma(1)} \cdots a_{k\sigma(k)} \cdots a_{n\sigma(n)}$$

$$= c \det A.$$

To verify Property P_2, notice first that whenever a permutation τ is obtained from a permutation σ by interchanging two adjacent entries in the value vector, then the number of inversions in τ is either one more or one less than the number of inversions in σ, and hence sign $\tau = -\text{sign } \sigma$. If τ is obtained from σ by interchanging two entries that are not adjacent, say the ith and the jth, then this is the same as interchanging the ith entry successively with all the entries between the ith and the jth (there are $|i - j| - 1$ of these), then interchanging the ith and the jth, and finally interchanging the jth with each of the intervening entries to move it to the ith position. The total number of interchanges of adjacent entries in this process is $2(|i - j| - 1) + 1$. Hence

$$\text{sign } \tau = (-1)^{2(|i-j|-1)+1} \text{sign } \sigma = -\text{sign } \sigma,$$

whenever τ is obtained from σ by interchanging two entries in the value vector. It follows that, if $B \in \mathfrak{M}_{n \times n}$ is obtained from $A \in \mathfrak{M}_{n \times n}$ by interchanging the ith row with the jth, then

$$\det B = \sum_{\sigma \in S_n} (\text{sign } \sigma) a_{1\sigma(1)} \cdots \overset{i\text{th spot}}{a_{j\sigma(i)}} \cdots \overset{j\text{th spot}}{a_{i\sigma(j)}} \cdots a_{n\sigma(n)}$$

$$= \sum_{\sigma \in S_n} (\text{sign } \sigma) a_{1\tau(1)} \cdots \overset{i\text{th spot}}{a_{j\tau(j)}} \cdots \overset{j\text{th spot}}{a_{i\tau(i)}} \cdots a_{n\tau(n)}$$

where, for each $\sigma \in S_n$, τ is the permutation obtained from σ by interchanging the ith and jth entries in the value vector. Since τ runs through S_n as σ does, since sign $\tau = -$sign σ, and since multiplication of real numbers is commutative, we see that

$$\det B = \sum_{\tau \in S_n} -(\text{sign } \tau) a_{1\tau(1)} \cdots \overset{\overset{i\text{th spot}}{\downarrow}}{a_{i\tau(i)}} \cdots \overset{\overset{j\text{th spot}}{\downarrow}}{a_{j\tau(j)}} \cdots a_{n\tau(n)}$$

$$= -\det A.$$

The verification of Properties P$_3$ and P$_4$ are left as exercises. ▨

EXERCISES

1. List the 24 permutations of $\{1, 2, 3, 4\}$ and calculate the sign of each.

2. Find the determinant of each of the following matrices, using expansion by minors along a column of your choice.

(a) $\begin{pmatrix} 3 & -1 & 4 \\ 1 & 5 & -9 \\ 2 & 8 & 1 \end{pmatrix}$ (b) $\begin{pmatrix} 2 & 1 & 8 \\ -3 & 7 & 2 \\ 4 & -2 & 1 \end{pmatrix}$

(c) $\begin{pmatrix} 1 & -4 & 1 \\ 2 & 1 & 3 \\ 6 & -5 & 2 \end{pmatrix}$ (d) $\begin{pmatrix} 1 & -7 & 3 \\ 2 & 0 & 5 \\ 8 & -1 & 2 \end{pmatrix}$

3. Use expansion by minors along an appropriate column to evaluate each of the following determinants.

(a) $\det \begin{pmatrix} 1 & 1 & 1 \\ x & 1 & 2 \\ x^2 & 1 & 4 \end{pmatrix}$ (b) $\det \begin{pmatrix} \lambda - 1 & 0 & -1 \\ 2 & \lambda & -2 \\ 1 & 0 & \lambda + 1 \end{pmatrix}$

(c) $\det \begin{pmatrix} -1 & -1 & -1 & a \\ 1 & 0 & 0 & b \\ 0 & 1 & 0 & c \\ 0 & 0 & 1 & d \end{pmatrix}$

4. Verify that if the determinant is defined by the formula

$$\det \begin{pmatrix} a_{11} & \cdots & a_{1n} \\ & \cdots & \\ a_{n1} & \cdots & a_{nn} \end{pmatrix} = \sum_{\sigma \in S_n} (\text{sign } \sigma) a_{1\sigma(1)} \cdots a_{n\sigma(n)}$$

then

(a) $\det B = \det A$ whenever B is obtained from A by adding a scalar multiple of one row to another.

(b) $\det \begin{pmatrix} 1 & 0 & \cdots & 0 \\ 0 & 1 & \cdots & 0 \\ & & \cdots & \\ 0 & 0 & \cdots & 1 \end{pmatrix} = 1.$

Orthogonality

6.1 ORTHONORMAL BASES

In Chapter 4, you learned that the choice of an ordered basis for a vector space leads to the important idea of a coordinatization of the vector space: given an ordered basis $\mathbf{B} = (\mathbf{v}_1, \ldots, \mathbf{v}_k)$ for V, we can attach to each vector $\mathbf{v} \in V$ its $\mathbf{B}$-coordinate k-tuple, $(c_1, \ldots, c_k) \in \mathbb{R}^k$, where the real numbers $c_1, \ldots, c_k$ are determined by the equation

$$\mathbf{v} = c_1 \mathbf{v}_1 + \cdots + c_k \mathbf{v}_k.$$

It is easy to see that if $\mathbf{c} = (c_1, \ldots, c_k)$ is the $\mathbf{B}$-coordinate k-tuple of $\mathbf{v}$ and if $\mathbf{d} = (d_1, \ldots, d_k)$ is the $\mathbf{B}$-coordinate k-tuple of $\mathbf{w}$, then $\mathbf{c} + \mathbf{d} = (c_1 + d_1, \ldots, c_k + d_k)$ is the $\mathbf{B}$-coordinate k-tuple of $\mathbf{v} + \mathbf{w}$. Also, for each real number a, $a\mathbf{c} = (ac_1, \ldots, ac_k)$ is the $\mathbf{B}$-coordinate k-tuple of $a\mathbf{v}$. Thus, by passing from a vector $\mathbf{v}$ to its $\mathbf{B}$-coordinate k-tuple $\mathbf{c}$ we can convert computations in V into computations in $\mathbb{R}^k$. The only difficulty in this process is that of finding the $\mathbf{B}$-coordinates of any given vector. This can be a formidable task. However, when V is a subspace of $\mathbb{R}^n$ for some n, we can use the dot product in $\mathbb{R}^n$ to identify special ordered bases for which the problem of finding coordinates becomes very easy. These special bases are the orthonormal ones.

A finite subset $\{\mathbf{v}_1, \ldots, \mathbf{v}_k\}$ of $\mathbb{R}^n$ is **orthogonal** if the vectors $\mathbf{v}_1, \ldots, \mathbf{v}_k$ are mutually perpendicular ($\mathbf{v}_i \cdot \mathbf{v}_j = 0$ whenever $i \neq j$). The set $\{\mathbf{v}_1, \ldots, \mathbf{v}_k\}$ is **orthonormal** if it is orthogonal and if, in addition, each $\mathbf{v}_i$ is a unit vector ($\|\mathbf{v}_i\| = 1$ for all i). An ordered basis $(\mathbf{v}_1, \ldots, \mathbf{v}_k)$ for a subspace V of $\mathbb{R}^n$ is called orthonormal if the set $\{\mathbf{v}_1, \ldots, \mathbf{v}_k\}$ is orthonormal.

Example 1. The standard basis $\{\mathbf{e}_1, \ldots, \mathbf{e}_n\}$ for $\mathbb{R}^n$ is orthonormal because $\mathbf{e}_i \cdot \mathbf{e}_j = 0$ whenever $i \neq j$ and $\|\mathbf{e}_i\| = 1$ for all i. ■

Example 2. The set $\{(1, 0, 1), (0, 1, 0)\}$ is an orthogonal set in $\mathbb{R}^3$, since $(1, 0, 1) \cdot (0, 1, 0) = 0$, but it is not orthonormal because $\|(1, 0, 1)\| = \sqrt{2} \neq 1$. ■

The following theorem describes the most important property of ortho-normal bases.

Theorem 1.   *Let V be a subspace of $\mathbb{R}^n$ and let $\mathbf{B} = (\mathbf{v}_1, \ldots, \mathbf{v}_k)$ be an ordered orthonormal basis for V. Then the $\mathbf{B}$-coordinate k-tuple of each vector $\mathbf{v} \in V$ is the k-tuple*

$$(\mathbf{v} \cdot \mathbf{v}_1, \ldots, \mathbf{v} \cdot \mathbf{v}_k).$$

Proof.   The $\mathbf{B}$-coordinate k-tuple of $\mathbf{v}$ is $(c_1, \ldots, c_k)$ where the real numbers $c_1, \ldots, c_k$ satisfy the equation

$$\mathbf{v} = c_1\mathbf{v}_1 + \cdots + c_k\mathbf{v}_k.$$

Taking the dot product of both sides of this equation with $\mathbf{v}_1$, we get

$$
\begin{aligned}
\mathbf{v} \cdot \mathbf{v}_1 &= (c_1\mathbf{v}_1 + c_2\mathbf{v}_2 + \cdots + c_n\mathbf{v}_n) \cdot \mathbf{v}_1 \\
&= c_1(\mathbf{v}_1 \cdot \mathbf{v}_1) + c_2(\mathbf{v}_2 \cdot \mathbf{v}_1) + \cdots + c_n(\mathbf{v}_n \cdot \mathbf{v}_1) \\
&= c_1\|\mathbf{v}_1\|^2 + c_2(0) + \cdots + c_n(0) \\
&= c_1,
\end{aligned}
$$

so $c_1 = \mathbf{v} \cdot \mathbf{v}_1$ as claimed. To see that $c_2 = \mathbf{v} \cdot \mathbf{v}_2$, simply take the dot product of both sides of the above equation with $\mathbf{v}_2$. Continuing in this way, we see that $c_i = \mathbf{v} \cdot \mathbf{v}_i$ for each i.   ∎

Example 3.   Let

$$\mathbf{v}_1 = \left(\frac{\sqrt{3}}{2}, \frac{1}{2}\right) \quad \text{and} \quad \mathbf{v}_2 = \left(-\frac{1}{2}, \frac{\sqrt{3}}{2}\right).$$

Because $\mathbf{B} = (\mathbf{v}_1, \mathbf{v}_2)$ is an ordered orthonormal basis for $\mathbb{R}^2$, the $\mathbf{B}$-coordinate pairs of vectors in $\mathbb{R}^2$ are easy to compute. For example, the $\mathbf{B}$-coordinate pair of $\mathbf{v} = (-2, 3)$ is

$$(\mathbf{v} \cdot \mathbf{v}_1, \mathbf{v} \cdot \mathbf{v}_2) = \left(-\sqrt{3} + \frac{3}{2}, 1 + \frac{3\sqrt{3}}{2}\right).$$

As a check, it is easy to verify that

$$\left(-\sqrt{3} + \frac{3}{2}\right)\mathbf{v}_1 + \left(1 + \frac{3\sqrt{3}}{2}\right)\mathbf{v}_2 = (-2, 3).   ∎$$

A nice feature of any orthonormal set is that it forms a basis for the subspace that it spans. This is a consequence of the following theorem.

Theorem 2. *Let* $\{\mathbf{v}_1, \ldots, \mathbf{v}_k\}$ *be an orthonormal set in* $\mathbb{R}^n$. *Then* $\{\mathbf{v}_1, \ldots, \mathbf{v}_k\}$ *is linearly independent.*

Proof. Suppose we have a dependence relation

$$c_1\mathbf{v}_1 + \cdots + c_k\mathbf{v}_k = \mathbf{0}.$$

Then

$$c_1 = \mathbf{v}_1 \bullet (c_1\mathbf{v}_1 + \cdots + c_k\mathbf{v}_k) = \mathbf{v}_1 \bullet \mathbf{0} = 0$$

$$\cdots$$

$$c_k = \mathbf{v}_k \bullet (c_1\mathbf{v}_1 + \cdots + c_k\mathbf{v}_k) = \mathbf{v}_k \bullet \mathbf{0} = 0$$

and hence $c_1 = \cdots = c_k = 0$ as was to be shown. ■

Corollary. *Let* $\{\mathbf{v}_1, \ldots, \mathbf{v}_k\}$ *be an orthonormal set in* $\mathbb{R}^n$. *Then* $\{\mathbf{v}_1, \ldots, \mathbf{v}_k\}$ *is a basis for* $\mathcal{L}(\mathbf{v}_1, \ldots, \mathbf{v}_k)$.

Example 4. Let

$$\mathbf{v}_1 = (\tfrac{1}{2}, \tfrac{1}{2}, -\tfrac{1}{2}, -\tfrac{1}{2})$$

$$\mathbf{v}_2 = (\tfrac{1}{2}, -\tfrac{1}{2}, \tfrac{1}{2}, -\tfrac{1}{2})$$

$$\mathbf{v}_3 = (-\tfrac{1}{2}, \tfrac{1}{2}, \tfrac{1}{2}, -\tfrac{1}{2}).$$

Then $\{\mathbf{v}_1, \mathbf{v}_2, \mathbf{v}_3\}$ is easily seen to be an orthonormal set in $\mathbb{R}^4$. The subspace $\mathcal{L}(\mathbf{v}_1, \mathbf{v}_2, \mathbf{v}_3)$ of $\mathbb{R}^4$ spanned by $\{\mathbf{v}_1, \mathbf{v}_2, \mathbf{v}_3\}$ is the hyperplane H in $\mathbb{R}^4$ defined by the equation $x_1 + x_2 + x_3 + x_4 = 0$. To see this, first notice that $\mathbf{v}_i \in H$ for each i, so $\{\mathbf{v}_1, \mathbf{v}_2, \mathbf{v}_3\}$ is a subset of H. Then observe that the set $\{\mathbf{v}_1, \mathbf{v}_2, \mathbf{v}_3\}$ is linearly independent because it is orthonormal. Finally, $\{\mathbf{v}_1, \mathbf{v}_2, \mathbf{v}_3\}$ must be a basis for H, since the dimension of H is 3.

The vector $\mathbf{v} = (-1, 4, -5, 2)$ is also in H. Its **B**-coordinate triple, where $\mathbf{B} = (\mathbf{v}_1, \mathbf{v}_2, \mathbf{v}_3)$, is

$$(\mathbf{v} \bullet \mathbf{v}_1, \mathbf{v} \bullet \mathbf{v}_2, \mathbf{v} \bullet \mathbf{v}_3) = (3, -6, -1).$$ ■

We mentioned in the opening paragraph of this section that the use of **B**-coordinates in a vector space V enables us to convert algebraic computations (addition, scalar multiplication, forming linear combinations) in V into corresponding computations in $\mathbb{R}^k$. When V is a subspace of $\mathbb{R}^n$ and the basis is orthonormal, we can also convert geometric computations (dot product, length) in V into corresponding computations in $\mathbb{R}^k$, as the following theorem shows.

Theorem 3. *Let* V *be a subspace of* $\mathbb{R}^n$ *and let* $\mathbf{B} = (\mathbf{v}_1, \ldots, \mathbf{v}_k)$ *be an ordered orthonormal basis for* V. *Suppose* $\mathbf{v} \in V$ *and* $\mathbf{w} \in V$ *have* **B**-*coordinate k-tuples*

$\mathbf{c} = (c_1, \ldots, c_k)$ *and* $\mathbf{d} = (d_1, \ldots, d_k)$ *respectively. Then*

$$\mathbf{v} \bullet \mathbf{w} = \mathbf{c} \bullet \mathbf{d} = c_1 d_1 + \cdots + c_k d_k$$

and

$$\|\mathbf{v}\| = \|\mathbf{c}\| = (c_1^2 + \cdots + c_k^2)^{\frac{1}{2}}.$$

Proof. Since

$$\mathbf{v} = c_1 \mathbf{v}_1 + \cdots + c_k \mathbf{v}_k = \sum_{i=1}^{k} c_i \mathbf{v}_i$$

and

$$\mathbf{w} = d_1 \mathbf{v}_1 + \cdots + d_k \mathbf{v}_k = \sum_{j=1}^{k} d_j \mathbf{v}_j$$

we find

$$\mathbf{v} \bullet \mathbf{w} = \left(\sum_{i=1}^{k} c_i \mathbf{v}_i \right) \bullet \left(\sum_{j=1}^{k} d_j \mathbf{v}_j \right)$$

$$= \sum_{i,j=1}^{k} c_i d_j (\mathbf{v}_i \bullet \mathbf{v}_j)$$

$$= \sum_{i=1}^{k} c_i d_i (\mathbf{v}_i \bullet \mathbf{v}_i) \quad \text{(since } \mathbf{v}_i \bullet \mathbf{v}_j = 0 \text{ for } i \neq j)$$

$$= \sum_{i=1}^{k} c_i d_i \quad \text{(since } \mathbf{v}_i \bullet \mathbf{v}_i = \|\mathbf{v}_i\|^2 = 1 \text{ for all } i)$$

$$= \mathbf{c} \bullet \mathbf{d}$$

and

$$\|\mathbf{v}\| = (\mathbf{v} \bullet \mathbf{v})^{\frac{1}{2}} = (\mathbf{c} \bullet \mathbf{c})^{\frac{1}{2}} = \|\mathbf{c}\|. \quad \blacksquare$$

We have seen some of the advantages of *using* orthonormal bases for subspaces of $\mathbb{R}^n$. Now let us consider the problem of how to *find* an orthonormal basis for a given subspace V of $\mathbb{R}^n$. Since we already know how to find bases for these subspaces (start with any linearly independent set and enlarge it to a basis), we need only describe how to replace a given basis by an orthonormal one.

If the dimension k of V is 1, then the given basis will consist of a single vector $\mathbf{v}$. To obtain an orthonormal basis we need only *normalize* $\mathbf{v}$; that is, divide $\mathbf{v}$ by its length. Thus, $\{\mathbf{v}/\|\mathbf{v}\|\}$ is an orthonormal basis for V.

If the dimension of V is 2, then the given basis will consist of two vectors, $\{\mathbf{v}_1, \mathbf{v}_2\}$. First, let $\mathbf{w}_1 = \mathbf{v}_1$. Then replace $\mathbf{v}_2$ by $\mathbf{w}_2 = \mathbf{v}_2 + c\mathbf{w}_1 \in V$, where c is chosen so that $\mathbf{w}_2$ is perpendicular to $\mathbf{w}_1$. It is easy to obtain a formula for c. We require that

$$0 = \mathbf{w}_2 \cdot \mathbf{w}_1 = (\mathbf{v}_2 + c\mathbf{w}_1) \cdot \mathbf{w}_1 = \mathbf{v}_2 \cdot \mathbf{w}_1 + c\mathbf{w}_1 \cdot \mathbf{w}_1$$

so

$$c = -\frac{\mathbf{v}_2 \cdot \mathbf{w}_1}{\mathbf{w}_1 \cdot \mathbf{w}_1}.$$

Thus $\{\mathbf{w}_1, \mathbf{w}_2\}$ is an orthogonal set in V, where

$$\mathbf{w}_1 = \mathbf{v}_1$$

$$\mathbf{w}_2 = \mathbf{v}_2 - \frac{\mathbf{v}_2 \cdot \mathbf{w}_1}{\mathbf{w}_1 \cdot \mathbf{w}_1}\mathbf{w}_1.$$

To obtain an ortho*normal* set, simply normalize $\mathbf{w}_1$ and $\mathbf{w}_2$ to get the orthonormal basis $\left\{\dfrac{\mathbf{w}_1}{\|\mathbf{w}_1\|}, \dfrac{\mathbf{w}_2}{\|\mathbf{w}_2\|}\right\}$ for V. (Note that $\mathbf{w}_2 \neq \mathbf{0}$ since $\{\mathbf{v}_1, \mathbf{v}_2\}$ is linearly independent, and that $\left\{\dfrac{\mathbf{w}_1}{\|\mathbf{w}_1\|}, \dfrac{\mathbf{w}_2}{\|\mathbf{w}_2\|}\right\}$ is a basis for V because it is linearly independent [it is orthonormal!] and the dimension of V is 2.)

If the dimension of V is 3 and $\{\mathbf{v}_1, \mathbf{v}_2, \mathbf{v}_3\}$ is a basis for V, the procedure is similar. First define $\mathbf{w}_1$ and $\mathbf{w}_2$ as before:

$$\mathbf{w}_1 = \mathbf{v}_1$$

$$\mathbf{w}_2 = \mathbf{v}_2 - \frac{\mathbf{v}_2 \cdot \mathbf{w}_1}{\mathbf{w}_1 \cdot \mathbf{w}_1}\mathbf{w}_1.$$

Then take

$$\mathbf{w}_3 = \mathbf{v}_3 + c_1\mathbf{w}_1 + c_2\mathbf{w}_2$$

where c_1 and c_2 are chosen so that $\mathbf{w}_3 \cdot \mathbf{w}_1 = \mathbf{w}_3 \cdot \mathbf{w}_2 = 0$. It is easy to check that

$$c_1 = -\frac{\mathbf{v}_3 \cdot \mathbf{w}_1}{\mathbf{w}_1 \cdot \mathbf{w}_1} \quad \text{and} \quad c_2 = -\frac{\mathbf{v}_3 \cdot \mathbf{w}_2}{\mathbf{w}_2 \cdot \mathbf{w}_2}$$

(see Exercise 5). Then $\{\mathbf{w}_1, \mathbf{w}_2, \mathbf{w}_3\}$ will be an orthogonal set in V and $\left\{\dfrac{\mathbf{w}_1}{\|\mathbf{w}_1\|}, \dfrac{\mathbf{w}_2}{\|\mathbf{w}_2\|}, \dfrac{\mathbf{w}_3}{\|\mathbf{w}_3\|}\right\}$ will be an orthonormal basis for V.

This process, called the **Gram-Schmidt orthogonalization process,** generalizes in a straightforward way to subspaces of $\mathbb{R}^n$ of any dimension, as follows.

Theorem 4.   *Suppose V is a subspace of $\mathbb{R}^n$ with basis $\{v_1, \ldots, v_k\}$. Define $w_1, \ldots, w_k \in V$ by*

$$w_1 = v_1$$

$$w_2 = v_2 - \frac{v_2 \cdot w_1}{w_1 \cdot w_1} w_1$$

$$w_3 = v_3 - \frac{v_3 \cdot w_1}{w_1 \cdot w_1} w_1 - \frac{v_3 \cdot w_2}{w_2 \cdot w_2} w_2$$

$$\ldots$$

$$w_k = v_k - \frac{v_k \cdot w_1}{w_1 \cdot w_1} - \cdots - \frac{v_k \cdot w_{k-1}}{w_{k-1} \cdot w_{k-1}} w_{k-1}.$$

Then $\left\{ \dfrac{w_1}{\|w_1\|}, \ldots, \dfrac{w_k}{\|w_k\|} \right\}$ is an orthonormal basis for V.

Remark.   If, in Theorem 4, we are given only a spanning set $\{v_1, \ldots, v_k\}$ for V (not necessarily a basis), then we can still apply the Gram-Schmidt process to obtain an orthonormal basis for V. The only change in the procedure is that we may find that one or more of the vectors w_i turns out to be zero. If we simply discard any such w_i and omit the terms $\dfrac{v_j \cdot w_i}{w_i \cdot w_i} w_i$ for those i from all subsequent computations, we will still end up with an orthonormal basis for V.

Example 5.   Let us apply the Gram-Schmidt process to find an orthonormal basis for the subspace $V = \mathcal{L}(v_1, v_2)$ of $\mathbb{R}^3$ where $v_1 = (1, -1, 2)$ and $v_2 = (2, 1, -1)$. We find

$$w_1 = v_1 = (1, -1, 2)$$

$$w_2 = v_2 - \frac{v_2 \cdot w_1}{w_1 \cdot w_1} w_1 = (2, 1, -1) - \frac{(-1)}{6}(1, -1, 2) = \frac{1}{6}(13, 5, -4)$$

and so

$$\left\{ \frac{w_1}{\|w_1\|}, \frac{w_2}{\|w_2\|} \right\} = \left\{ \frac{1}{\sqrt{6}}(1, -1, 2), \frac{1}{\sqrt{210}}(13, 5, -4) \right\}$$

is the required orthonormal basis for V.   ∎

Example 6.   Applying the Gram-Schmidt process to find an orthonormal basis for the subspace $\mathcal{L}(v_1, v_2, v_3)$ of $\mathbb{R}^4$, where $v_1 = (2, 0, 1, 2)$, $v_2 = (-3, 1, 2, 0)$, and $v_3 = (1, 1, 4, 4)$, we get

$$\mathbf{w}_1 = \mathbf{v}_1 = (2, 0, 1, 2)$$

$$\mathbf{w}_2 = \mathbf{v}_2 - \frac{\mathbf{v}_2 \cdot \mathbf{w}_1}{\mathbf{w}_1 \cdot \mathbf{w}_1} \mathbf{w}_1 = (-3, 1, 2, 0) - \frac{(-4)}{9}(2, 0, 1, 2)$$

$$= \tfrac{1}{9}(-19, 9, 22, 8)$$

$$\mathbf{w}_3 = \mathbf{v}_3 - \frac{\mathbf{v}_3 \cdot \mathbf{w}_1}{\mathbf{w}_1 \cdot \mathbf{w}_1} \mathbf{w}_1 - \frac{\mathbf{v}_3 \cdot \mathbf{w}_2}{\mathbf{w}_2 \cdot \mathbf{w}_2} \mathbf{w}_2$$

$$= (1, 1, 4, 4) - \tfrac{14}{9}(2, 0, 1, 2) - \frac{(110/81)}{(990/81)}(-19, 9, 22, 8)$$

$$= (1, 1, 4, 4) - \tfrac{1}{9}(28, 0, 14, 28) - \tfrac{1}{9}(-19, 9, 22, 8)$$

$$= (0, 0, 0, 0).$$

Since $\mathbf{w}_3 = \mathbf{0}$ we discard it. Thus

$$\left\{ \frac{\mathbf{w}_1}{\|\mathbf{w}_1\|}, \frac{\mathbf{w}_2}{\|\mathbf{w}_2\|} \right\} = \left\{ \frac{1}{3}(2, 0, 1, 2), \frac{1}{3\sqrt{110}}(-19, 9, 22, 8) \right\}$$

is an orthonormal basis for V. Note that the dimension of V is 2. The given spanning set was linearly dependent! ■

The formulas that occur in Theorem 4 have nice geometric interpretations. Consider the formula for $\mathbf{w}_2$:

$$\mathbf{w}_2 = \mathbf{v}_2 - \frac{\mathbf{v}_2 \cdot \mathbf{w}_1}{\mathbf{w}_1 \cdot \mathbf{w}_1} \mathbf{w}_1.$$

Since $\mathbf{v}_2 \cdot \mathbf{w}_1 = \|\mathbf{v}_2\| \, \|\mathbf{w}_1\| \cos \theta$, where θ is the angle between the vectors $\mathbf{v}_2$ and $\mathbf{w}_1$, and since $\mathbf{w}_1 \cdot \mathbf{w}_1 = \|\mathbf{w}_1\|^2$, we can rewrite this formula as

$$\mathbf{w}_2 = \mathbf{v}_2 - (\|\mathbf{v}_2\| \cos \theta) \frac{\mathbf{w}_1}{\|\mathbf{w}_1\|}.$$

The geometric interpretation of this formula can now be seen in Figure 6.1. The

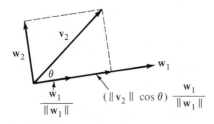

FIGURE 6.1

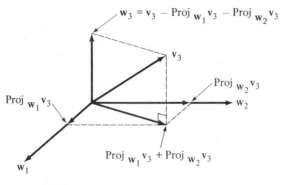

FIGURE 6.2

vector $\mathbf{w}_1/\|\mathbf{w}_1\|$ is the unit vector pointing in the direction of $\mathbf{w}_1$; hence

$$(\|\mathbf{v}_2\| \cos \theta)\frac{\mathbf{w}_1}{\|\mathbf{w}_1\|} = \text{Proj}_{\mathbf{w}_1}\mathbf{v}_2$$

is the **projection** of $\mathbf{v}_2$ along $\mathbf{w}_1$ (see Exercise 20 in Section 1.4). The vector $\mathbf{w}_2$ is obtained by subtracting from $\mathbf{v}_2$ its projection along $\mathbf{w}_1$:

$$\mathbf{w}_2 = \mathbf{v}_2 - \text{Proj}_{\mathbf{w}_1}\mathbf{v}_2.$$

Similarly, each of the vectors $\mathbf{w}_i$ is obtained by subtracting from the given vector $\mathbf{v}_i$ its projections along the vectors $\mathbf{w}_1, \ldots, \mathbf{w}_{i-1}$:

$$\mathbf{w}_i = \mathbf{v}_i - \text{Proj}_{\mathbf{w}_1}\mathbf{v}_i - \text{Proj}_{\mathbf{w}_2}\mathbf{v}_i - \cdots - \text{Proj}_{\mathbf{w}_{i-1}}\mathbf{v}_i$$

(see Figure 6.2).

Given any two vectors $\mathbf{v}$ and $\mathbf{w}$ in $\mathbb{R}^n$, with $\mathbf{w} \neq \mathbf{0}$, the vector

$$\text{Proj}_{\mathbf{w}}\mathbf{v} = \frac{\mathbf{v} \cdot \mathbf{w}}{\mathbf{w} \cdot \mathbf{w}}\mathbf{w}$$

is sometimes called the **component of v parallel to w.** The vector

$$\mathbf{v} - \text{Proj}_{\mathbf{w}}\mathbf{v} = \mathbf{v} - \frac{\mathbf{v} \cdot \mathbf{w}}{\mathbf{w} \cdot \mathbf{w}}\mathbf{w}$$

is called **the component of v perpendicular to w.** Notice that $\mathbf{v}$ is the sum of these two components (see Figure 6.3).

Given any subspace W of $\mathbb{R}^n$, we can similarly decompose each $\mathbf{v} \in \mathbb{R}^n$ into a sum of two *components,* one *parallel* to W and one *perpendicular* to W (see Figure 6.4). This is the content of the next theorem.

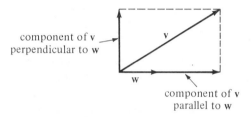

FIGURE 6.3

Theorem 5.   *Let W be a subspace of* $\mathbb{R}^n$. *Then each vector* $\mathbf{v} \in \mathbb{R}^n$ *can be decomposed in one and only one way into a sum*

$$\mathbf{v} = \mathbf{v}_1 + \mathbf{v}_2$$

where $\mathbf{v}_1 \in W$ *and* $\mathbf{v}_2$ *is perpendicular to* W.

Proof.   Let $\{\mathbf{w}_1, \ldots, \mathbf{w}_k\}$ be an orthonormal basis for W. The vector $\mathbf{v}_1$, if it exists, must be a linear combination of the basis vectors $\mathbf{w}_1, \ldots, \mathbf{w}_k$. In fact, since these basis vectors are orthonormal and since we require that $\mathbf{v}_2 \cdot \mathbf{w}_i = 0$ for all i, we must have

$$\begin{aligned} \mathbf{v}_1 &= (\mathbf{v}_1 \cdot \mathbf{w}_1)\mathbf{w}_1 + \cdots + (\mathbf{v}_1 \cdot \mathbf{w}_k)\mathbf{w}_k \\ &= ((\mathbf{v} - \mathbf{v}_2) \cdot \mathbf{w}_1)\mathbf{w}_1 + \cdots + ((\mathbf{v} - \mathbf{v}_2) \cdot \mathbf{w}_k)\mathbf{w}_k \\ &= (\mathbf{v} \cdot \mathbf{w}_1)\mathbf{w}_1 + \cdots + (\mathbf{v} \cdot \mathbf{w}_k)\mathbf{w}_k. \end{aligned}$$

This formula shows that there is at most one possible choice for $\mathbf{v}_1$, namely,

$$\mathbf{v}_1 = (\mathbf{v} \cdot \mathbf{w}_1)\mathbf{w}_1 + \cdots + (\mathbf{v} \cdot \mathbf{w}_k)\mathbf{w}_k,$$

and at most one possible choice for $\mathbf{v}_2$, namely

$$\mathbf{v}_2 = \mathbf{v} - \mathbf{v}_1.$$

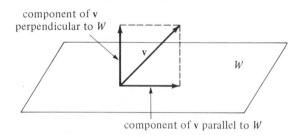

FIGURE 6.4

So it remains only to show that if we define v_1 and v_2 by the above two formulas, then $v_1 \in W$, v_2 is perpendicular to W, and $v = v_1 + v_2$. The first and third of these conditions are clear. For the second, simply note that

$$v_2 \cdot w_i = (v - v_1) \cdot w_i = v \cdot w_i - v_1 \cdot w_i = v \cdot w_i - v \cdot w_i = 0$$

for all i, so v_2 is perpendicular to each element of the basis $\{w_1, \dots, w_k\}$ for W and hence is perpendicular to every vector in W. ■

The component of $v \in \mathbb{R}^n$ parallel to the subspace W is often called the **orthogonal projection** $\mathrm{Proj}_W v$ of v onto W. From the proof of Theorem 5, we see that this orthogonal projection is given by the formula

$$\mathrm{Proj}_W v = (v \cdot w_1)w_1 + \cdots + (v \cdot w_k)w_k,$$

where $\{w_1, \dots, w_k\}$ is any orthonormal basis for W. If $\{w_1, \dots, w_k\}$ is an orthogonal, but not necessarily orthonormal, basis for W, this formula becomes

$$\mathrm{Proj}_W v = \frac{v \cdot w_1}{w_1 \cdot w_1} w_1 + \cdots + \frac{v \cdot w_k}{w_k \cdot w_k} w_k.$$

EXERCISES

1. Which of the following sets are orthogonal? Which are orthonormal?
   (a) $\{(1, 0, 1, 0), (-1, 0, 1, 0), (-1, 0, -1, 0)\}$.
   (b) $\{(1, 0, 1, 0), (-1, 0, 1, 1), (1, 0, -1, 2)\}$.
   (c) $\left\{ \left(\frac{1}{\sqrt{3}}, \frac{1}{\sqrt{3}}, \frac{1}{\sqrt{3}} \right), \left(-\frac{1}{\sqrt{6}}, \frac{2}{\sqrt{6}}, -\frac{1}{\sqrt{6}} \right) \right\}$.
   (d) $\left\{ \left(\frac{1}{2}, -\frac{1}{2}, \frac{1}{2}, \frac{1}{2} \right), \left(\frac{1}{2\sqrt{3}}, -\frac{1}{2\sqrt{3}}, -\frac{3}{2\sqrt{3}}, \frac{1}{2\sqrt{3}} \right), \left(-\frac{1}{\sqrt{2}}, 0, 0, \frac{1}{\sqrt{2}} \right) \right\}$
   (e) $\{(12, 5, 9, -4), (1, 1, -1, 2), (10, -13, -7, -2)\}$

2. Verify that $\mathbf{B} = \left(\left(\frac{2}{3}, \frac{2}{3}, \frac{1}{3} \right), \left(\frac{\sqrt{2}}{2}, -\frac{\sqrt{2}}{2}, 0 \right), \left(\frac{\sqrt{2}}{6}, \frac{\sqrt{2}}{6}, -\frac{2\sqrt{2}}{3} \right) \right)$ is an ordered orthonormal basis for $\mathbb{R}^3$, and find the **B**-coordinate triple of each of the following vectors.
   (a) $\left(\frac{\sqrt{2}}{2}, -\frac{\sqrt{2}}{2}, 0 \right)$ (d) $(1, -1, 2)$

   (b) $(1, 1, -1)$ (e) $\left(\frac{2}{3} \sqrt{2}, -\frac{1}{3} \sqrt{2}, -\frac{2}{3} \sqrt{2} \right)$

   (c) $(1, 0, 0)$

3. Find an orthonormal basis for each of the following spaces:
   (a) $\mathcal{L}((1, 1, 0), (-1, 1, 0))$
   (b) $\mathcal{L}((1, 1, 1), (-1, 1, 1))$
   (c) $\mathcal{L}((1, 1, 1, 1), (-2, -1, 1, 2), (1, 0, -1, 2), (0, 0, 1, 5))$

(d) $\mathcal{L}((0, 1, 0, \ldots, 0), (0, 1, 2, 0, \ldots, 0), (0, 1, 2, 3, 0, \ldots, 0), \ldots,$
     $(0, 1, 2, \ldots, n))$

4. Find an orthonormal basis for each of the following spaces:
    (a) The plane $3x_1 + x_2 - x_3 = 0$ in $\mathbb{R}^3$
    (b) The hyperplane $2x_1 - x_2 + 3x_3 - x_4 = 0$ in $\mathbb{R}^4$
    (c) The solution space of the system

$$x_1 + x_2 + x_3 + x_4 = 0$$
$$-x_1 + x_2 \qquad + x_4 = 0$$

5. Verify that if $\{\mathbf{w}_1, \mathbf{w}_2\}$ is an orthogonal set in $\mathbb{R}^n$ with $\mathbf{w}_1$ and $\mathbf{w}_2$ both nonzero, and if $\mathbf{v}_3 \in \mathbb{R}^n$, then

$$\mathbf{w}_3 = \mathbf{v}_3 + c_1\mathbf{w}_1 + c_2\mathbf{w}_2$$

is perpendicular to both $\mathbf{w}_1$ and $\mathbf{w}_2$ if and only if

$$c_1 = -\frac{\mathbf{v}_3 \cdot \mathbf{w}_1}{\mathbf{w}_1 \cdot \mathbf{w}_1} \quad \text{and} \quad c_2 = -\frac{\mathbf{v}_3 \cdot \mathbf{w}_2}{\mathbf{w}_2 \cdot \mathbf{w}_2}.$$

6. Let $\mathbf{w}_1 = (\frac{3}{5}, \frac{2}{5}, \frac{2}{5}, \frac{2}{5}, \frac{2}{5})$, $\mathbf{w}_2 = (0, \frac{1}{2}, -\frac{1}{2}, \frac{1}{2}, -\frac{1}{2})$, and $\mathbf{w}_3 = (0, \frac{1}{2}, \frac{1}{2}, -\frac{1}{2}, -\frac{1}{2})$.
    (a) Show that $\{\mathbf{w}_1, \mathbf{w}_2, \mathbf{w}_3\}$ is an orthonormal set in $\mathbb{R}^5$.
    (b) Find the orthogonal projection of $\mathbf{v} = (1, 0, -2, 1, 3)$ onto $W = \mathcal{L}(\mathbf{w}_1, \mathbf{w}_2, \mathbf{w}_3)$.
    (c) Find the component of $\mathbf{v}$ perpendicular to W ($\mathbf{v}$ and W as in part (b)).

7. Let $\mathbf{v}_1 = (1, -1, 1)$, $\mathbf{v}_2 = (1, 1, 0)$, and $W = \mathcal{L}(\mathbf{v}_1, \mathbf{v}_2)$.
    (a) Find an orthonormal basis for W.
    (b) Find the orthogonal projection of $\mathbf{v} = (5, 7, -3)$ onto W.
    (c) Find the component of $\mathbf{v} = (5, 7, -3)$ perpendicular to W.

8. Let $\mathbf{v} \in \mathbb{R}^n$, let $\mathbf{d}$ be a unit vector in $\mathbb{R}^n$, and let H be the hyperplane $\mathbf{d} \cdot \mathbf{x} = 0$. Show that the orthogonal projection of $\mathbf{v}$ onto H is given by the formula

$$\text{Proj}_H\mathbf{v} = \mathbf{v} - (\mathbf{v} \cdot \mathbf{d})\mathbf{d}.$$

[Hint: Use Theorem 5.]

9. Let W be a subspace of $\mathbb{R}^n$ and let $\mathbf{v} \in \mathbb{R}^n$. Show that the orthogonal projection of $\mathbf{v}$ onto W is the vector in W closest to $\mathbf{v}$ by showing that, for each $\mathbf{w} \in W$,

$$\|\mathbf{w} - \mathbf{v}\| \geq \|\text{Proj}_W\mathbf{v} - \mathbf{v}\|$$

and equality holds if and only if $\mathbf{w} = \text{Proj}_W\mathbf{v}$. [Hint: Let $\mathbf{a} = \mathbf{w} - \text{Proj}_W\mathbf{v}$ so that $\mathbf{w} = \mathbf{a} + \text{Proj}_W\mathbf{v}$. Then verify that $\|\mathbf{w} - \mathbf{v}\|^2 = \|\mathbf{a}\|^2 + \|\text{Proj}_W\mathbf{v} - \mathbf{v}\|^2$.]

6.2 LEAST SQUARES APPROXIMATIONS

Often, in engineering, science, or social science, theory predicts that there is a linear relation $y = mx + b$ between two variables x and y, but the constants m and b are not known. In order to find m and b, a sequence of measurements is taken that gives, for several values $(x_1, x_2, \ldots, x_n)$ of x, corresponding experimental values $(y_1, y_2, \ldots, y_n)$ of y. If the points $(x_1, y_1), \ldots, (x_n, y_n)$ all lie on a line in $\mathbb{R}^2$, then it is easy to determine the values of m and b predicted by the experiment: they are just the slope and y-intercept of the line through these n points. But

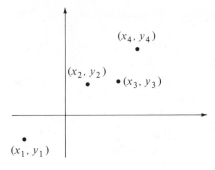

FIGURE 6.5

usually, due to experimental error, the points $(x_1, y_1), \ldots, (x_n, y_n)$ do not lie on a common line (see Figure 6.5), so the problem is then to find the line that in some sense comes closest to passing through the n points $(x_1, y_1), \ldots, (x_n, y_n)$.

There are various ways to measure how close a point is to a line. Geometrically, the perpendicular distance from the point to the line is probably the best measure. But, from the point of view of analyzing data, vertical distance is usually preferred. If the vertical distance $|y_i - (mx_i + b)|$ is small, then the experimental value y_i of y will be close to the predicted value $mx_i + b$.

A standard measure of the closeness of n points $(x_1, y_1), \ldots, (x_n, y_n)$ in $\mathbb{R}^2$ to the line $y = mx + b$ is the sum S of the squares of the vertical distances from the points to the line:

$$S = \sum_{i=1}^{n} (y_i - (mx_i + b))^2$$

(see Figure 6.6). The reason that the sum of the squares of the distances is used rather than simply the sum of the distances is that large errors are less tolerable than small errors. Squaring magnifies the contribution to S of large errors.

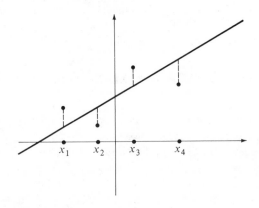

FIGURE 6.6

The **least squares linear approximation** to the points $(x_1, y_1), \ldots, (x_n, y_n)$ is defined to be the line $y = mx + b$ for which S is minimum. Finding m and b to make S minimum is a minimization problem for the function S of the two variables m and b. This problem could be solved using calculus, but we shall instead translate the problem into a problem in linear algebra.

Consider first the case where the n points $(x_1, y_1), \ldots, (x_n, y_n)$ are collinear. Then the slope m and y-intercept b of the line through these points must satisfy the linear system

$$mx_1 + b = y_1$$
$$mx_2 + b = y_2$$
$$\ldots$$
$$mx_n + b = y_n.$$

This system is equivalent to the vector equation

$$mx + b1 = y$$

where $x = (x_1, \ldots, x_n)$, $1 = (1, 1, \ldots, 1)$, and $y = (y_1, \ldots, y_n)$. We see, then, that the n given points lie on a common line ℓ in $\mathbb{R}^2$ if and only if y is a linear combination of x and 1. The coefficients of this linear combination are the slope and y-intercept of ℓ, respectively.

If the points $(x_1, y_1), \ldots, (x_n, y_n)$ are not collinear, then it is not possible to find m and b such that

$$mx + b1 = y.$$

So we seek m and b so that $mx + b1$ is as close to y as possible; that is, we want to find the values of m and b that make $\| y - (mx + b1) \|$ as small as possible. But a positive function is smallest when its square is smallest. Since

$$\| y - (mx + b1) \|^2 = \sum_{i=1}^{n} (y_i - (mx_i + b))^2 = S,$$

we see that finding m and b to minimize $\| y - (mx + b1) \|$ is exactly the same as finding the least squares linear approximation to the points $(x_1, y_1), \ldots, (x_n, y_n)$.

The geometry of the problem is now clear. The equation $mx + b1 = y$ has a solution if and only if the vector y is in the subspace $W = \mathcal{L}(x, 1)$ of $\mathbb{R}^n$ spanned by $\{x, 1\}$. If y is not in this subspace (see Figure 6.7) then the equation $mx + b1 = y$ has no solution so we try to find m and b so that $\| y - (mx + b1) \|$ is as small as possible. Since the vector in W closest to y is $\text{Proj}_W y$ (see Exercise 9 of the previous section for an outline of an analytic proof of this geometrically evident fact), the solution to the least squares approximation problem will be the solution of the vector equation

$$mx + b1 = \text{Proj}_W y.$$

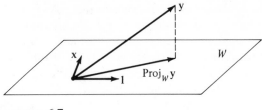

FIGURE 6.7

We could compute $\text{Proj}_W\mathbf{y}$ and solve this system directly for m and b (see Exercise 7), but instead we shall take a shortcut. Since the vector

$$\mathbf{y} - \text{Proj}_W\mathbf{y} = \mathbf{y} - (m\mathbf{x} + b\mathbf{1})$$

is perpendicular to W (Figure 6.7), and since $\mathbf{x} \in W$ and $\mathbf{1} \in W$, we must have

$$\mathbf{x} \cdot [\mathbf{y} - (m\mathbf{x} + b\mathbf{1})] = 0$$

and

$$\mathbf{1} \cdot [\mathbf{y} - (m\mathbf{x} + b\mathbf{1})] = 0.$$

Hence (m, b) must satisfy the linear system

$$(\mathbf{x} \cdot \mathbf{x})m + (\mathbf{x} \cdot \mathbf{1})b = \mathbf{x} \cdot \mathbf{y}$$
$$(\mathbf{1} \cdot \mathbf{x})m + (\mathbf{1} \cdot \mathbf{1})b = \mathbf{1} \cdot \mathbf{y}.$$

The solution of this system is easily seen to be

$$m = \frac{(\mathbf{1} \cdot \mathbf{1})(\mathbf{x} \cdot \mathbf{y}) - (\mathbf{1} \cdot \mathbf{x})(\mathbf{1} \cdot \mathbf{y})}{(\mathbf{1} \cdot \mathbf{1})(\mathbf{x} \cdot \mathbf{x}) - (\mathbf{1} \cdot \mathbf{x})^2}$$

$$b = \frac{(\mathbf{x} \cdot \mathbf{x})(\mathbf{1} \cdot \mathbf{y}) - (\mathbf{x} \cdot \mathbf{y})(\mathbf{1} \cdot \mathbf{x})}{(\mathbf{1} \cdot \mathbf{1})(\mathbf{x} \cdot \mathbf{x}) - (\mathbf{1} \cdot \mathbf{x})^2}.$$

Notice that the denominator is equal to

$$(\mathbf{1} \cdot \mathbf{1})(\mathbf{x} \cdot \mathbf{x}) - (\mathbf{1} \cdot \mathbf{x})^2 = \|\mathbf{1}\|^2 \|\mathbf{x}\|^2 - \|\mathbf{1}\|^2 \|\mathbf{x}\|^2 \cos^2 \theta$$
$$= \|\mathbf{1}\|^2 \|\mathbf{x}\|^2 \sin^2 \theta$$

where θ is the angle between $\mathbf{1}$ and $\mathbf{x}$. This is the area of the parallelogram spanned by $\mathbf{1}$ and $\mathbf{x}$. It is zero only when $\mathbf{x}$ is a scalar multiple of $\mathbf{1}$; that is, when $x_1 = x_2 = \cdots = x_n$. In this case the given points are collinear (they all lie on a vertical line).

Thus we have proved the following theorem.

Theorem 1. *The least squares linear approximation to a set* $(x_1, y_1), \ldots, (x_n, y_n)$
of noncollinear points in $\mathbb{R}^2$ *is the line* $y = mx + b$ *where*

$$m = \frac{(\mathbf{1} \cdot \mathbf{1})(\mathbf{x} \cdot \mathbf{y}) - (\mathbf{1} \cdot \mathbf{x})(\mathbf{1} \cdot \mathbf{y})}{(\mathbf{1} \cdot \mathbf{1})(\mathbf{x} \cdot \mathbf{x}) - (\mathbf{1} \cdot \mathbf{x})^2} \qquad \mathbf{x} = (x_1, \ldots, x_n)$$

$$\mathbf{y} = (y_1, \ldots, y_n)$$

$$b = \frac{(\mathbf{x} \cdot \mathbf{x})(\mathbf{1} \cdot \mathbf{y}) - (\mathbf{x} \cdot \mathbf{y})(\mathbf{1} \cdot \mathbf{x})}{(\mathbf{1} \cdot \mathbf{1})(\mathbf{x} \cdot \mathbf{x}) - (\mathbf{1} \cdot \mathbf{x})^2} \qquad \mathbf{1} = (1, 1, \ldots, 1).$$

Example 1. To find the least squares linear approximation to the set $\{(-1, 1),$
$(1, -1), (3, -4), (5, -4)\}$, we put $\mathbf{x} = (-1, 1, 3, 5)$ and $\mathbf{y} = (1, -1, -4, -4)$,
and compute

$$m = \frac{(4)(-34) - (8)(-8)}{(4)(36) - (8)^2} = -\frac{9}{10}$$

$$b = \frac{(36)(-8) - (-34)(8)}{(4)(36) - (8)^2} = -\frac{1}{5}$$

The line is $y = -\frac{9}{10}x - \frac{1}{5}$ (see Figure 6.8). ■

Example 2. The population of a certain small town in Maine has been growing
slowly over a period of 20 years, according to the following table:

YEAR	POPULATION
1960	2000
1970	2050
1980	2080

Let us try to predict the town's population in the year 2000.

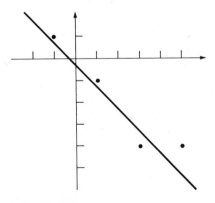

FIGURE 6.8

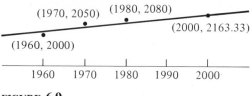

(1970, 2050) (1980, 2080)

(2000, 2163.33)

(1960, 2000)

1960   1970   1980   1990   2000

FIGURE 6.9

Without any additional information available to suggest otherwise, it is reasonable to approximate the population graph by a line (see Figure 6.9). We shall find the least squares linear approximation to the data and use it to predict the population in the year 2000. Setting x = (1960, 1970, 1980), y = (2000, 2050, 2080), 1 = (1, 1, 1), and using the formulas in Theorem 1, we find that $m = 4$ and, to two decimal places, $b = -5836.67$. So the least squares approximation to the data is

$$y = 4x - 5836.67.$$

Plugging in $x = 2000$ we obtain $y = 2163.33$. We therefore predict that the population in the year 2000 will be approximately 2163.   ■

Least squares linear approximations can also be used to fit certain nonlinear functions to experimental data. The most common examples are power functions and exponentials. If there is a theoretical relationship between positive variables x and y of the form

$$y = ax^b$$

then, taking the logarithm of both sides, we obtain a linear relation

$$\ln y = b \ln x + \ln a$$

between the variables $\ln x$ and $\ln y$, so a least squares linear approximation can be used to determine experimental values of b and $\ln a$, hence of a and b. Similarly, suppose there is a theoretical relationship between x and y of the form

$$y = ae^{bx},$$

where $y > 0$ and $a > 0$. Then, taking the logarithm of both sides, we obtain a linear relation

$$\ln y = bx + \ln a$$

between the variables x and $\ln y$, so again we can use a least squares linear approximation to determine experimental values of b and $\ln a$, hence of a and b.

Example 3.   If we plot the points (1, 4), (2, 2), and (3, 1) we see that they seem to lie on a curve shaped like the graph of $y = \dfrac{c}{x^k} = cx^{-k}$ for some c and k (see

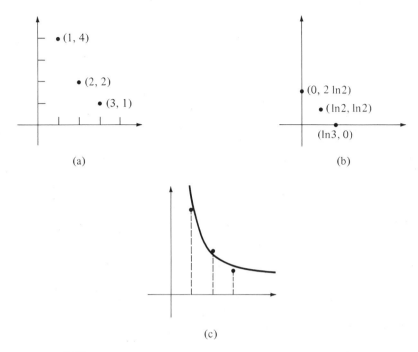

FIGURE 6.10

Figure 6.10a). Let us use the least squares formula to find values of c and k that yield a good fit to these data. Taking the logarithm of both sides of the equation $y = cx^{-k}$ we get

$$\ln y = -k \ln x + \ln c = m \ln x + b$$

where $m = -k$ and $b = \ln c$. We want to find m and b that yield the least squares linear approximation to the set $\{(\ln 1, \ln 4), (\ln 2, \ln 2), (\ln 3, \ln 1)\} = \{(0, 2 \ln 2),$ $(\ln 2, \ln 2), (\ln 3, 0)\}$ (see Figure 6.10b). Setting $\mathbf{x} = (0, \ln 2, \ln 3)$, $\mathbf{y} = (2 \ln 2,$ $\ln 2, 0)$, $\mathbf{1} = (1, 1, 1)$, and applying the formula of Theorem 1, we find that, to two decimal places,

$$m = \frac{3(\ln 2)^2 - (\ln 2 + \ln 3)(3 \ln 2)}{3((\ln 2)^2 + (\ln 3)^2) - (\ln 2 + \ln 3)^2} = -1.23$$

$$b = \frac{((\ln 2)^2 + (\ln 3)^2)(3 \ln 2) - (\ln 2)^2(\ln 2 + \ln 3)}{3((\ln 2)^2 + (\ln 3)^2) - (\ln 2 + \ln 3)^2} = 1.43.$$

Thus, $k = -m = 1.23$, $c = e^b = 4.18$, and the curve we seek is $y = \dfrac{4.18}{x^{1.23}}$. If we compute the values of y as x runs through the values 1, 2, 3 we find that y runs through the values 4.18, 1.78, 1.08, in fairly good agreement with the given data (see Figure 6.10c). ■

Example 4. The uninhibited growth of a bacteria population is described by an equation of the form $N = N_0 e^{kt}$ (remember the differential equation $\dfrac{dN}{dt} = kN$?) where N is the number of bacteria at time t, N_0 is the number at time $t = 0$, and k is a constant. Population counts of a particular bacteria colony yield the following data:

t	N
3	4.5×10^7
4	7.5×10^7
5	1.2×10^8

We shall use a least squares approximation to determine approximate values of N_0 and k from these data, and we shall use these approximate values to predict the number of bacteria when $t = 10$.

Taking the logarithm to the base 10 of both sides of the equation $N = N_0 e^{kt}$ yields the linear relation

$$\log_{10} N = kt \log_{10} e + \log_{10} N_0,$$

or

$$\log_{10} N = (k \log_{10} e)t + \log_{10} N_0,$$

between the variables t and $\log_{10} N$. Taking the $\log_{10}$ of the data in the N column, we get

t	$\log_{10} N$
3	7.653
4	7.875
5	8.079

Setting $\mathbf{x} = (3, 4, 5)$, $\mathbf{y} = (7.653, 7.875, 8.079)$, $\mathbf{1} = (1, 1, 1)$, and using the formulas in Theorem 1, we find that

$$\left. \begin{array}{l} m(= k \log_{10} e) = 0.213 \\[2mm] b(= \log_{10} N_0) = 7.017 \end{array} \right\} \quad \text{so} \quad \left\{ \begin{array}{l} k = \dfrac{m}{\log_{10} e} = 0.49 \\[2mm] N_0 = 10^b = 1.04 \times 10^7. \end{array} \right.$$

Hence the least squares approximation is

$$N = (1.04 \times 10^7)e^{0.49t}.$$

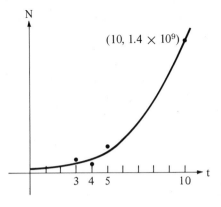

FIGURE 6.11

When $t = 10$ we get, for the predicted value of N,

$$N|_{t=10} = (1.04 \times 10^7)e^{4.9} = 1.4 \times 10^9.$$

(See Figure 6.11.)   ∎

The least squares method can be extended to a method for finding approximate solutions for inconsistent systems of linear equations.

Consider the linear system

$$a_{11}x_1 + \cdots + a_{1n}x_n = b_1$$

(∗)
$$\cdots$$

$$a_{m1}x_1 + \cdots + a_{mn}x_n = b_m.$$

This system is equivalent to the vector equation

$$x_1\mathbf{a}_1 + \cdots + x_n\mathbf{a}_n = \mathbf{b}$$

where $\mathbf{a}_1, \ldots, \mathbf{a}_n, \mathbf{b}$ are the column vectors of the matrix

$$\begin{pmatrix} a_{11} & \cdots & a_{1n} & b_1 \\ & \cdots & & \\ a_{m1} & \cdots & a_{mn} & b_m \end{pmatrix}.$$

As we discussed in Section 2.3 (see Theorem 1), it is evident from this vector equation that the system (∗) is consistent (has a solution) if and only if $\mathbf{b}$ is in the column space $W = \mathcal{L}(\mathbf{a}_1, \ldots, \mathbf{a}_n)$ of the coefficient matrix

$$\begin{pmatrix} a_{11} & \cdots & a_{1n} \\ & \cdots & \\ a_{m1} & \cdots & a_{mn} \end{pmatrix}$$

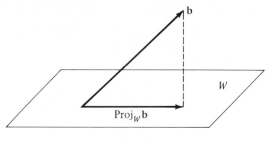

FIGURE 6.12

If $\mathbf{b} \notin W$ (see Figure 6.12) then the system has no solution. In this case, it is often useful to find an approximate solution. A ***least squares approximate solution*** of $(*)$ is a vector $(x_1, \ldots, x_n)$ in $\mathbb{R}^n$ with the property that $x_1\mathbf{a}_1 + \cdots + x_n\mathbf{a}_n$ is as close as possible to $\mathbf{b}$ (that is, such that $\|\mathbf{b} - (x_1\mathbf{a}_1 + \cdots + x_n\mathbf{a}_n)\|$ is as small as possible). Since the vector in $W = \mathcal{L}(\mathbf{a}_1, \ldots, \mathbf{a}_n)$ closest to $\mathbf{b}$ is the vector $\mathrm{Proj}_W\mathbf{b}$ (Figure 6.12), *the least squares approximate solutions of the system*

$$x_1\mathbf{a}_1 + \cdots + x_n\mathbf{a}_n = \mathbf{b}$$

are the solutions of the system

$$x_1\mathbf{a}_1 + \cdots + x_n\mathbf{a}_n = \mathrm{Proj}_W\mathbf{b}.$$

We shall derive, in Section 7.3, an efficient algorithm for finding these approximate solutions. For now, we shall be content with a simple example in which a direct attack on the problem is feasible.

Example 5. Let us find a least squares approximate solution of the system

$$
\begin{aligned}
x_1 + x_2 &= 9 \\
3x_1 - x_2 &= 1 \\
x_1 + 2x_2 &= 10.
\end{aligned}
$$

Row reduction of

$$
\begin{pmatrix}
1 & 1 & 9 \\
3 & -1 & 1 \\
1 & 2 & 10
\end{pmatrix}
\quad \text{yields} \quad
\begin{pmatrix}
1 & 0 & 0 \\
0 & 1 & 0 \\
0 & 0 & 1
\end{pmatrix}
$$

so this system has no solution. The vector $(x_1, x_2) \in \mathbb{R}^2$ is a least squares approximate solution if it satisfies the equation

$$x_1(1, 3, 1) + x_2(1, -1, 2) = \mathrm{Proj}_W(9, 1, 10),$$

where $W = \mathcal{L}((1, 3, 1), (1, -1, 2))$. Since $\{(1, 3, 1), (1, -1, 2)\}$ is an orthogonal basis for W, it is easy to compute the approximate solution in this case: (x_1, x_2) is just the coordinate pair of $\text{Proj}_W(9, 1, 10)$ relative to the ordered orthogonal basis $((1, 3, 1), (1, -1, 2))$ for W. Since

$$\text{Proj}_W(9, 1, 10) = \frac{(9, 1, 10) \cdot (1, 3, 1)}{(1, 3, 1) \cdot (1, 3, 1)}(1, 3, 1)$$

$$+ \frac{(9, 1, 10) \cdot (1, -1, 2)}{(1, -1, 2) \cdot (1, -1, 2)}(1, -1, 2)$$

$$= 2(1, 3, 1) + \tfrac{14}{3}(1, -1, 2),$$

the approximate solution we seek is $(2, \tfrac{14}{3})$. ■

EXERCISES

[Note: Use of a hand calculator is recommended for many of these exercises.]

1. Find the least squares linear approximation $y = mx + b$ to the following sets of points.
    (a) $(-1, 0), (1, 3), (3, 4)$
    (b) $(1, 5), (2, 3), (3, 0)$
    (c) $(-1, 0), (0, -1), (1, 1), (2, -1)$
    (d) $(0, -1), (1, 1), (3, 5), (4, 7)$
    (e) $(-1, 3), (0, 1), (1, -1), (2, -2)$
    (f) $(1, 5), (2, 6), (3, 4), (4, 5)$
    (g) $(-1, -4), (1, -3), (2, 0), (3, 0)$

2. Use an appropriate least squares linear approximation to find a curve $y = ax^b$ that provides a good fit with the given set of points.
    (a) $(1, 3), (2, 11), (3, 25)$
    (b) $(1, 2), (2, 1), (3, 1/2)$
    (c) $(1, 100), (2, 3), (3, 1/2)$
    (d) $(1, 3), (2, 4), (3, 5)$
    (e) $(1, 10), (2, 75), (3, 260)$

3. Use an appropriate least squares linear approximation to find an exponential curve $y = ce^{kt}$ to fit the following points.
    (a) $(0, 4), (2, 13), (4, 35)$
    (b) $(0, 10), (1, 50), (2, 250)$
    (c) $(-1, 1/2), (0, 1), (1, 3/2)$
    (d) $(-2, 5), (-1, 2), (1, 1)$
    (e) $(1, 2), (2, 4), (3, 8)$
    (f) $(0, 8), (1, 20), (2, 60), (3, 200)$

4. A certain population appears to be growing exponentially with time t. The (t, N) data are $(0, 50)$, $(1, 350)$, $(2, 2400)$. Use the method of least squares to find an exponential function $N = ce^{kt}$ to approximate these data. Then predict the approximate value of N when $t = 3$.

5. Workers on an assembly line are observed to spend a total of 800 worker-hours assembling the first of a new model of automobile, 560 worker-hours assembling the second, 460 assembling the third, and 405 assembling the fourth. Assuming that the total number y of worker-hours required to assemble the xth automobile is given by an equation of the form $y = ax^b$, use the method of least squares to determine values of a and b that provide a good fit to the given data. Then predict the number of worker-hours required to assemble the 100th automobile.

6. Find least squares approximate solutions to the following linear systems.

(a) $2x_1 + x_2 = 3$
     $x_1 + 3x_2 = 4$
     $5x_1 - x_2 = 5$

(b) $-3x_1 + x_2 = 0$
     $2x_1 + x_2 = 6$
     $x_1 + x_2 = 5$

(c) $x_1 + x_2 - x_3 = 1$
     $x_1 \qquad - x_3 = 0$
     $x_1 - x_2 - x_3 = -1$
     $x_1 \qquad + 3x_3 = -2$

7. Suppose $(x_1, y_1), \ldots, (x_n, y_n)$ is a set of points in $\mathbb{R}^2$ with $x_1 + \cdots + x_n = 0$ (so that $\mathbf{1} \cdot \mathbf{x} = 0$, where $\mathbf{1} = (1, 1, \ldots, 1)$ and $\mathbf{x} = (x_1, \ldots, x_n)$). Find the least squares linear approximation $y = mx + b$ to this set of points by calculating $\text{Proj}_W \mathbf{y}$, where $W = \mathcal{L}(\mathbf{1}, \mathbf{x})$ and $\mathbf{y} = (y_1, \ldots, y_n)$, and reading off the coefficients of the linear combination $\text{Proj}_W \mathbf{y} = m\mathbf{x} + b\mathbf{1}$. Check that your answer is consistent with the formulas in Theorem 1.

8. Suppose we know that the data $(x_1, y_1), \ldots, (x_n, y_n)$ should describe a linear relationship of the form $y = mx$. Then we should look, not for the line in $\mathbb{R}^2$ that comes closest to passing through these points, but instead for the line *through the origin* that comes closest to passing through these points. Thus we want to choose m to minimize $\|\mathbf{y} - m\mathbf{x}\|$, where $\mathbf{x} = (x_1, \ldots, x_n)$ and $\mathbf{y} = (y_1, \ldots, y_n)$. Show that this m is given by the formula $m = \mathbf{x} \cdot \mathbf{y}/\mathbf{x} \cdot \mathbf{x}$.

9. Suppose we want to find the *horizontal* line $y = b$ that provides the best approximation to the points $(x_1, y_1), \ldots, (x_n, y_n)$. Then we want to choose b to minimize $\|\mathbf{y} - b\mathbf{1}\|$, where $\mathbf{y} = (y_1, \ldots, y_n)$ and $\mathbf{1} = (1, \ldots, 1)$. Show that b is given by the formula $b = (y_1 + \cdots + y_n)/n$.

10. (a) Suppose we want to find a least squares *quadratic* approximation $y = ax^2 + bx + c$ to the set of points $(x_1, y_1), \ldots, (x_n, y_n)$ in $\mathbb{R}^2$. Then we seek a, b, and c that minimize

$$\sum_{i=1}^{n} [y_i - (ax_i^2 + bx_i + c)]^2$$

(see Figure 6.13.) Show that $(a, b, c) \in \mathbb{R}^3$ is a solution of this least squares quadratic approximation problem if and only if (a, b, c) is a least squares

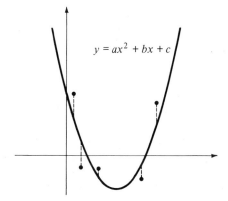

FIGURE 6.13

approximate solution of the linear system with matrix

$$\begin{pmatrix} x_1^2 & x_1 & 1 & y_1 \\ \vdots & \vdots & \vdots & \vdots \\ x_n^2 & x_n & 1 & y_n \end{pmatrix}.$$

(b) Generalize part (a), replacing "quadratic approximation $y = ax^2 + bx + c$" by "kth degree polynomial approximation $y = a_k x^k + \cdots + a_1 x + a_0$."

6.3 INNER PRODUCTS

In Section 6.1 it was shown that the computation of **B**-coordinates, where **B** is an ordered basis for a subspace of $\mathbb{R}^n$, is simplest when the basis **B** is orthonormal. It would certainly be helpful if orthonormal bases were available in a larger class of vector spaces than just subspaces of $\mathbb{R}^n$. In order to define orthonormal bases in a vector space, we need to be able to compute dot products of vectors. For spaces other than subspaces of $\mathbb{R}^n$, a "dot product" is called an "inner product," and the inner product of two vectors **v** and **w** is denoted by the symbol $\langle \mathbf{v}, \mathbf{w} \rangle$, rather than by $\mathbf{v} \cdot \mathbf{w}$.

In this section we shall study inner products on vector spaces and describe an application of inner products to Fourier series.

An ***inner product*** on a vector space V is a rule that assigns to each pair of vectors **v**, **w** in V a real number $\langle \mathbf{v}, \mathbf{w} \rangle$. This rule is required to have the following properties: for each **u**, **v**, and $\mathbf{w} \in V$ and $c \in \mathbb{R}$,

(P_1) $$\langle \mathbf{u}, \mathbf{v} \rangle = \langle \mathbf{v}, \mathbf{u} \rangle$$

(P_2) $$\langle \mathbf{u} + \mathbf{v}, \mathbf{w} \rangle = \langle \mathbf{u}, \mathbf{w} \rangle + \langle \mathbf{v}, \mathbf{w} \rangle$$

(P_3) $$\langle c\mathbf{u}, \mathbf{v} \rangle = c\langle \mathbf{u}, \mathbf{v} \rangle$$

(P_4) $$\langle \mathbf{u}, \mathbf{u} \rangle \geq 0, \text{ and } \langle \mathbf{u}, \mathbf{u} \rangle = 0 \text{ if and only if } \mathbf{u} = \mathbf{0}.$$

Example 1. We saw in Section 1.4 that the dot product on $\mathbb{R}^n$ has these four properties. Thus setting $\langle \mathbf{v}, \mathbf{w} \rangle = \mathbf{v} \cdot \mathbf{w}$ defines an inner product on $\mathbb{R}^n$. ■

Example 2. Let I be the closed interval $I_{[a,b]}$. For $f, g \in \mathcal{C}(I)$, define

$$\langle f, g \rangle = \int_a^b f(x)g(x)\, dx.$$

The four properties of an inner product are easy to verify: for f, g and $h \in \mathcal{C}(I)$ and $c \in \mathbb{R}$,

(P_1) $$\langle f, g \rangle = \int_a^b f(x)g(x)\, dx = \int_a^b g(x)f(x)\, dx = \langle g, f \rangle$$

(P_2) $$\langle f + g, h \rangle = \int_a^b (f(x) + g(x))h(x)\, dx$$

$$= \int_a^b (f(x)h(x) + g(x)h(x))\, dx$$

$$= \int_a^b f(x)h(x)\, dx + \int_a^b g(x)h(x)\, dx$$

$$= \langle f, h \rangle + \langle g, h \rangle$$

(P_3) $$\langle cf, g \rangle = \int_a^b cf(x)g(x)\, dx = c \int_a^b f(x)g(x)\, dx = c\langle f, g \rangle$$

(P_4) $$\langle f, f \rangle = \int_a^b (f(x))^2\, dx \geq 0, \text{ and}$$

$$\langle f, f \rangle = \int_a^b (f(x))^2\, dx = 0 \text{ if and only if } f(x) = 0$$
$$\text{for all } x \in I \text{ (since } f \text{ is continuous)}.$$

Thus the formula $\langle f, g \rangle = \int_a^b f(x)g(x)\, dx$ defines an inner product $\mathcal{C}(I)$. ■

Example 3. Let V be the subset of $\mathbb{R}^\infty$ consisting of all $\mathbf{a} = (a_1, a_2, a_3, \ldots)$ such that the infinite series $\sum_{k=1}^{\infty} |a_k|$ converges. Then V is a subspace of $\mathbb{R}^\infty$ because:

(i) If $\mathbf{a} = (a_1, a_2, \ldots) \in V$ and $\mathbf{b} = (b_1, b_2, \ldots) \in V$, then $\mathbf{a} + \mathbf{b} = (a_1 + b_1, a_2 + b_2, \ldots) \in V$ because $\sum_{k=1}^{\infty} |a_k + b_k|$ converges by comparison with the convergent series $\sum_{k=1}^{\infty} (|a_k| + |b_k|)$.

(ii) If $\mathbf{a} = (a_1, a_2, \ldots) \in V$ and $c \in \mathbb{R}$, then $c\mathbf{a} = (ca_1, ca_2, \ldots) \in V$ because

$$\sum_{k=1}^{\infty} |ca_k| = |c| \sum_{k=1}^{\infty} |a_k| \text{ converges.}$$

We can define an inner product on V by the formula

$$\langle \mathbf{a}, \mathbf{b} \rangle = \sum_{k=1}^{\infty} a_k b_k.$$

Notice that the infinite series $\Sigma a_k b_k$ converges by comparison with the series $\displaystyle\sum_{k=1}^{\infty} |a_k|$. Indeed, $|b_k| < 1$ for sufficiently large k so $|a_k b_k| < |a_k|$ for sufficiently large k. Furthermore, if $\mathbf{a} = (a_1, a_2, a_3, \ldots)$, $\mathbf{b} = (b_1, b_2, b_3, \ldots)$ and $\mathbf{c} = (c_1, c_2, c_3, \ldots)$ are three vectors in V and $t \in \mathbb{R}$, then

(P_1) $\qquad \langle \mathbf{a}, \mathbf{b} \rangle = \displaystyle\sum_{k=1}^{\infty} a_k b_k = \sum_{k=1}^{\infty} b_k a_k = \langle \mathbf{b}, \mathbf{a} \rangle$

(P_2) $\langle \mathbf{a} + \mathbf{b}, \mathbf{c} \rangle = \displaystyle\sum_{k=1}^{\infty} (a_k + b_k)c_k = \sum_{k=1}^{\infty} (a_k c_k + b_k c_k) = \sum_{k=1}^{\infty} a_k c_k + \sum_{k=1}^{\infty} b_k c_k$

$$= \langle \mathbf{a}, \mathbf{c} \rangle + \langle \mathbf{b}, \mathbf{c} \rangle$$

(P_3) $\qquad \langle t\mathbf{a}, \mathbf{b} \rangle = \displaystyle\sum_{k=1}^{\infty} (ta_k)b_k = t \sum_{k=1}^{\infty} a_k b_k = t\langle \mathbf{a}, \mathbf{b} \rangle$

(P_4) $\qquad \langle \mathbf{a}, \mathbf{a} \rangle = \displaystyle\sum_{k=1}^{\infty} a_k^2 \geq 0$, and $\langle \mathbf{a}, \mathbf{a} \rangle = 0$ if and only if $\mathbf{a} = \mathbf{0}$.

Hence the rule that assigns to each pair $\mathbf{a}, \mathbf{b}$ of vectors in V the real number $\langle \mathbf{a}, \mathbf{b} \rangle$ does have the properties required of an inner product. ■

Example 4. Let V be a vector space, let $\langle \ , \ \rangle$ be an inner product on V, and let W be a subspace of V. Then the rule that assigns to each pair $\mathbf{v}, \mathbf{w}$ of vectors in W the real number $\langle \mathbf{v}, \mathbf{w} \rangle$ is an inner product on W. The required properties are automatically satisfied in W because they are satisfied in V. Thus, for example, if W is the subspace of $\mathcal{C}(I_{[0,1]})$ spanned by $\{1, x, x^2\}$, the formula

$$\langle f, g \rangle = \int_0^1 f(x)g(x)\, dx$$

defines an inner product on W. ■

Given an inner product on a vector space V, we define the **length** $\|\mathbf{v}\|$ of a vector $\mathbf{v} \in V$ by

$$\|\mathbf{v}\| = \langle \mathbf{v}, \mathbf{v} \rangle^{\frac{1}{2}}.$$

We say that vectors $\mathbf{v}$ and $\mathbf{w} \in V$ are **perpendicular** if $\langle \mathbf{v}, \mathbf{w} \rangle = 0$. A finite set $\{\mathbf{v}_1, \ldots, \mathbf{v}_n\}$ in V is **orthogonal** if $\langle \mathbf{v}_i, \mathbf{v}_j \rangle = 0$ whenever $i \neq j$, and it is **orthonormal** if, in addition, $\|\mathbf{v}_i\| = 1$ for all i. With these definitions, all of the theorems of Section 6.1 are valid for any vector space with an inner product. Simply replace the dot product every time it appears by the inner product.

Example 5. Let $I = I_{[0,1]}$ and let V be the subspace of $\mathcal{C}(I)$ spanned by $\{1, x, x^2\}$. Using the inner product $\langle f, g \rangle = \int_0^1 f(x)g(x)\, dx$, we shall apply the Gram-Schmidt process to find an orthonormal basis for V. We find

$$\mathbf{w}_1 = 1$$

$$\mathbf{w}_2 = x - \frac{\langle x, 1 \rangle}{\langle 1, 1 \rangle} 1 = x - \frac{\int_0^1 x \cdot 1 \, dx}{\int_0^1 1 \cdot 1 \, dx} 1 = x - \tfrac{1}{2}$$

$$\mathbf{w}_3 = x^2 - \frac{\langle x^2, 1 \rangle}{\langle 1, 1 \rangle} 1 - \frac{\langle x^2, x - \tfrac{1}{2} \rangle}{\langle x - \tfrac{1}{2}, x - \tfrac{1}{2} \rangle}(x - \tfrac{1}{2})$$

$$= x^2 - \frac{\int_0^1 x^2 \cdot 1 \, dx}{\int_0^1 1 \cdot 1 \, dx} 1 - \frac{\int_0^1 x^2 \cdot (x - \tfrac{1}{2}) \, dx}{\int_0^1 (x - \tfrac{1}{2})^2 \, dx}(x - \tfrac{1}{2})$$

$$= x^2 - \tfrac{1}{3} - \frac{\tfrac{1}{12}}{\tfrac{1}{12}}(x - \tfrac{1}{2})$$

$$= x^2 - x + \tfrac{1}{6}.$$

Since

$$\|\mathbf{w}_1\| = \langle \mathbf{w}_1, \mathbf{w}_1 \rangle^{\frac{1}{2}} = \left(\int_0^1 1 \cdot 1 \, dx \right)^{\frac{1}{2}} = 1,$$

$$\|\mathbf{w}_2\| = \langle \mathbf{w}_2, \mathbf{w}_2 \rangle^{\frac{1}{2}} = \left(\int_0^1 (x - \tfrac{1}{2})^2 \, dx \right)^{\frac{1}{2}} = \frac{1}{2\sqrt{3}},$$

and

$$\|\mathbf{w}_3\| = \langle \mathbf{w}_3, \mathbf{w}_3 \rangle^{\frac{1}{2}} = \left(\int_0^1 (x^2 - x + \tfrac{1}{6})^2 \, dx \right)^{\frac{1}{2}} = \frac{1}{6\sqrt{5}}$$

we conclude that

$$\left\{ \frac{\mathbf{w}_1}{\|\mathbf{w}_1\|}, \frac{\mathbf{w}_2}{\|\mathbf{w}_2\|}, \frac{\mathbf{w}_3}{\|\mathbf{w}_3\|} \right\} = \{1, 2\sqrt{3}(x - \tfrac{1}{2}), 6\sqrt{5}(x^2 - x + \tfrac{1}{6})\}$$

is an orthonormal basis for V. ∎

Example 6. Let $V = \mathcal{L}(1, x, x^2) \subset \mathcal{C}(I_{[0,1]})$ and $\langle f, g \rangle = \int_0^1 f(x)g(x)\,dx$ as in Example 5. Let us calculate the **B**-coordinates of x^2, where **B** is the ordered basis $(1, \ 2\sqrt{3}(x - \frac{1}{2}), \ 6\sqrt{5}(x^2 - x + \frac{1}{6}))$. Since this basis is orthonormal, we find that

$$x^2 = c_1 \cdot 1 + c_2(2\sqrt{3}(x - \tfrac{1}{2})) + c_3(6\sqrt{5}(x^2 - x + \tfrac{1}{6}))$$

where

$$c_1 = \langle x^2, 1 \rangle = \int_0^1 x^2 \cdot 1 \, dx = \tfrac{1}{3}$$

$$c_2 = \langle x^2, 2\sqrt{3}(x - \tfrac{1}{2}) \rangle = \int_0^1 2\sqrt{3}(x^3 - \tfrac{1}{2}x^2)\,dx = \frac{\sqrt{3}}{6}$$

$$c_3 = \langle x^2, 6\sqrt{5}(x^2 - x + \tfrac{1}{6}) \rangle = \int_0^1 6\sqrt{5}(x^4 - x^3 + \tfrac{1}{6}x^2) = \frac{\sqrt{5}}{30}$$

so the **B**-coordinate triple of x^2 is $\left(\dfrac{1}{3}, \dfrac{\sqrt{3}}{6}, \dfrac{\sqrt{5}}{30}\right)$. ∎

Example 7. Let $V = \mathcal{C}(I_{[0, 2\pi]})$. Then the formula

$$\langle f, g \rangle = \frac{1}{\pi} \int_0^{2\pi} f(x)g(x)\,dx$$

defines an inner product on V. The familiar integral formulas

$$\frac{1}{\pi} \int_0^{2\pi} \sin mx \sin nx \, dx = 0 \quad \text{if } m \neq n$$

$$\frac{1}{\pi} \int_0^{2\pi} \sin mx \cos nx \, dx = 0$$

$$\frac{1}{\pi} \int_0^{2\pi} \cos mx \cos nx \, dx = 0 \quad \text{if } m \neq n,$$

which are valid whenever m and n are nonnegative integers, state that the set

$$\{1, \cos x, \sin x, \cos 2x, \sin 2x, \ldots, \cos nx, \sin nx\}$$

(recall that $1 = \cos 0x$) is orthogonal, for each n. Since

$$\frac{1}{\pi} \int_0^{2\pi} 1 \, dx = 2$$

and

$$\frac{1}{\pi} \int_0^{2\pi} \cos^2 nx \, dx = \frac{1}{\pi} \int_0^{2\pi} \sin^2 nx \, dx = 1 \quad \text{for } n \geq 1,$$

we can conclude that the set

$$\left\{ \frac{1}{\sqrt{2}}, \cos x, \sin x, \cos 2x, \sin 2x, \ldots, \cos nx, \sin nx \right\}$$

is orthonormal with respect to the inner product

$$\langle f, g \rangle = \frac{1}{\pi} \int_0^{2\pi} f(x)g(x)\, dx \quad \text{on } \mathcal{C}(I_{[0,2\pi]}),$$

for each positive integer n. ∎

Example 7 is the key to the theory of Fourier series. For any function $f \in \mathcal{C}(I_{[0,2\pi]})$, we can use the fact that

$$\left(\frac{1}{\sqrt{2}}, \cos x, \sin x, \ldots, \cos nx, \sin nx \right)$$

is an ordered orthonormal basis for the subspace

$$\mathcal{F}^n = \mathcal{L}(1, \cos x, \sin x, \ldots, \cos nx, \sin nx)$$

to compute the projection of f onto this subspace. We find that

$$\text{Proj}_{\mathcal{F}^n} f = c \cdot \frac{1}{\sqrt{2}} + a_1 \cos x + b_1 \sin x + \cdots + a_n \cos nx + b_n \sin nx$$

where

$$c = \left\langle f(x), \frac{1}{\sqrt{2}} \right\rangle, \quad a_1 = \langle f(x), \cos x \rangle, b_1 = \langle f(x), \sin x \rangle, \ldots$$

$$a_n = \langle f(x), \cos nx \rangle, \quad b_n = \langle f(x), \sin nx \rangle.$$

If we set $a_0 = \langle f(x), 1 \rangle = \sqrt{2}c$, we see that:

$$\text{Proj}_{\mathcal{F}^n} f = \frac{a_0}{2} + a_1 \cos x + b_1 \sin x + \cdots + a_n \cos nx + b_n \sin nx$$

where, for each $k \geq 0$,

$$a_k = \frac{1}{\pi} \int_0^{2\pi} f(x) \cos kx\, dx$$

and

$$b_k = \frac{1}{\pi} \int_0^{2\pi} f(x) \sin kx\, dx.$$

The numbers $a_0, a_1, a_2, \ldots$ and $b_1, b_2, \ldots$ are called the **Fourier coefficients** of f. These numbers solve the following "least squares" approximation problem: find $a_0, a_1, \ldots, a_n, b_1, b_2, \ldots, b_n$ so that

$$\| f - (\tfrac{1}{2}a_0 + a_1 \cos x + \cdots + a_n \cos nx + b_1 \sin x + \cdots + b_n \sin nx)\|$$

is as small as possible. If we increase n then we increase the size of the subspace $\mathcal{F}^n$, so $\mathrm{Proj}_{\mathcal{F}^n} f$ gets closer to f as n gets larger. In fact, it can be shown that

$$\lim_{n \to \infty} \|(\mathrm{Proj}_{\mathcal{F}^n} f) - f\| = 0.$$

Furthermore, it can be shown that, if f is continuous on $I_{[0, 2\pi]}$ and differentiable on $I_{(0, 2\pi)}$, then, for each x in the open interval $I_{(0, 2\pi)}$,

$$\lim_{n \to \infty} \left(\frac{a_0}{2} + \sum_{k=1}^{n} (a_k \cos kx + b_k \sin kx) \right) = f(x).$$

Since this limit must have the same value for $x = 2\pi$ as for $x = 0$, we cannot expect that it equals $f(x)$ at the end points. It turns out that when $x = 0$ and when $x = 2\pi$, the above limit is equal to $\tfrac{1}{2}(f(0) + f(2\pi))$, provided f is differentiable from the right at $x = 0$ and is differentiable from the left at $x = 2\pi$.

The infinite series

$$\frac{a_0}{2} + \sum_{k=1}^{\infty} (a_k \cos kx + b_k \sin kx)$$

is called the **Fourier series** of f.

Example 8. Let us find the Fourier series of $f(x) = x$. The Fourier coefficients of f are

$$a_0 = \frac{1}{\pi} \int_0^{2\pi} x \, dx = 2\pi$$

$$a_k = \frac{1}{\pi} \int_0^{2\pi} x \cos kx \, dx = 0 \quad \text{for } k > 0$$

$$b_k = \frac{1}{\pi} \int_0^{2\pi} x \sin kx \, dx = -2/k.$$

Hence the Fourier series for $f(x) = x$ is

$$\pi - \sum_{k=1}^{\infty} \frac{2}{k} \sin kx = \pi - 2\left(\sin x + \frac{1}{2} \sin 2x + \frac{1}{3} \sin 3x + \cdots \right). \quad \blacksquare$$

EXERCISES

1. Which of the following formulas define inner products on $\mathcal{C}(I_{[0,1]})$?

   (a) $\langle f, g \rangle = f(0)g(0)$

   (d) $\langle f, g \rangle = \int_0^1 |f(x)g(x)|\, dx$

   (b) $\langle f, g \rangle = \int_0^1 (f(x) + g(x))\, dx$

   (e) $\langle f, g \rangle = \int_0^1 e^x f(x)g(x)\, dx$

   (c) $\langle f, g \rangle = \int_0^1 f(x)g(1 - x)\, dx$

   (f) $\langle f, g \rangle = \int_0^{\frac{1}{2}} f(x)g(x)\, dx$

2. Let $\mathcal{I}(I_{[a,b]})$ denote the vector space of all integrable functions with domain $I_{[a,b]}$. Show that the formula

$$\langle f, g \rangle = \int_a^b f(x)g(x)\, dx$$

   does not define an inner product on $\mathcal{I}(I_{[a,b]})$.

3. Consider the vector space $\mathcal{C}(I_{[-1,1]})$ with inner product $\langle f, g \rangle = \int_{-1}^1 f(x)g(x)\, dx$. Which of the following subsets of $\mathcal{C}(I_{[-1,1]})$ are orthogonal?

   (a) $\{1, \sin \pi x, \cos \pi x\}$
   (b) $\{1, x, x^2\}$
   (c) $\{3 - 5x^2, x, x^2\}$
   (d) $\{1, e^x, e^{2x}\}$
   (e) $\{1, \sin \pi x, \sin 2\pi x\}$

4. Find the **B**-coordinates of each of the following polynomials, where **B** is the ordered orthonormal basis for the subspace $\mathcal{L}(1, x, x^2)$ of $\mathcal{C}(I_{[0,1]})$ found in Example 5.

   (a) $1 + x + x^2$ (c) $x - 1$
   (b) x (d) $(x - 1)^2$

5. (a) Find an orthonormal basis for the subspace $\mathcal{L}(1, x, x^2)$ of $\mathcal{C}(I_{[-1,1]})$, where the inner product on $\mathcal{C}(I_{[-1,1]})$ is defined by $\langle f, g \rangle = \int_{-1}^1 f(x)g(x)\, dx$.

   (b) Find an orthonormal basis for $\mathcal{L}(1, x, x^2, x^3) \subset \mathcal{C}(I_{[-1,1]})$ where $\langle f, g \rangle = \int_{-1}^1 f(x)g(x)\, dx$.

6. Find an orthonormal basis for the subspace of $\mathcal{C}(I_{[0,1]})$ spanned by $\{x, x^2\}$, where the inner product on $\mathcal{C}(I_{[0,1]})$ is given by $\langle f, g \rangle = \int_0^1 f(x)g(x)\, dx$.

7. Find an orthonormal basis for the subspace of $\mathcal{C}(I_{[0,2\pi]})$ spanned by $\{1, \cos x, \cos 2x, \sin x, \sin 2x\}$, where the inner product on $\mathcal{C}(I_{[0,2\pi]})$ is given by $\langle f, g \rangle = \int_0^{2\pi} f(x)g(x)\, dx$.

8. Let $V = \{(a_1, a_2, \ldots) \in \mathbb{R}^\infty | \sum\limits_{k=1}^{\infty} |a_k| \text{ converges}\}$ and let

$$\langle (a_1, a_2, \ldots), (b_1, b_2, \ldots) \rangle = \sum_{k=1}^{\infty} a_k b_k,$$

as in Example 3.

(a) Find $\|(1, a, a^2, a^3, \ldots)\|$ where $0 < a < 1$.

(b) Find an orthogonal basis for the subspace of V spanned by $\{(1, \frac{1}{2}, \frac{1}{4}, \frac{1}{8}, \ldots), (1, \frac{1}{3}, \frac{1}{9}, \frac{1}{27}, \ldots)\}$.

9. Let $W = \left\{ (a_1, a_2, a_3, \ldots) \in \mathbb{R}^\infty | \sum\limits_{k=1}^{\infty} a_k^2 \text{ converges} \right\}$.

(a) Show that W is a subspace of $\mathbb{R}^\infty$.

(b) Show that the formula $\langle (a_1, a_2, a_3, \ldots), (b_1, b_2, b_3, \ldots) \rangle = \sum\limits_{k=1}^{\infty} a_k b_k$ defines an inner product on W. [Hint: To see that $\sum\limits_{k=1}^{\infty} a_k b_k$ converges, use the fact that $|\mathbf{a} \cdot \mathbf{b}| \le \|\mathbf{a}\| \, \|\mathbf{b}\|$ for $\mathbf{a}, \mathbf{b} \in \mathbb{R}^n$.]

(c) Show that $W \supseteq V$, where V is as in Exercise 8.

(d) Show that W is strictly larger than the space V of Exercise 8 by finding a vector $\mathbf{a} \in \mathbb{R}^\infty$ that is in W but is not in V.

10. Find the Fourier series for each of the following functions $f \in \mathcal{C}(I_{[0, 2\pi]})$.

(a) $f(x) = x^2$ (c) $f(x) = \cos^2 x$

(b) $f(x) = \sin^2 x$ (d) $f(x) = \pi - x$

11. (a) Show that the set

$$\left\{ \frac{1}{\sqrt{2}}, \cos\frac{2\pi x}{l}, \sin\frac{2\pi x}{l}, \cos\frac{4\pi x}{l}, \ldots, \cos\frac{2n\pi x}{l}, \sin\frac{2n\pi x}{l} \right\}$$

is an orthonormal set in $\mathcal{C}(I_{[0,l]})$, where the inner product is given by $\langle f, g \rangle = \frac{2}{l} \int_0^l f(x)g(x)\,dx, \ l > 0$.

(b) Let $f \in \mathcal{C}(I_{[0,l]})$. Find the coefficients $a_0, \ldots, a_n, b_1, \ldots, b_n$ such that

$$\text{Proj}_W f = \frac{a_0}{2} + a_1 \cos\frac{2\pi x}{l} + \cdots + a_n \cos\frac{2n\pi x}{l}$$

$$+ b_1 \sin\frac{2\pi x}{l} + \cdots + b_n \sin\frac{2n\pi x}{l},$$

where W is the subspace of $\mathcal{C}(I_{[0,l]})$ spanned by the set in part (a).

12. Let V be the subset of $\mathcal{C}(I_{[0,\infty]})$ consisting of those functions f that are bounded; i.e., for each $f \in V$ there is a real number $K > 0$ such that $|f(x)| \le K$ for all $x \ge 0$.

(a) Show that V is a subspace of $\mathcal{C}(I_{[0,\infty]})$.

(b) Show that the formula

$$\langle f, g \rangle = \int_0^\infty e^{-x} f(x)g(x)\,dx$$

defines an inner product on V.

(c) Explain why an inner product on V cannot be defined by the formula

$$\langle f, g \rangle = \int_0^\infty f(x)g(x)\,dx.$$

13. Let V be a finite dimensional vector space with basis $\{v_1, \ldots, v_n\}$. Show that there is an inner product on V with respect to which the basis $\{v_1, \ldots, v_n\}$ is orthonormal. [Hint: If $\langle \quad , \quad \rangle$ is such an inner product, what must

$$\left\langle \sum_{i=1}^n a_i v_i, \sum_{j=1}^n b_j v_j \right\rangle$$

be?]

Linear Maps

SEVEN

7.1 LINEAR MAPS

In this chapter we shall study functions defined on vector spaces. The most interesting functions from the point of view of linear algebra are those that respect the vector operations of addition and scalar multiplication. These functions are called *linear* functions, or linear maps.

Let V and W be vector spaces and let $\phi: V \to W$ be a function; that is, ϕ is a rule that assigns to each vector $\mathbf{v} \in V$ a vector $\phi(\mathbf{v}) \in W$. The function ϕ is called a ***linear map*** if, for each $\mathbf{u}, \mathbf{v} \in V$ and $c \in \mathbb{R}$,

(i)
$$\phi(\mathbf{u} + \mathbf{v}) = \phi(\mathbf{u}) + \phi(\mathbf{v})$$

and

(ii)
$$\phi(c\mathbf{v}) = c\phi(\mathbf{v}).$$

Thus, by definition, a function $\phi: V \to W$ is a linear map if it preserves vector addition and scalar multiplication. It is easy to check that linear maps also preserve vector subtraction and linear combinations:

$$\phi(\mathbf{u} - \mathbf{v}) = \phi(\mathbf{u} + (-1)\mathbf{v}) = \phi(\mathbf{u}) + \phi((-1)\mathbf{v})$$
$$= \phi(\mathbf{u}) + (-1)\phi(\mathbf{v}) = \phi(\mathbf{u}) - \phi(\mathbf{v})$$

and

$$\phi(c_1\mathbf{v}_1 + \cdots + c_k\mathbf{v}_k) = \phi(c_1\mathbf{v}_1) + \cdots + \phi(c_k\mathbf{v}_k)$$
$$= c_1\phi(\mathbf{v}_1) + \cdots + c_k\phi(\mathbf{v}_k).$$

Example 1. Let $D: \mathcal{D} \to \mathcal{F}(\mathbb{R})$ denote differentiation. Thus D is the rule that assigns to each differentiable function f its derivative. D is a linear map because

(i) the derivative of a sum is the sum of the derivatives; $D(f + g) = D(f) + D(g)$, and

(ii) the derivative of a constant times a function is equal to that constant times the derivative of the function; $D(cf) = cD(f)$. ∎

Example 2. Let $\phi: \mathcal{C}(I_{[a,b]}) \to \mathbb{R}$ be defined by

$$\phi(f) = \int_a^b f.$$

Thus ϕ is the rule that assigns to each continuous function on the closed interval $I_{[a,b]}$ its integral over that interval. ϕ is a linear map because

(i) $\phi(f + g) = \int_a^b (f + g) = \int_a^b f + \int_a^b g = \phi(f) + \phi(g)$

(the integral of the sum is the sum of the integrals) and

(ii) $\phi(cf) = \int_a^b (cf) = c \int_a^b f = c\phi(f)$

(the integral of a constant times a function is equal to that constant times the integral of the function). ∎

Example 3. Let $\phi: \mathcal{C}(I_{[a,b]}) \to \mathcal{C}(I_{[a,b]})$ be defined by

$$[\phi(f)](x) = \int_a^x f.$$

The verification that ϕ is a linear map is the same as in Example 2. ∎

Example 4. For each $a \in \mathbb{R}$, let $\phi_a: \mathbb{R}^n \to \mathbb{R}^n$ be defined by

$$\phi_a(\mathbf{v}) = a\mathbf{v}.$$

Then ϕ is a linear map because

(i) $\phi(\mathbf{u} + \mathbf{v}) = a(\mathbf{u} + \mathbf{v}) = a\mathbf{u} + a\mathbf{v} = \phi(\mathbf{u}) + \phi(\mathbf{v})$

and

(ii) $\phi(c\mathbf{v}) = ac\mathbf{v} = c(a\mathbf{v}) = c\phi(\mathbf{v}).$

This map $\phi_a: \mathbb{R}^n \to \mathbb{R}^n$ is called "expansion by a factor of a" if $a > 1$ and is called "contraction by a factor of a" if $0 < a < 1$. ∎

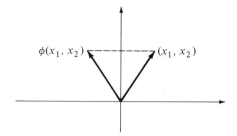

FIGURE 7.1

Example 5.   Let $\phi: \mathbb{R}^2 \to \mathbb{R}^2$ be defined by

$$\phi(x_1, x_2) = (-x_1, x_2).$$

This map ϕ is usually called "reflection in the x_2-axis" (see Figure 7.1). It is linear because

(i) $$\phi((x_1, x_2) + (y_1, y_2)) = \phi(x_1 + y_1, x_2 + y_2)$$
$$= (-(x_1 + y_1), (x_2 + y_2))$$
$$= (-x_1, x_2) + (-y_1, y_2)$$
$$= \phi(x_1, x_2) + \phi(y_1, y_2) \qquad \text{(see Figure 7.2)}$$

and

(ii) $$\phi(c(x_1, x_2)) = \phi(cx_1, cx_2) = (-cx_1, cx_2)$$
$$= c(-x_1, x_2) = c\phi(x_1, x_2) \qquad \text{(see Figure 7.3).} \quad \blacksquare$$

Example 6.   For each $\theta \in \mathbb{R}$, define $r_\theta: \mathbb{R}^2 \to \mathbb{R}^2$ as follows. Given $\mathbf{v} \in \mathbb{R}^2$, let $r_\theta(\mathbf{v})$ be the vector in $\mathbb{R}^2$ obtained by rotating $\mathbf{v}$ counterclockwise through an angle

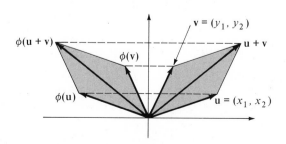

FIGURE 7.2

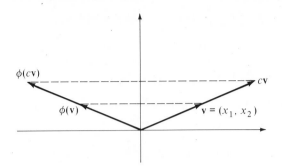

FIGURE 7.3

of θ radians (see Figure 7.4). r_θ is called "rotation through the angle θ." To see that r_θ is a linear map, simply observe that

(i)   the parallelogram that describes the vector sum $r_\theta(\mathbf{u}) + r_\theta(\mathbf{v})$ is obtained by rotating through the angle θ the parallelogram describing the vector sum $\mathbf{u} + \mathbf{v}$ (see Figure 7.5), so $r_\theta(\mathbf{u} + \mathbf{v}) = r_\theta(\mathbf{u}) + r_\theta(\mathbf{v})$, and

(ii)   the vector obtained by rotating $c\mathbf{v}$ through the angle θ is just c times the vector obtained by rotating $\mathbf{v}$ through the angle θ (see Figure 7.6), so $r_\theta(c\mathbf{v}) = cr_\theta(\mathbf{v})$. ∎

Example 7.   For $\mathbf{a} \in \mathbb{R}^n$, define $\phi_{\mathbf{a}}: \mathbb{R}^n \to \mathbb{R}$ by

$$\phi_{\mathbf{a}}(\mathbf{v}) = \mathbf{a} \cdot \mathbf{v}.$$

Then $\phi_{\mathbf{a}}$ is a linear map because

(i) $$\phi_{\mathbf{a}}(\mathbf{u} + \mathbf{v}) = \mathbf{a} \cdot (\mathbf{u} + \mathbf{v}) = \mathbf{a} \cdot \mathbf{u} + \mathbf{a} \cdot \mathbf{v} = \phi_{\mathbf{a}}(\mathbf{u}) + \phi_{\mathbf{a}}(\mathbf{v})$$

and

(ii) $$\phi_{\mathbf{a}}(c\mathbf{v}) = \mathbf{a} \cdot (c\mathbf{v}) = c(\mathbf{a} \cdot \mathbf{v}) = c\phi_{\mathbf{a}}(\mathbf{v}).$$ ∎

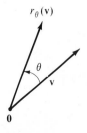

FIGURE 7.4

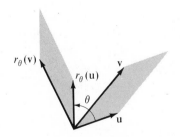

FIGURE 7.5

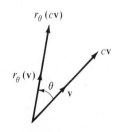

FIGURE 7.6

Example 8. For $A \in \mathfrak{M}_{m \times n}$, define $\phi_A \colon \mathbb{R}^n \to \mathbb{R}^m$ by

$$\phi_A(\mathbf{v}) = (\mathbf{a}_1 \bullet \mathbf{v}, \ldots, \mathbf{a}_m \bullet \mathbf{v})$$

where $\mathbf{a}_1, \ldots, \mathbf{a}_m \in \mathbb{R}^n$ are the row vectors of A. Then ϕ_A is a linear map because

(i)
$$\begin{aligned}
\phi_A(\mathbf{u} + \mathbf{v}) &= (\mathbf{a}_1 \bullet (\mathbf{u} + \mathbf{v}), \ldots, \mathbf{a}_m \bullet (\mathbf{u} + \mathbf{v})) \\
&= (\mathbf{a}_1 \bullet \mathbf{u} + \mathbf{a}_1 \bullet \mathbf{v}, \ldots, \mathbf{a}_m \bullet \mathbf{u} + \mathbf{a}_m \bullet \mathbf{v}) \\
&= (\mathbf{a}_1 \bullet \mathbf{u}, \ldots, \mathbf{a}_m \bullet \mathbf{u}) + (\mathbf{a}_1 \bullet \mathbf{v}, \ldots, \mathbf{a}_m \bullet \mathbf{v}) \\
&= \phi_A(\mathbf{u}) + \phi_A(\mathbf{v})
\end{aligned}$$

and

(ii)
$$\begin{aligned}
\phi_A(c\mathbf{v}) &= (\mathbf{a}_1 \bullet (c\mathbf{v}), \ldots, \mathbf{a}_m \bullet (c\mathbf{v})) \\
&= (c\mathbf{a}_1 \bullet \mathbf{v}, \ldots, c\mathbf{a}_m \bullet \mathbf{v}) \\
&= c(\mathbf{a}_1 \bullet \mathbf{v}, \ldots, \mathbf{a}_m \bullet \mathbf{v}) \\
&= c\phi_A(\mathbf{v}).
\end{aligned}$$

This map $\phi_A \colon \mathbb{R}^n \to \mathbb{R}^m$ is called *the linear map associated with the matrix* A. ■

Example 9. Let $\phi \colon \mathbb{R}^3 \to \mathbb{R}^2$ be defined by

$$\phi(x_1, x_2, x_3) = (2x_1 + 3x_2 - x_3, -x_1 + 4x_2 + 6x_3).$$

Then ϕ is a linear map because, in fact, $\phi = \phi_A$ where

$$A = \begin{pmatrix} 2 & 3 & -1 \\ -1 & 4 & 6 \end{pmatrix}$$

(see Example 8). ■

Example 10. Given any two vector spaces V and W, there is an important linear map $\Theta \colon V \to W$, called the *zero map*, defined by $\Theta(\mathbf{v}) = \mathbf{0}$ for all $\mathbf{v} \in V$. There is another important linear map $i \colon V \to V$, which maps any vector space V to itself, called the *identity map*, defined by $i(\mathbf{v}) = \mathbf{v}$ for all $\mathbf{v} \in V$. Note that the zero map $\Theta \colon \mathbb{R}^n \to \mathbb{R}^m$ is just ϕ_O where O is the zero matrix in $\mathfrak{M}_{m \times n}$, and the identity map $i \colon \mathbb{R}^n \to \mathbb{R}^n$ is just ϕ_I where $I \in \mathfrak{M}_{n \times n}$ is the *identity matrix*

$$I = \begin{pmatrix} 1 & 0 & \cdots & 0 \\ 0 & 1 & \cdots & 0 \\ & & \cdots & \\ 0 & 0 & \cdots & 1 \end{pmatrix}$$

in $\mathfrak{M}_{n \times n}$. ■

The linear maps ϕ_A associated with $m \times n$ matrices A provide us with a large collection of examples of linear maps from $\mathbb{R}^n$ to $\mathbb{R}^m$. It turns out that, in fact, *every* linear map from $\mathbb{R}^n$ to $\mathbb{R}^m$ is ϕ_A for some $A \in \mathfrak{M}_{m \times n}$. This is a consequence of the following theorem, which asserts that a linear map is completely determined by what it does to the elements of any spanning set.

Theorem 1. *Let V be a vector space with spanning set $\{v_1, \ldots, v_k\}$. Suppose $\phi \colon V \to W$ and $\psi \colon V \to W$ are linear maps such that*

$$\phi(v_1) = \psi(v_1), \quad \phi(v_2) = \psi(v_2), \quad \ldots, \quad \phi(v_k) = \psi(v_k).$$

Then $\phi = \psi$.

Proof. For each $v \in V$ we can find $c_1, \ldots, c_k \in \mathbb{R}$ such that

$$v = c_1 v_1 + \cdots + c_k v_k = \sum_{i=1}^{k} c_i v_i.$$

Then

$$\phi(v) = \phi\left(\sum_{i=1}^{k} c_i v_i \right) = \sum_{i=1}^{k} c_i \phi(v_i)$$

$$= \sum_{i=1}^{k} c_i \psi(v_i) = \psi\left(\sum_{i=1}^{k} c_i v_i \right) = \psi(v).$$

Thus $\phi(v) = \psi(v)$ for all $v \in V$; that is, $\phi = \psi$. ∎

Theorem 2. *Let $\phi \colon \mathbb{R}^n \to \mathbb{R}^m$ be a linear map. Then $\phi = \phi_A$, where A is the matrix whose column vectors are the vectors $\phi(e_1), \ldots, \phi(e_n)$, $\{e_1, \ldots, e_n\}$ being the standard basis for $\mathbb{R}^n$.*

Proof. Let

$$A = \begin{pmatrix} a_{11} & \cdots & a_{1n} \\ & \cdots & \\ a_{m1} & \cdots & a_{mn} \end{pmatrix}$$

and let a_i be the ith row vector of A. Then

$$\phi_A(e_j) = (a_1 \bullet e_j, a_2 \bullet e_j, \ldots, a_m \bullet e_j)$$

$$= (a_{1j}, a_{2j}, \ldots, a_{mj})$$

$$= j\text{th column vector of } A.$$

Thus, if A is the matrix whose column vectors are the vectors $\phi(\mathbf{e}_1), \ldots, \phi(\mathbf{e}_n)$, then we have, for each j $(1 \leq j \leq n)$,

$$\phi_A(\mathbf{e}_j) = j\text{th column vector of } A = \phi(\mathbf{e}_j)$$

so $\phi_A = \phi$ by Theorem 1. ■

The matrix A such that $\phi = \phi_A$, where $\phi \colon \mathbb{R}^n \to \mathbb{R}^m$, is called *the matrix of* ϕ.

Theorem 2 can be used to find algebraic descriptions of linear maps that are described geometrically.

Example 11. For $\theta \in \mathbb{R}$, let $r_\theta \colon \mathbb{R}^2 \to \mathbb{R}^2$ be rotation through the angle θ, as in Example 6. Then the matrix of r_θ is the 2×2 matrix A whose columns are the vectors $r_\theta(\mathbf{e}_1)$ and $r_\theta(\mathbf{e}_2)$. But from Figure 7.7 we see easily that

$$r_\theta(\mathbf{e}_1) = (\cos \theta, \sin \theta)$$
$$r_\theta(\mathbf{e}_2) = (-\sin \theta, \cos \theta)$$

so the matrix of r_θ is

$$A = \begin{pmatrix} \cos \theta & -\sin \theta \\ \sin \theta & \cos \theta \end{pmatrix}.$$

For each $(x_1, x_2) \in \mathbb{R}^2$, we can now easily compute $r_\theta(x_1, x_2)$:

$$r_\theta(x_1, x_2) = \phi_A(x_1, x_2) = ((\cos \theta)x_1 - (\sin \theta)x_2, (\sin \theta)x_1 + (\cos \theta)x_2)$$

or

$$r_\theta(x_1, x_2) = (x_1 \cos \theta - x_2 \sin \theta, x_1 \sin \theta + x_2 \cos \theta).$$ ■

According to Theorem 1, a linear map is completely determined by what it does to a spanning set. If $\{\mathbf{v}_1, \ldots, \mathbf{v}_k\}$ is a spanning set for V, then each linear map ϕ with domain V is completely determined once the values $\phi(\mathbf{v}_1), \ldots, \phi(\mathbf{v}_k)$ of ϕ on the elements of the spanning set are specified. However, these values cannot be

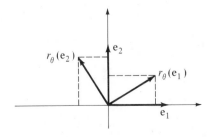

FIGURE 7.7

specified arbitrarily. Indeed, if the spanning set $\{v_1, \ldots, v_k\}$ is linearly dependent, then the image set $\{\phi(v_1), \ldots, \phi(v_k)\}$ must also be linearly dependent, since each dependence relation

$$c_1 v_1 + \cdots + c_k v_k = \mathbf{0}$$

gives rise to a dependence relation

$$c_1 \phi(v_1) + \cdots + c_k \phi(v_k) = \phi(c_1 v_1 + \cdots + c_k v_k) = \phi(\mathbf{0}) = \phi(0\mathbf{0}) = 0\phi(\mathbf{0}) = \mathbf{0}.$$

If, however, the spanning set $\{v_1, \ldots, v_k\}$ is a *basis* for V, then for each set $\{w_1, \ldots, w_k\}$ of vectors in a vector space W there does exist a linear map ϕ such that $\phi(v_i) = w_i$ for each i.

Theorem 3. *Let $\{v_1, \ldots, v_k\}$ be a basis for a vector space V and let $w_1, \ldots, w_k$ be any vectors in some vector space W. Then there exists a unique linear map $\phi: V \to W$ such that*

$$\phi(v_1) = w_1, \quad \ldots, \quad \phi(v_k) = w_k.$$

Proof: Each vector $v \in V$ can be expressed in one and only one way as $v = c_1 v_1 + \cdots + c_k v_k$, where $c_1, \ldots, c_k \in \mathbb{R}$. Define $\phi(v) = c_1 w_1 + \cdots + c_k w_k$. Then ϕ is a linear map, because if $u = b_1 v_1 + \cdots + b_k v_k$, $v = c_1 v_1 + \cdots + c_k v_k$, and $c \in \mathbb{R}$ then

(i)
$$\phi(u + v) = \phi((b_1 + c_1)v_1 + \cdots + (b_k + c_k)v_k)$$
$$= (b_1 + c_1)w_1 + \cdots + (b_k + c_k)w_k$$
$$= (b_1 w_1 + \cdots + b_k w_k) + (c_1 w_1 + \cdots + c_k w_k)$$
$$= \phi(u) + \phi(v)$$

and

(ii)
$$\phi(cv) = \phi((cc_1)v_1 + \cdots + (cc_k)v_k)$$
$$= (cc_1)w_1 + \cdots + (cc_k)w_k$$
$$= c(c_1 w_1 + \cdots + c_k w_k)$$
$$= c\phi(v)$$

Moreover, since $v_j = 0v_1 + \cdots + 0v_{j-1} + 1v_j + 0v_{j+1} + \cdots + 0v_n$, we have

$$\phi(v_j) = 0w_1 + \cdots + 0w_{j-1} + 1w_j + 0w_{j+1} + \cdots + 0w_n$$
$$= w_j$$

for each j, as promised. ∎

Example 12. Theorem 3 tells us that there is a unique linear map ϕ: $\mathbb{R}^3 \to \mathbb{R}^3$ with $\phi(e_1) = (2, 3, 4)$, $\phi(e_2) = (-2, 3, -1)$, $\phi(e_3) = (1, 2, 1)$. Following the proof of Theorem 3, we can describe this map explicitly:

$$\phi(x_1, x_2, x_3) = \phi(x_1 e_1 + x_2 e_2 + x_3 e_3) = x_1 \phi(e_1) + x_2 \phi(e_2) + x_3 \phi(e_3)$$
$$= x_1(2, 3, 4) + x_2(-2, 3, -1) + x_3(1, 2, 1)$$
$$= (2x_1 - 2x_2 + x_3, \, 3x_1 + 3x_2 + 2x_3, \, 4x_1 - x_2 + x_3).$$

In other words $\phi = \phi_A$, where

$$A = \begin{pmatrix} 2 & -2 & 1 \\ 3 & 3 & 2 \\ 4 & -1 & 1 \end{pmatrix},$$

in agreement with Theorem 2. ∎

We close this section with a theorem that describes some important properties of linear maps.

Theorem 4. *Let ϕ: $V \to W$ be a linear map. Then*

(i) $\phi(0) = 0$,

(ii) *if ℓ is a line in V, then $\phi(\ell) = \{\phi(v) | v \in \ell\}$ is either a line in W or a single point, and*

(iii) *if S is a subspace of V, then $\phi(S) = \{\phi(v) | v \in S\}$ is a subspace of W.*

Proof. (i) $\phi(0) = \phi(0 + 0) = \phi(0) + \phi(0)$. Subtracting $\phi(0)$ from both sides yields $0 = \phi(0)$.

(ii) Let **a** and **b** be in ℓ. Then each vector in ℓ is of the form $v = a + t(b - a)$ for some $t \in \mathbb{R}$. Since

$$\phi(v) = \phi(a + t(b - a)) = \phi(a) + t\phi(b - a)$$
$$= \phi(a) + t(\phi(b) - \phi(a)),$$

we see that $\phi(\ell)$ is a line in W unless $\phi(a) = \phi(b)$, in which case $\phi(\ell)$ is the single point $\phi(a)$. Moreover, if $\phi(a) \neq \phi(b)$ *then ϕ maps the line in V through* **a** *and* **b** *onto the line in W through $\phi(a)$ and $\phi(b)$.*

(iii) We must check the three conditions for a subspace:

(1) $0 \in \phi(S)$ because $0 = \phi(0)$ by (i).

(2) If $w_1 = \phi(v_1) \in \phi(S)$ and $w_2 = \phi(v_2) \in \phi(S)$ then $w_1 + w_2 = \phi(v_1) + \phi(v_2) = \phi(v_1 + v_2) \in \phi(S)$, and

(3) If $w = \phi(v) \in \phi(S)$ then $cw = c\phi(v) = \phi(cv) \in \phi(S)$, for all $c \in \mathbb{R}$. ∎

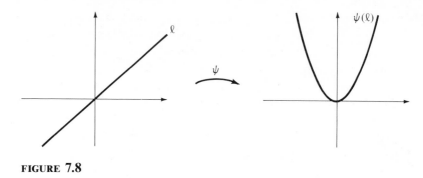

FIGURE 7.8

Theorem 4 can be useful in showing that certain functions are *not* linear maps.

Example 13. The function $\phi\colon \mathbb{R}^2 \to \mathbb{R}^2$ defined by

$$\phi(x_1, x_2) = (x_1, x_2 + 1)$$

is *not* a linear map because

$$\phi(\mathbf{0}) = \phi(0, 0) = (0, 1) \neq \mathbf{0}.$$

The function $\psi\colon \mathbb{R}^2 \to \mathbb{R}^2$ defined by

$$\psi(x_1, x_2) = (x_1, x_2^2)$$

is *not* a linear map because, for example, if ℓ is the line $\{(t, t) \mid t \in \mathbb{R}\}$ then $\psi(\ell)$ is the parabola $\{(t, t^2) \mid t \in \mathbb{R}\}$, which is neither a line nor a point (see Figure 7.8). ■

EXERCISES

1. Which of the following are linear maps?
    (a) $\phi\colon \mathbb{R} \to \mathbb{R}$, $\phi(x) = ax + b$, where a and b are fixed real numbers.
    (b) $\phi\colon \mathbb{R}^2 \to \mathbb{R}$, $\phi(x_1, x_2) = x_1 + x_2$
    (c) $\phi\colon \mathbb{R}^2 \to \mathbb{R}$, $\phi(x_1, x_2) = x_1 x_2$
    (d) $\phi\colon \mathbb{R} \to \mathbb{R}^2$, $\phi(x) = (x, x^2)$
    (e) $\phi\colon \mathbb{R}^2 \to \mathbb{R}^2$, $\phi(x_1, x_2) = (x_1 + x_2 + 1, x_1 - x_2)$
    (f) $\phi\colon \mathbb{R}^4 \to \mathbb{R}^5$, $\phi(x_1, x_2, x_3, x_4) = (x_1 - x_2, x_3, x_4, x_3, x_4)$
    (g) $\phi\colon \mathbb{R}^3 \to \mathbb{R}^3$, $\phi(\mathbf{x}) = \mathbf{a} \times \mathbf{x}$, where $\mathbf{a}$ is a fixed vector in $\mathbb{R}^3$
2. Which of the following are linear maps?
    (a) $\phi\colon \mathbb{R}^n \to \mathbb{R}$, $\phi(x_1, \ldots, x_n) = x_1$
    (b) $\phi\colon \mathbb{R}^n \to \mathbb{R}$, $\phi(x_1, \ldots, x_n) = x_1 + x_2 + \cdots + x_n$
    (c) $\phi\colon \mathbb{R} \to \mathbb{R}^n$, $\phi(x) = (x, x, \ldots, x)$

(d) $\phi: \mathbb{R}^n \to \mathbb{R}$, $\phi(\mathbf{x}) = \|\mathbf{x}\|$

(e) $\phi: \mathbb{R}^n \to \mathbb{R}^n$, $\phi(\mathbf{x}) = \text{Proj}_\mathbf{a}\mathbf{x}$, where $\mathbf{a}$ is some fixed vector in $\mathbb{R}^n$

(f) $\phi: \mathbb{R}^n \to W$, $\phi(\mathbf{x}) = \text{Proj}_W\mathbf{x}$, where W is some fixed subspace of $\mathbb{R}^n$

3. Let V be the vector space of all twice differentiable functions with domain $\mathbb{R}$. Which of the following functions $\phi: V \to \mathcal{F}(\mathbb{R})$ are linear maps?

(a) $\phi(y) = y''$ (d) $\phi(y) = -y$

(b) $\phi(y) = y'' + y$ (e) $\phi(y) = y^2$

(c) $\phi(y) = y'' + y + 1$

4. Which of the following functions $\phi: \mathcal{C}(I_{[0,1]}) \to \mathbb{R}$ are linear maps?

(a) $\phi(f) = f(\frac{1}{2})$ (d) $\phi(f) = \int_0^1 e^x f(x)\, dx$

(b) $\phi(f) = |f(\frac{1}{2})|$ (e) $\phi(f) = \int_0^1 (f(x))^2\, dx$

(c) $\phi(f) = \frac{1}{2}[f(0) + f(1)]$ (f) $\phi(f) = \max\{f(x) | 0 \le x \le 1\}$

5. Let $\phi_A: \mathbb{R}^4 \to \mathbb{R}^3$ be the linear map associated with the matrix

$$A = \begin{pmatrix} 2 & -1 & 5 & 3 \\ -1 & 7 & 6 & 2 \\ 1 & 4 & 2 & 9 \end{pmatrix}.$$

Find

(a) $\phi_A(3, -1, 2, 4)$ (d) $\phi_A(1, 0, 0, 0)$

(b) $\phi_A(-1, 1, -1, 1)$ (e) $\phi_A(0, 1, 0, 0)$

(c) $\phi_A(2, 2, -2, 3)$ (f) $\phi_A(0, 0, 0, 1)$

6. Find an explicit formula for $\phi_A(\mathbf{x})$ where

(a) $A = \begin{pmatrix} 3 & -1 & 2 \\ 2 & 4 & 3 \end{pmatrix}$, $\mathbf{x} = (x_1, x_2, x_3)$

(b) $A = (3 \quad 2 \quad 1)$, $\mathbf{x} = (x_1, x_2, x_3)$

(c) $A = \begin{pmatrix} 1 \\ 2 \\ 3 \end{pmatrix}$, $\mathbf{x} = (x)$

(d) $A = \begin{pmatrix} \dfrac{1}{\sqrt{2}} & -\dfrac{1}{\sqrt{2}} \\ \dfrac{1}{\sqrt{2}} & \dfrac{1}{\sqrt{2}} \end{pmatrix}$, $\mathbf{x} = (x_1, x_2)$

7. Find the matrix of each of the following linear maps.

(a) $\phi: \mathbb{R}^2 \to \mathbb{R}^2$, $\phi(x_1, x_2) = (2x_1 - 3x_2, 5x_1 + 4x_2)$

(b) $\phi: \mathbb{R}^2 \to \mathbb{R}^3$, $\phi(x_1, x_2) = (x_1 + x_2, x_1 - x_2, 2x_1 + 3x_2)$

(c) $\phi: \mathbb{R}^3 \to \mathbb{R}^2$, $\phi(x_1, x_2, x_3) = (x_1 - x_2 + x_3, x_1 + 2x_2 + x_3)$

(d) $\phi: \mathbb{R}^3 \to \mathbb{R}^3$, $\phi(x_1, x_2, x_3) = (2x_1, 3x_2, -x_3)$

(e) $\phi: \mathbb{R}^3 \to \mathbb{R}^3$, $\phi(x_1, x_2, x_3) = (x_2, x_3, x_1)$

(f) $\phi: \mathbb{R}^3 \to \mathbb{R}^1$, $\phi(x_1, x_2, x_3) = x_2$

(g) $\phi: \mathbb{R}^1 \to \mathbb{R}^3$, $\phi(x) = (x, -x, 0)$

8. Find the matrix of each of the following linear maps:

(a) $\phi: \mathbb{R}^2 \to \mathbb{R}^2$, rotation through the angle $\pi/2$

(b) $\phi: \mathbb{R}^2 \to \mathbb{R}^2$, reflection in the x_2-axis

(c) $\phi: \mathbb{R}^2 \to \mathbb{R}^2$, reflection in the line $x_2 = x_1$

(d) $\phi\colon \mathbb{R}^3 \to \mathbb{R}^3$, $\phi(\mathbf{x}) = \mathrm{Proj}_{\mathbf{e}_2}(\mathbf{x})$

(e) $\phi\colon \mathbb{R}^3 \to \mathbb{R}^3$, $\phi(\mathbf{x}) = \mathrm{Proj}_{\mathbf{a}}(\mathbf{x})$ where $\mathbf{a} = (a_1, a_2, a_3)$ is any unit vector

(f) $\phi\colon \mathbb{R}^3 \to \mathbb{R}^3$, $\phi(\mathbf{x}) = \mathbf{a} \times \mathbf{x}$, where $\mathbf{a} = (a_1, a_2, a_3)$

9. Let $\phi\colon \mathbb{R}^3 \to \mathbb{R}^2$ be the linear map with the properties that $\phi(\mathbf{e}_1) = (2, -1)$, $\phi(\mathbf{e}_2) = (1, 1)$, $\phi(\mathbf{e}_3) = (-1, 4)$. Find

(a) $\phi(2, -1, 1)$

(b) $\phi(1, 1, 1)$

(c) $\phi(1, 2, 3)$

10. Let $\phi\colon \mathbb{R}^3 \to \mathbb{R}^2$ be the linear map such that $\phi(1, 0, 1) = (2, -1)$, $\phi(0, 1, 1) = (1, 1)$, $\phi(1, 1, 0) = (-1, 4)$. Find

(a) $\phi(2, -1, 1)$

(b) $\phi(1, 1, 1)$

(c) $\phi(1, 2, 3)$

11. Let $\phi\colon \mathcal{P}^3 \to \mathcal{D}$ be the linear map such that $\phi(1) = \cos 2x$, $\phi(x) = \cos^2 x$, $\phi(x^2) = \sin^2 x$. Find

(a) $\phi(x + x^2)$ (d) $\phi(1 - 2x)$

(b) $\phi(x - x^2)$ (e) $\phi(1 + 2x^2)$

(c) $\phi(1 - x + x^2)$

12. Let $\mathbf{d}$ be a unit vector in $\mathbb{R}^n$ and let H be the hyperplane $\mathbf{d} \cdot \mathbf{x} = 0$. Show that the function $\phi\colon \mathbb{R}^n \to \mathbb{R}^n$ defined by $\phi(\mathbf{v}) = \mathbf{v} - 2(\mathbf{v} \cdot \mathbf{d})\mathbf{d}$ is a linear map. (ϕ is called **reflection in the hyperplane** H; see Figure 7.9.)

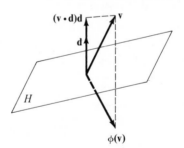

FIGURE 7.9

13. Let $\mathbf{d}$ be a unit vector in $\mathbb{R}^3$ and let $\theta \in \mathbb{R}$. Show that the function $\phi\colon \mathbb{R}^3 \to \mathbb{R}^3$ defined by

$$\phi(\mathbf{v}) = (\cos \theta)\mathbf{v} + (\sin \theta)\mathbf{d} \times \mathbf{v}$$

is a linear map. (ϕ is called **rotation about d through the angle** θ; see Figure 7.10.)

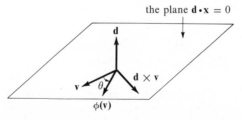

FIGURE 7.10

14. Let $\phi: \mathbb{R}^3 \to \mathbb{R}^3$ be a linear map and let P be a plane in $\mathbb{R}^3$. Show that

$$\phi(P) = \{\mathbf{w} \in \mathbb{R}^3 \,|\, \mathbf{w} = \phi(\mathbf{v}) \quad \text{for some } \mathbf{v} \in P\}$$

is either a plane, a line, or a single point in $\mathbb{R}^3$. [*Hint:* If $\mathbf{a} \in P$ and if $\mathbf{v}_1$ and $\mathbf{v}_2$ are parallel to P with $\{\mathbf{v}_1, \mathbf{v}_2\}$ linearly independent, then each vector $\mathbf{v} \in P$ is of the form

$$\mathbf{v} = \mathbf{a} + c_1\mathbf{v}_1 + c_2\mathbf{v}_2$$

for some $c_1, c_2 \in \mathbb{R}$. Compute $\phi(\mathbf{v})$.]

7.2 THE ALGEBRA OF LINEAR MAPS

We have seen that each linear map $\phi: \mathbb{R}^n \to \mathbb{R}^m$ corresponds to a matrix: $\phi = \phi_A$ for some $A \in \mathfrak{M}_{m \times n}$. In this section we shall see that two linear maps ϕ and $\psi: \mathbb{R}^n \to \mathbb{R}^m$ can be added, and that the matrix corresponding to $\phi + \psi$ is just $A + B$ where $\phi = \phi_A$ and $\psi = \phi_B$. Similarly, we shall see that $c\phi: V \to W$ is a linear map and that the matrix for $c\phi$ is just cA. Then we shall study the composition $\psi \circ \phi$ of linear maps $\phi: \mathbb{R}^n \to \mathbb{R}^k$ and $\psi: \mathbb{R}^k \to \mathbb{R}^m$ and shall see that $\psi \circ \phi$ is a linear map whose matrix is the "matrix product" BA, where $\psi = \phi_B$ and $\phi = \phi_A$.

Let ϕ and ψ be linear maps from a vector space V to a vector space W. We define the **sum** of ϕ and ψ to be the function $\phi + \psi: V \to W$ defined by

$$(\phi + \psi)(\mathbf{v}) = \phi(\mathbf{v}) + \psi(\mathbf{v}) \text{ for all } \mathbf{v} \in V.$$

Then $\phi + \psi$ is a linear map because if $\mathbf{u}, \mathbf{v} \in V$ and $c \in \mathbb{R}$, then

(i)
$$\begin{aligned}
(\phi + \psi)(\mathbf{u} + \mathbf{v}) &= \phi(\mathbf{u} + \mathbf{v}) + \psi(\mathbf{u} + \mathbf{v}) \\
&= \phi(\mathbf{u}) + \phi(\mathbf{v}) + \psi(\mathbf{u}) + \psi(\mathbf{v}) \\
&= \phi(\mathbf{u}) + \psi(\mathbf{u}) + \phi(\mathbf{v}) + \psi(\mathbf{v}) \\
&= (\phi + \psi)(\mathbf{u}) + (\phi + \psi)(\mathbf{v}),
\end{aligned}$$

and

(ii)
$$\begin{aligned}
(\phi + \psi)(c\mathbf{v}) &= \phi(c\mathbf{v}) + \psi(c\mathbf{v}) \\
&= c\phi(\mathbf{v}) + c\psi(\mathbf{v}) \\
&= c(\phi(\mathbf{v}) + \psi(\mathbf{v})) \\
&= c(\phi + \psi)(\mathbf{v}).
\end{aligned}$$

Similarly, if $c \in \mathbb{R}$ and $\phi: V \to W$ is a linear map, we define the linear map $c\phi: V \to W$ by

$$(c\phi)(\mathbf{v}) = c\phi(\mathbf{v}) \text{ for all } \mathbf{v} \in V.$$

The verification that $c\phi$ is a linear map is straightforward (see Exercise 9).

Recall that each linear map ϕ: $\mathbb{R}^n \to \mathbb{R}^m$ is associated with a matrix: $\phi = \phi_A$ where $A \in \mathfrak{M}_{m \times n}$ is the matrix of ϕ. It is natural to ask "What is the relationship between the matrix of $\phi + \psi$ and the matrices of ϕ and of ψ?" and "What is the relationship between the matrix of ϕ and the matrix of $c\phi$?" The next theorem answers these questions.

Theorem 1.   *Let ϕ: $\mathbb{R}^n \to \mathbb{R}^m$ and ψ: $\mathbb{R}^n \to \mathbb{R}^m$ be linear maps and let $c \in \mathbb{R}$. Suppose the matrix of ϕ is A and the matrix of ψ is B. Then*

    (i)   *the matrix of $\phi + \psi$ is $A + B$*

    (ii) *the matrix of $c\phi$ is cA.*

Proof.   Let the row vectors of A and B, respectively, be $\mathbf{a}_1, \ldots, \mathbf{a}_m$ and $\mathbf{b}_1, \ldots, \mathbf{b}_m$, and let $\mathbf{v} \in \mathbb{R}^n$. Then $\mathbf{a}_1 + \mathbf{b}_1, \ldots, \mathbf{a}_m + \mathbf{b}_m$ are the row vectors of $A + B$ and

$$
\begin{aligned}
(\phi + \psi)(\mathbf{v}) &= \phi(\mathbf{v}) + \psi(\mathbf{v}), \text{ by definition of } \phi + \psi \\
&= \phi_A(\mathbf{v}) + \phi_B(\mathbf{v}), \text{ since } \phi = \phi_A \text{ and } \psi = \phi_B \\
&= (\mathbf{a}_1 \cdot \mathbf{v}, \ldots, \mathbf{a}_m \cdot \mathbf{v}) + (\mathbf{b}_1 \cdot \mathbf{v}, \ldots, \mathbf{b}_m \cdot \mathbf{v}), \\
&\qquad \text{by definition of } \phi_A, \phi_B \\
&= (\mathbf{a}_1 \cdot \mathbf{v} + \mathbf{b}_1 \cdot \mathbf{v}, \ldots, \mathbf{a}_m \cdot \mathbf{v} + \mathbf{b}_m \cdot \mathbf{v}) \\
&= ((\mathbf{a}_1 + \mathbf{b}_1) \cdot \mathbf{v}, \ldots, (\mathbf{a}_m + \mathbf{b}_m) \cdot \mathbf{v}) \\
&= \phi_{A+B}(\mathbf{v}).
\end{aligned}
$$

Hence

$$
\phi + \psi = \phi_{A+B},
$$

proving (i).

The verification of (ii) is left as an exercise.   ∎

Recall that if ϕ: $U \to V$ and ψ: $V \to W$ are functions, then the ***composition*** $\psi \circ \phi$: $U \to W$ is the function defined by

$$
\psi \circ \phi(\mathbf{u}) = \psi(\phi(\mathbf{u})) \text{ for all } \mathbf{u} \in U.
$$

It is easy to verify that if ϕ and ψ are linear maps then so is $\psi \circ \phi$ (see Exercise 12). In particular, if ϕ: $\mathbb{R}^n \to \mathbb{R}^k$ and ψ: $\mathbb{R}^k \to \mathbb{R}^m$ are linear maps, then so is $\psi \circ \phi$: $\mathbb{R}^n \to \mathbb{R}^m$. Our next goal is to describe the relationship between the matrix of $\psi \circ \phi$ and the matrices of ψ and of ϕ.

To do this we shall apply Theorem 2 of the previous section. Let $\{\mathbf{e}_1, \ldots, \mathbf{e}_n\}$ be the standard basis for $\mathbb{R}^n$. Then the matrix of $\psi \circ \phi$ is the matrix C whose column vectors are the vectors $\psi \circ \phi(\mathbf{e}_1), \ldots, \psi \circ \phi(\mathbf{e}_n)$. In particular, the (i, j)-entry

c_{ij} of C, the number that appears in the ith row and jth column of C, is the ith entry in the vector

$$\psi \circ \phi(\mathbf{e}_j) = \phi_A \circ \phi_B(\mathbf{e}_j) = \phi_A(\phi_B(\mathbf{e}_j))$$
$$= (\mathbf{a}_1 \cdot \phi_B(\mathbf{e}_j), \quad \ldots, \quad \mathbf{a}_m \cdot \phi_B(\mathbf{e}_j))$$

where $A \in \mathfrak{M}_{m \times k}$ is the matrix of ψ, $B \in \mathfrak{M}_{k \times n}$ is the matrix of ϕ, and $\mathbf{a}_1, \ldots, \mathbf{a}_m$ are the row vectors of A. Thus

$$c_{ij} = \mathbf{a}_i \cdot \phi_B(\mathbf{e}_j).$$

But $\phi_B(\mathbf{e}_j)$ is just the jth column vector of B, so $c_{ij} = $ (the ith row vector of A) $\cdot$ (the jth column vector of B).

The matrix $C \in \mathfrak{M}_{m \times n}$ whose (i, j)-entry is equal to the dot product of the ith row of $A \in \mathfrak{M}_{m \times k}$ and the jth column of $B \in \mathfrak{M}_{k \times n}$ is called the **product** of the matrices A and B, and we write $C = AB$. Thus the (i, j) entry in the matrix AB is obtained by matching the entries in the ith row of A with the entries in the jth column of B, multiplying corresponding entries, and adding the results.

The above computation proves the following theorem.

Theorem 2. *Let $\phi: \mathbb{R}^n \to \mathbb{R}^k$ and $\psi: \mathbb{R}^k \to \mathbb{R}^m$ be linear maps. Suppose the matrix of ψ is A and the matrix of ϕ is B. Then the matrix of $\psi \circ \phi$ is AB.*

Remarks. In order that the product AB of two matrices A and B be defined, *the number of columns in A must be equal to the number of rows in B.* Notice also that the number of rows in the product matrix AB is the same as the number of rows in A, and the number of columns in AB is the same as the number of columns in B.

Example 1. Let $A = \begin{pmatrix} 1 & -1 & 3 \\ 2 & 3 & 4 \end{pmatrix}$ and $B = \begin{pmatrix} 1 & 3 \\ 0 & -1 \\ 1 & 2 \end{pmatrix}$.

Then

$$AB = \begin{pmatrix} 1 & -1 & 3 \\ 2 & 3 & 4 \end{pmatrix} \begin{pmatrix} 1 & 3 \\ 0 & -1 \\ 1 & 2 \end{pmatrix}$$

$$= \begin{pmatrix} (1)(1) + (-1)(0) + (3)(1) & (1)(3) + (-1)(-1) + (3)(2) \\ (2)(1) + (3)(0) + (4)(1) & (2)(3) + (3)(-1) + (4)(2) \end{pmatrix}$$

$$= \begin{pmatrix} 4 & 10 \\ 6 & 11 \end{pmatrix}.$$

Also

$$BA = \begin{pmatrix} 1 & 3 \\ 0 & -1 \\ 1 & 2 \end{pmatrix} \begin{pmatrix} 1 & -1 & 3 \\ 2 & 3 & 4 \end{pmatrix} = \begin{pmatrix} 1+6 & -1+9 & 3+12 \\ 0-2 & 0-3 & 0-4 \\ 1+4 & -1+6 & 3+8 \end{pmatrix}$$

$$= \begin{pmatrix} 7 & 8 & 15 \\ -2 & -3 & -4 \\ 5 & 5 & 11 \end{pmatrix}.$$

Notice that AB and BA are not of the same size. In particular, $AB \neq BA$. ■

Example 2.   Let $A = \begin{pmatrix} 1 & -1 \\ 2 & 3 \end{pmatrix}$, and $B = \begin{pmatrix} 1 & 0 \\ -1 & 1 \end{pmatrix}$.

Then

$$AB = \begin{pmatrix} 2 & -1 \\ -1 & 3 \end{pmatrix} \quad \text{and} \quad BA = \begin{pmatrix} 1 & -1 \\ 1 & 4 \end{pmatrix}.$$

Notice that even though the 2×2 matrices AB and BA are both defined and are of the same size, they are still not equal. This observation is equivalent to the statement that $\phi_A \circ \phi_B \neq \phi_B \circ \phi_A$. Matrix multiplication, like composition of functions, is not a commutative operation! ■

Example 3.   Let $\phi \colon \mathbb{R}^3 \to \mathbb{R}^2$ and $\psi \colon \mathbb{R}^2 \to \mathbb{R}^4$ be defined by

$$\phi(x_1, x_2, x_3) = (x_1 + x_2, -x_1 + x_3)$$

and

$$\psi(x_1, x_2) = (x_1 + 2x_2, -x_1 + x_2, x_2, 3x_1 - x_2).$$

We can use Theorem 2 to compute $\psi \circ \phi \colon \mathbb{R}^3 \to \mathbb{R}^4$. Note that the matrices A of ψ and B of ϕ are

$$A = \begin{pmatrix} 1 & 2 \\ -1 & 1 \\ 0 & 1 \\ 3 & -1 \end{pmatrix} \quad \text{and} \quad B = \begin{pmatrix} 1 & 1 & 0 \\ -1 & 0 & 1 \end{pmatrix}.$$

The matrix C of $\psi \circ \phi$ is

$$C = AB = \begin{pmatrix} -1 & 1 & 2 \\ -2 & -1 & 1 \\ -1 & 0 & 1 \\ 4 & 3 & -1 \end{pmatrix}.$$

Hence

$$\psi \circ \phi(x_1, x_2, x_3)$$
$$= (-x_1 + x_2 + 2x_3, -2x_1 - x_2 + x_3, -x_1 + x_3, 4x_1 + 3x_2 - x_3). \quad \blacksquare$$

Example 4. We can use Theorem 2 to give an easy proof of the familiar trigonometric identities for the sine and cosine of a sum. For α and $\beta \in \mathbb{R}$, consider the linear maps $r_\alpha \colon \mathbb{R}^2 \to \mathbb{R}^2$ (rotation through the angle α) and $r_\beta \colon \mathbb{R}^2 \to \mathbb{R}^2$ (rotation through the angle β) as in Example 6 of the previous section. Clearly a rotation through the angle β followed by a rotation through the angle α results in a rotation through the angle $\alpha + \beta$. Thus

$$r_\alpha \circ r_\beta = r_{\alpha+\beta}.$$

It follows that

$$AB = C$$

where A, B, and C are the matrices of r_α, r_β, and $r_{\alpha+\beta}$ respectively. But (see Example 11 of the previous section),

$$A = \begin{pmatrix} \cos \alpha & -\sin \alpha \\ \sin \alpha & \cos \alpha \end{pmatrix}, \quad B = \begin{pmatrix} \cos \beta & -\sin \beta \\ \sin \beta & \cos \beta \end{pmatrix},$$

and hence

$$\begin{pmatrix} \cos(\alpha + \beta) & -\sin(\alpha + \beta) \\ \sin(\alpha + \beta) & \cos(\alpha + \beta) \end{pmatrix} = C = AB$$

$$= \begin{pmatrix} \cos \alpha \cos \beta - \sin \alpha \sin \beta & -(\sin \alpha \cos \beta + \cos \alpha \sin \beta) \\ \sin \alpha \cos \beta + \cos \alpha \sin \beta & \cos \alpha \cos \beta - \sin \alpha \sin \beta \end{pmatrix}.$$

Comparing entries, we see immediately that

$$\cos(\alpha + \beta) = \cos \alpha \cos \beta - \sin \alpha \sin \beta$$

and

$$\sin(\alpha + \beta) = \sin \alpha \cos \beta + \cos \alpha \sin \beta. \quad \blacksquare$$

Matrix multiplication can also be used to rewrite systems of linear equations in compact form. Indeed, you can easily check that the system of m equations

$$a_{11}x_1 + a_{12}x_2 + \cdots + a_{1n}x_n = b_1$$
$$a_{21}x_1 + a_{22}x_2 + \cdots + a_{2n}x_n = b_2$$
$$\cdots$$
$$a_{m1}x_1 + a_{m2}x_2 + \cdots + a_{mn}x_n = b_m$$

is equivalent to the single matrix equation

$$\begin{pmatrix} a_{11} & a_{12} & \cdots & a_{1n} \\ a_{21} & a_{22} & \cdots & a_{2n} \\ & & \cdots & \\ a_{m1} & a_{m2} & \cdots & a_{mn} \end{pmatrix} \begin{pmatrix} x_1 \\ x_2 \\ \vdots \\ x_n \end{pmatrix} = \begin{pmatrix} b_1 \\ b_2 \\ \vdots \\ b_m \end{pmatrix}.$$

Notice also that this system of equations is equivalent to the single vector equation

$$\phi_A(x_1, \ldots, x_n) = (b_1, \ldots, b_m)$$

where

$$A = \begin{pmatrix} a_{11} & \cdots & a_{1n} \\ & \cdots & \\ a_{m1} & \cdots & a_{mn} \end{pmatrix}.$$

EXERCISES

1. Let $A = \begin{pmatrix} -1 & 3 & 2 \\ 2 & -1 & 1 \end{pmatrix}$, $B = \begin{pmatrix} 1 & -1 \\ -1 & 2 \\ 3 & -3 \end{pmatrix}$, $C = (1 \ -1 \ 1)$, $D = \begin{pmatrix} 2 \\ -1 \\ 2 \end{pmatrix}$.
   Compute

   (a) AB (d) CB
   (b) AD (e) CD
   (c) BA (f) DC

2. Check that the six products listed in Exercise 1 are the only pairwise products of A, B, C, D that are defined. (There are 16 possibilities!)

3. Compute the following matrix products:

   (a) $\begin{pmatrix} -1 & 1 \\ 3 & 2 \end{pmatrix}\begin{pmatrix} 5 & 7 \\ 9 & -3 \end{pmatrix}$ (b) $\begin{pmatrix} 2 & -3 \\ -3 & 2 \end{pmatrix}\begin{pmatrix} 4 \\ 1 \end{pmatrix}$

   (c) $\begin{pmatrix} -1 & 3 & 2 \\ 1 & 4 & -5 \\ 3 & 6 & 9 \end{pmatrix}\begin{pmatrix} -1 \\ 2 \\ 1 \end{pmatrix}$ (d) $\begin{pmatrix} -1 & 1 & 2 \\ 0 & 3 & -2 \\ 1 & 1 & 1 \end{pmatrix}\begin{pmatrix} 3 & 7 & 1 \\ -5 & 0 & 0 \\ 1 & -3 & -1 \end{pmatrix}$

   (e) $(-1 \ \ 2 \ \ 1)\begin{pmatrix} -1 & 3 & 2 \\ 1 & 4 & 5 \\ 3 & 6 & 9 \end{pmatrix}$ (f) $\begin{pmatrix} 1 & 3 & 4 \\ 2 & 5 & 3 \\ 0 & 1 & 4 \end{pmatrix}\begin{pmatrix} 17 & -8 & -11 \\ -8 & 4 & 5 \\ 2 & -1 & -1 \end{pmatrix}$

4. Compute $A^2 = AA$ and $A^3 = A^2A$ for

   (a) $A = \begin{pmatrix} 0 & a & b \\ 0 & 0 & c \\ 0 & 0 & 0 \end{pmatrix}$ (b) $A = \begin{pmatrix} 1 & 0 & a \\ 0 & 1 & 0 \\ 0 & 0 & 1 \end{pmatrix}$

5. Let $A = \begin{pmatrix} 1 & 1 \\ 0 & 1 \end{pmatrix}$.

    (a) Find A^2, A^3, and A^4.

    (b) Find A^k for all positive integers k.

    (c) Find a square root of A; that is, find a matrix B such that $B^2 = A$.

6. Compute A^k for $k = 2, 3, 4$ when

    $$A = \begin{pmatrix} 0 & 1 & 0 & 0 \\ 0 & 0 & 1 & 0 \\ 0 & 0 & 0 & 1 \\ 0 & 0 & 0 & 0 \end{pmatrix}.$$

7. For each of the following pairs of linear maps ϕ and ψ, (i) find the matrices A of ψ and B of ϕ, (ii) compute the matrix product AB, and (iii) write a formula for $\psi \circ \phi(x_1, \ldots, x_n)$.

    (a) $\phi(x_1, x_2) = (2x_1 - x_2, 3x_1 + 4x_2)$
         $\psi(x_1, x_2) = (-3x_1 + x_2, x_1 + x_2)$ $(n = 2)$

    (b) $\phi(x_1, x_2) = (x_1 + x_2, x_1 - x_2, 2x_1)$
         $\psi(x_1, x_2, x_3) = (x_3, x_2, x_1)$ $(n = 2)$

    (c) $\phi(x_1, x_2, x_3) = (x_1 + x_2 + x_3, x_1 + x_2, x_1)$
         $\psi(x_1, x_2, x_3) = (x_1 - x_2, x_2 - x_3, x_3 - x_1)$ $(n = 3)$

    (d) $\phi(x_1, x_2, x_3) = (x_1 + x_2, x_2 + x_3)$
         $\psi(x_1, x_2) = x_1 - x_2$ $(n = 3)$

    (e) $\phi(x) = (x, -x, 2x, 3x)$
         $\psi(x_1, x_2, x_3, x_4) = x_1 + x_2 + x_3 + x_4$ $(n = 1)$

8. Use the method of Example 4 to derive trigonometric identities for the following:

    (a) $\sin(\alpha - \beta)$ and $\cos(\alpha - \beta)$

    (b) $\sin 2\alpha$ and $\cos 2\alpha$

9. Let $\phi: V \to W$ be a linear map and let $c \in \mathbb{R}$. Show that $c\phi: V \to W$ is a linear map.

10. Let V and W be vector spaces and let $\mathcal{L}(V, W)$ denote the set of all linear maps from V to W. Show that, with the operations of addition and scalar multiplication defined in this section, $\mathcal{L}(V, W)$ is a vector space.

11. Verify that if $\phi: \mathbb{R}^n \to \mathbb{R}^m$ is a linear map and $c \in \mathbb{R}$ then the matrix of $c\phi$ is cA where A is the matrix of ϕ.

12. Show that if $\phi: U \to V$ and $\psi: V \to W$ are linear maps then $\psi \circ \phi: U \to W$ is also a linear map.

13. (a) Let $\phi_1, \phi_2: U \to V$ and $\psi: V \to W$ be linear maps. Show that $\psi \circ (\phi_1 + \phi_2) = \psi \circ \phi_1 + \psi \circ \phi_2$.

    (b) Let $\phi_1: V \to W$, $\phi_2: U \to V$, $\phi_3: T \to U$ be linear maps. Show that
         $$\phi_1 \circ (\phi_2 \circ \phi_3) = (\phi_1 \circ \phi_2) \circ \phi_3.$$

    (c) Let $A \in \mathcal{M}_{m \times k}$ and $B, C \in \mathcal{M}_{k \times n}$. Use part (a) of this exercise and Theorems 1 and 2 to prove that
         $$A(B + C) = AB + AC.$$

    (d) Let $A \in \mathcal{M}_{k \times l}$, $B \in \mathcal{M}_{l \times m}$, and $C \in \mathcal{M}_{m \times n}$. Use part (b) of this exercise and Theorem 2 to prove that
         $$A(BC) = (AB)C.$$

14. Let σ be a permutation of the integers $\{1, \ldots, n\}$ and let M_σ be the matrix of σ as in Section 5.3.

   (a) Show that the linear map $\phi_\sigma : \mathbb{R}^n \to \mathbb{R}^n$ associated with the matrix M_σ is given by the formula

   $$\phi_\sigma(x_1, \ldots, x_n) = (x_{\sigma(1)}, \ldots, x_{\sigma(n)}).$$

   (b) Show that if $A = \begin{pmatrix} \mathbf{a}_1 \\ \vdots \\ \mathbf{a}_n \end{pmatrix} \in \mathfrak{M}_{n \times k}$, then $M_\sigma A = \begin{pmatrix} \mathbf{a}_{\sigma(1)} \\ \vdots \\ \mathbf{a}_{\sigma(n)} \end{pmatrix}$.

15. (a) Show that if $A \in \mathfrak{M}_{m \times n}$ and

   $$E = \begin{pmatrix} \mathbf{e}_1 \\ \cdots \\ c\mathbf{e}_i \\ \cdots \\ \mathbf{e}_m \end{pmatrix} \in \mathfrak{M}_{m \times m}$$

   then EA is the matrix obtained from A by multiplying the ith row by c.

   (b) Show that if $A \in \mathfrak{M}_{m \times n}$ and

   $$E = \begin{pmatrix} \mathbf{e}_1 \\ \cdots \\ \mathbf{e}_j \\ \cdots \\ \mathbf{e}_i \\ \cdots \\ \mathbf{e}_m \end{pmatrix} \quad \substack{i\text{th row} \\ \\ j\text{th row}} \in \mathfrak{M}_{m \times m}$$

   then EA is the matrix obtained from A by interchanging the ith and the jth rows.

   (c) Show that if $A \in \mathfrak{M}_{m \times n}$ and

   $$E = \begin{pmatrix} \mathbf{e}_1 \\ \cdots \\ \mathbf{e}_i + c\mathbf{e}_j \\ \cdots \\ \mathbf{e}_m \end{pmatrix} \quad i\text{th row}$$

   then EA is the matrix obtained from A by adding c times the jth row to the ith row.

   Remark. Matrices of the above three types are called **elementary matrices.**

   (d) Conclude that for each matrix $A \in \mathfrak{M}_{m \times n}$ there is a sequence $E_1, \ldots, E_k$ of elementary matrices such that $E_k \cdots E_1 A$ is the row echelon matrix of A.

   (e) Show that if $A \in \mathfrak{M}_{n \times n}$ has rank n, then A is a product of elementary matrices.

16. (a) Show that if $E \in \mathfrak{M}_{n \times n}$ is any elementary matrix (see Exercise 15), then $\det(EA) = (\det E)(\det A)$ for all $A \in \mathfrak{M}_{n \times n}$.

   (b) Show that, if A and B are any two $n \times n$ matrices, then

   $$\det(AB) = (\det A)(\det B).$$

   [Hint: If A has rank n, write A as a product of elementary matrices as in Exercise 15(e). If A has rank $< n$, argue that both sides of this equation must be equal to zero.]

7.3 INVERSE MAPS

Let V and W be vector spaces. Linear maps $\phi: V \to W$ and $\psi: W \to V$ are said to be *inverse* to each other if

$$\psi \circ \phi = i_V \quad \text{and} \quad \phi \circ \psi = i_W$$

where $i_V: V \to V$ and $i_W: W \to W$ are the identity maps of V and W respectively. Thus $\phi: V \to W$ and $\psi: W \to V$ are inverse to each other if each reverses the effect of the other:

(i)   if $\phi(\mathbf{v}) = \mathbf{w}$, then $\psi(\mathbf{w}) = \mathbf{v}$

and

(ii)   if $\psi(\mathbf{w}) = \mathbf{v}$, then $\phi(\mathbf{v}) = \mathbf{w}$.

Example 1.   (a) The linear maps $\phi(\mathbf{v}) = 2\mathbf{v}$ and $\psi(\mathbf{v}) = \frac{1}{2}\mathbf{v}$ ($\phi, \psi: V \to V$) are inverse to each other.

(b) If $r_\theta: \mathbb{R}^2 \to \mathbb{R}^2$ denotes rotation through the angle θ, then r_θ and $r_{-\theta}$ are inverse to each other.

(c) If $\phi: \mathbb{R}^4 \to \mathfrak{M}_{2\times 2}$ and $\psi: \mathfrak{M}_{2\times 2} \to \mathbb{R}^4$ are defined by

$$\phi(x_1, x_2, x_3, x_4) = \begin{pmatrix} x_1 & x_2 \\ x_3 & x_4 \end{pmatrix} \quad \text{and} \quad \psi \begin{pmatrix} a & b \\ c & d \end{pmatrix} = (a, b, c, d)$$

then ϕ and ψ are inverse to each other.

(d) Let $V = \mathcal{C}(\mathbb{R})$ and let W be the vector space of all continuously differentiable functions in $\mathfrak{F}(\mathbb{R})$. If $\phi: V \to W$ and $\psi: W \to V$ are defined by

$$\phi(f)(x) = \int_0^x f \quad \text{and} \quad \psi(f) = f'$$

then $\psi \circ \phi = i_V$. But $\phi \circ \psi \neq i_W$ since, for example,

$$\phi(\psi(\cos x)) = \phi(-\sin x) = \cos x - 1,$$

so these linear maps are *not* inverse to each other. ∎

Given a linear map $\phi: V \to W$, there is at most one linear map $\psi: W \to V$ such that ϕ and ψ are inverse to each other. For suppose the linear maps $\psi_1: W \to V$ and $\psi_2: W \to V$ both satisfy

$$\psi_i \circ \phi = i_V \quad \text{and} \quad \phi \circ \psi_i = i_W \quad (i = 1, 2).$$

Then

$$\psi_2 = i_V \circ \psi_2 = (\psi_1 \circ \phi) \circ \psi_2 = \psi_1 \circ (\phi \circ \psi_2) = \psi_1 \circ i_W = \psi_1.$$

A linear map $\phi\colon V \to W$ is called **invertible** if there exists a linear map $\psi\colon W \to V$ such that ϕ and ψ are inverse to each other. The map ψ, if it exists, is called the **inverse** ϕ^{-1} of ϕ.

The next theorem shows that a linear map $\phi\colon V \to W$ cannot possibly have an inverse unless $\dim V = \dim W$. Then Theorem 2 shows that, if $\dim V = \dim W$, each of the conditions $\psi \circ \phi = i_V$ and $\phi \circ \psi = i_W$ implies the other. Therefore, in order to check that $\phi\colon V \to W$ and $\psi\colon W \to V$ are inverse to each other, it suffices to check only one of these conditions.

Theorem 1. *Let V and W be finite dimensional vector spaces. If there exists a linear map from V to W that is invertible, then V and W must have the same dimension.*

Proof. Suppose there exists an invertible linear map $\phi\colon V \to W$. Let $\{\mathbf{v}_1, \ldots, \mathbf{v}_n\}$ be a basis for V. Then:

(i) we shall use the fact that $\phi \circ \psi = i_W$ for some linear map $\psi\colon W \to V$ to prove that $\{\phi(\mathbf{v}_1), \ldots, \phi(\mathbf{v}_n)\}$ spans W, and

(ii) we shall use the fact that $\psi \circ \phi = i_V$ for some linear map $\psi\colon W \to V$ to prove that $\{\phi(\mathbf{v}_1), \ldots, \phi(\mathbf{v}_n)\}$ is linearly independent.

From (i) and (ii) we can then conclude that $\{\phi(\mathbf{v}_1), \ldots, \phi(\mathbf{v}_n)\}$ is a basis for W and hence that $\dim W = n = \dim V$, as asserted.

First we prove (i): Let $\mathbf{w} \in W$. Then $\psi(\mathbf{w}) \in V$ so

$$\psi(\mathbf{w}) = c_1\mathbf{v}_1 + \cdots + c_n\mathbf{v}_n$$

for some $c_1, \ldots, c_n \in \mathbb{R}$. Hence

$$\mathbf{w} = i_W(\mathbf{w}) = \phi \circ \psi(\mathbf{w}) = \phi(c_1\mathbf{v}_1 + \cdots + c_n\mathbf{v}_n)$$

or

$$\mathbf{w} = c_1\phi(\mathbf{v}_1) + \cdots + c_n\phi(\mathbf{v}_n).$$

Thus each $\mathbf{w} \in W$ is a linear combination of the vectors $\phi(\mathbf{v}_1), \ldots, \phi(\mathbf{v}_n)$ and so $\{\phi(\mathbf{v}_1), \ldots, \phi(\mathbf{v}_n)\}$ does span W.

Now we prove (ii): Suppose

$$c_1\phi(\mathbf{v}_1) + \cdots + c_n\phi(\mathbf{v}_n) = \mathbf{0}$$

for some $c_1, \ldots, c_n \in \mathbb{R}$. We must show that $c_1 = \cdots = c_n = 0$. Applying ψ to both sides of the above equation yields

$$\psi(c_1\phi(\mathbf{v}_1) + \cdots + c_n\phi(\mathbf{v}_n)) = \psi(\mathbf{0}).$$

Since ψ is linear, this equation becomes

$$c_1\psi(\phi(\mathbf{v}_1)) + \cdots + c_n\psi(\phi(\mathbf{v}_n)) = \mathbf{0}.$$

Finally, since $\psi \circ \phi = i_V$, we get

$$c_1\mathbf{v}_1 + \cdots + c_n\mathbf{v}_n = \mathbf{0}.$$

But this implies that

$$c_1 = c_2 = \cdots = c_n = 0$$

since $\{\mathbf{v}_1, \ldots, \mathbf{v}_n\}$ is linearly independent. Thus $\{\phi(\mathbf{v}_1), \ldots, \phi(\mathbf{v}_n)\}$ is linearly independent, as was to be shown. ∎

Theorem 2. *Suppose V and W are two finite dimensional vector spaces of the same dimension and suppose $\phi\colon V \to W$ and $\psi\colon W \to V$ are linear maps. Then $\psi \circ \phi = i_V$ if and only if $\phi \circ \psi = i_W$. In particular, either one of these conditions is sufficient, when $\dim V = \dim W$, to guarantee that ϕ is invertible and that $\psi = \phi^{-1}$.*

Proof. Let $\{\mathbf{v}_1, \ldots, \mathbf{v}_n\}$ be a basis for V.

If $\psi \circ \phi = i_V$ then, by (ii) in the proof of Theorem 1, we know that $\{\phi(\mathbf{v}_1), \ldots, \phi(\mathbf{v}_n)\}$ is a linearly independent set in W, hence a basis for W (Theorem 2 of Section 4.4). But $\phi \circ \psi$ and i_W agree on this basis:

$$\phi \circ \psi(\phi(\mathbf{v}_i)) = \phi(\psi \circ \phi(\mathbf{v}_i)) = \phi(\mathbf{v}_i) = i_W(\phi(\mathbf{v}_i)).$$

By Theorem 1 of Section 7.1, this implies that $\phi \circ \psi = i_W$.

On the other hand, if $\phi \circ \psi = i_W$ then, by (i) in the proof of Theorem 1, $\{\phi(\mathbf{v}_1), \ldots, \phi(\mathbf{v}_n)\}$ spans W, and hence is a basis for W (Theorem 2 of Section 4.4). Let $\tilde{\psi}\colon W \to V$ be the linear map such that $\tilde{\psi}(\phi(\mathbf{v}_i)) = \mathbf{v}_i$ for each i. Then $\tilde{\psi} \circ \phi = i_V$. Hence $\tilde{\psi} = \tilde{\psi} \circ i_W = \tilde{\psi} \circ \phi \circ \psi = i_V \circ \psi = \psi$. Thus

$$\psi \circ \phi = \tilde{\psi} \circ \phi = i_V$$

as asserted. ∎

We now consider the question: for which matrices $A \in \mathfrak{M}_{m \times n}$ are the linear maps $\phi_A\colon \mathbb{R}^n \to \mathbb{R}^m$ invertible? By Theorem 1, invertibility of ϕ_A implies that $m = n$, so A must be a square matrix. Furthermore, if $\phi_A\colon \mathbb{R}^n \to \mathbb{R}^n$ is invertible, then $\phi_A^{-1} = \phi_B$ for some $B \in \mathfrak{M}_{n \times n}$. Since the identity map $i\colon \mathbb{R}^n \to \mathbb{R}^n$ is ϕ_I, where $I \in \mathfrak{M}_{n \times n}$ is the identity matrix, the equations

$$\phi_A \circ \phi_B = i \quad \text{and} \quad \phi_B \circ \phi_A = i$$

are equivalent to the matrix equations

$$AB = I \quad \text{and} \quad BA = I.$$

Hence we conclude that $\phi_A \colon \mathbb{R}^n \to \mathbb{R}^n$ is invertible if and only if there is a matrix $B \in \mathfrak{M}_{n \times n}$ such that $AB = BA = I$. A matrix $A \in \mathfrak{M}_{n \times n}$ with this property is called an **invertible matrix.** The matrix $B \in \mathfrak{M}_{n \times n}$ such that $AB = BA = I$ is called the **inverse** A^{-1} of A.

Remark. Since, by Theorem 2, $\phi_A \circ \phi_B = i$ if and only if $\phi_B \circ \phi_A = i$, we see that $AB = I$ if and only if $BA = I$. Thus, to find an inverse for a square matrix A we need only find a matrix B such that *either $AB = I$ or $BA = I$*. The other equation follows automatically.

The next theorem is an immediate consequence of the above discussion.

Theorem 3.   *Let $\phi \colon \mathbb{R}^n \to \mathbb{R}^n$ be a linear map and let $A \in \mathfrak{M}_{n \times n}$ be its matrix. Then ϕ is invertible if and only if A is invertible. If ϕ is invertible, then the matrix of ϕ^{-1} is A^{-1}.*

Example 2.   It is geometrically clear that the linear map $r_\theta \colon \mathbb{R}^2 \to \mathbb{R}^2$ (rotation through the angle θ) is invertible and that its inverse is rotation through the angle $-\theta \colon r_\theta^{-1} = r_{-\theta}$. We know that $r_\theta = \phi_A$, where

$$A = \begin{pmatrix} \cos\theta & -\sin\theta \\ \sin\theta & \cos\theta \end{pmatrix}.$$

It therefore follows that

$$A^{-1} = \begin{pmatrix} \cos(-\theta) & -\sin(-\theta) \\ \sin(-\theta) & \cos(-\theta) \end{pmatrix}$$

or

$$A^{-1} = \begin{pmatrix} \cos\theta & \sin\theta \\ -\sin\theta & \cos\theta \end{pmatrix}.$$

You should verify directly that $AA^{-1} = A^{-1}A = I$. ■

In the preceding example, it was easy to find A^{-1}, because we knew the linear map ϕ_A^{-1}. We now develop an algorithm for finding A^{-1} where A is any invertible matrix. This procedure will also tell us *whether* a given square matrix is invertible.

Let $A \in \mathfrak{M}_{n \times n}$. We would like to find (if there is one) a matrix $B \in \mathfrak{M}_{n \times n}$ such that $AB = I$. Let $\mathbf{a}_1, \ldots, \mathbf{a}_n$ be the *row* vectors of A, and let $\mathbf{b}_1, \ldots, \mathbf{b}_n$ be the *column* vectors of B. The condition $AB = I$ is precisely the statement that

$$\mathbf{a}_i \cdot \mathbf{b}_j = \begin{cases} 1, & \text{if } i = j, \\ 0, & \text{if } i \neq j. \end{cases}$$

In other words, to find $\mathbf{b}_j$, we need to solve the system of linear equations

$$\mathbf{a}_1 \cdot \mathbf{x} = 0$$

$$\cdots$$

$$\mathbf{a}_j \cdot \mathbf{x} = 1$$

$$\cdots$$

$$\mathbf{a}_n \cdot \mathbf{x} = 0.$$

The solution vector of this system will be the jth column of B. This system has a unique solution if and only if the coefficient matrix A has rank n; that is, if and only if A can be row reduced to the identity matrix I. Suppose A does have rank n. The matrix of the system then reduces as follows:

$$(A \,|\, \mathbf{e}_j) \rightarrow (I \,|\, \mathbf{b}_j).$$

Since we need to carry out the row reduction for each $j = 1, \ldots, n$, we might as well do it all at once:

$$(A \,|\, I) = (A \,|\, \mathbf{e}_1, \ldots, \mathbf{e}_n) \rightarrow (I \,|\, \mathbf{b}_1, \ldots, \mathbf{b}_n) = (I \,|\, B).$$

Thus we have proved the following theorem.

Theorem 4. *Let A be an $n \times n$ matrix. Then A is invertible if and only if A has rank n. The inverse matrix A^{-1} of an invertible matrix A can be found by reducing the matrix $(A \,|\, I)$ to row echelon form:*

$$(A \,|\, I) \xrightarrow[\text{reduction}]{\text{row}} (I \,|\, A^{-1}).$$

Example 3. If

$$A = \begin{pmatrix} 2 & 3 & 1 \\ -1 & 1 & 0 \\ 1 & 0 & 1 \end{pmatrix},$$

then row reduction of $(A \,|\, I)$ yields

$$\left(\begin{array}{ccc|ccc} 2 & 3 & 1 & 1 & 0 & 0 \\ -1 & 1 & 0 & 0 & 1 & 0 \\ 1 & 0 & 1 & 0 & 0 & 1 \end{array} \right) \rightarrow \left(\begin{array}{ccc|ccc} 1 & 0 & 1 & 0 & 0 & 1 \\ -1 & 1 & 0 & 0 & 1 & 0 \\ 2 & 3 & 1 & 1 & 0 & 0 \end{array} \right)$$

$$\rightarrow \left(\begin{array}{ccc|ccc} 1 & 0 & 1 & 0 & 0 & 1 \\ 0 & 1 & 1 & 0 & 1 & 1 \\ 0 & 3 & -1 & 1 & 0 & -2 \end{array} \right) \rightarrow \left(\begin{array}{ccc|ccc} 1 & 0 & 1 & 0 & 0 & 1 \\ 0 & 1 & 1 & 0 & 1 & 1 \\ 0 & 0 & -4 & 1 & -3 & -5 \end{array} \right)$$

$$\rightarrow \left(\begin{array}{ccc|ccc} 1 & 0 & 1 & 0 & 0 & 1 \\ 0 & 1 & 1 & 0 & 1 & 1 \\ 0 & 0 & 1 & -\frac{1}{4} & \frac{3}{4} & \frac{5}{4} \end{array} \right) \rightarrow \left(\begin{array}{ccc|ccc} 1 & 0 & 0 & \frac{1}{4} & -\frac{3}{4} & -\frac{1}{4} \\ 0 & 1 & 0 & \frac{1}{4} & \frac{1}{4} & -\frac{1}{4} \\ 0 & 0 & 1 & -\frac{1}{4} & \frac{3}{4} & \frac{5}{4} \end{array} \right).$$

Hence A is invertible and $A^{-1} = \begin{pmatrix} \frac{1}{4} & -\frac{3}{4} & -\frac{1}{4} \\ \frac{1}{4} & \frac{1}{4} & -\frac{1}{4} \\ -\frac{1}{4} & \frac{3}{4} & \frac{5}{4} \end{pmatrix}$.   ■

Example 4.   If $A = \begin{pmatrix} 1 & 2 \\ 2 & 4 \end{pmatrix}$, then row reduction of $(A \mid I)$ yields

$$\begin{pmatrix} 1 & 2 & 1 & 0 \\ 2 & 4 & 0 & 1 \end{pmatrix} \rightarrow \begin{pmatrix} 1 & 2 & 1 & 0 \\ 0 & 0 & -2 & 1 \end{pmatrix},$$

at which point it is clear that rank $A \neq 2$. Thus A is not invertible.   ■

Example 5.   We can use the algorithm for matrix inversion to find the inverse of any invertible linear map $\phi \colon \mathbb{R}^n \to \mathbb{R}^n$. Suppose, for example, that $\phi \colon \mathbb{R}^2 \to \mathbb{R}^2$ is given by

$$\phi(x_1, x_2) = (2x_1 - 3x_2, -x_1 + x_2).$$

Then $\phi = \phi_A$ where

$$A = \begin{pmatrix} 2 & -3 \\ -1 & 1 \end{pmatrix}.$$

Since

$$\begin{pmatrix} 2 & -3 & 1 & 0 \\ -1 & 1 & 0 & 1 \end{pmatrix} \rightarrow \begin{pmatrix} 1 & 0 & -1 & -3 \\ 0 & 1 & -1 & -2 \end{pmatrix},$$

we find that

$$A^{-1} = \begin{pmatrix} -1 & -3 \\ -1 & -2 \end{pmatrix}$$

and hence

$$\phi^{-1}(x_1, x_2) = \phi_{A^{-1}}(x_1, x_2) = (-x_1 - 3x_2, -x_1 - 2x_2).   ■$$

We can use matrix inverses to neatly describe solutions of $n \times n$ linear systems when the coefficient matrix has rank n. Such a system can be written as

$$A\mathbf{x} = \mathbf{b}$$

where $A \in \mathfrak{M}_{n \times n}$. If A has rank n, then A is invertible so the unique solution of $A\mathbf{x} = \mathbf{b}$ is

$$\mathbf{x} = A^{-1}\mathbf{b}.$$

This formula is especially useful when we are required to solve $A\mathbf{x} = \mathbf{b}$ for many different vectors $\mathbf{b}$.

Example 6.   Let us solve the following linear systems:

(a)
$$3x_1 + 5x_2 = 1$$
$$4x_1 + 7x_2 = 2$$

(b)
$$3x_1 + 5x_2 = -3$$
$$4x_1 + 7x_2 = 8$$

(c)
$$3x_1 + 5x_2 = \pi$$
$$4x_1 + 7x_2 = -\pi$$

All these systems are of the form $A\mathbf{x} = \mathbf{b}$ where $A = \begin{pmatrix} 3 & 5 \\ 4 & 7 \end{pmatrix}$. Since $A^{-1} = \begin{pmatrix} 7 & -5 \\ -4 & 3 \end{pmatrix}$, we can rapidly compute the solutions $\mathbf{x} = A^{-1}\mathbf{b}$:

$$\text{(a)}\quad \mathbf{x} = \begin{pmatrix} 7 & -5 \\ -4 & 3 \end{pmatrix}\begin{pmatrix} 1 \\ 2 \end{pmatrix} = \begin{pmatrix} -3 \\ 2 \end{pmatrix}$$

$$\text{(b)}\quad \mathbf{x} = \begin{pmatrix} 7 & -5 \\ -4 & 3 \end{pmatrix}\begin{pmatrix} -3 \\ 8 \end{pmatrix} = \begin{pmatrix} -61 \\ 36 \end{pmatrix}$$

$$\text{(c)}\quad \mathbf{x} = \begin{pmatrix} 7 & -5 \\ -4 & 3 \end{pmatrix}\begin{pmatrix} \pi \\ -\pi \end{pmatrix} = \begin{pmatrix} 12\pi \\ -7\pi \end{pmatrix}. \quad \blacksquare$$

We can use matrix multiplication and matrix inverses to describe least squares solutions of inconsistent linear systems.

Consider the linear system

$$a_{11}x_1 + \cdots + a_{1n}x_n = b_1$$
$$\cdots$$
$$a_{m1}x_1 + \cdots + a_{mn}x_n = b_m.$$

We can rewrite this system in matrix form as

$$A\mathbf{x} = \mathbf{b}$$

where

$$A = \begin{pmatrix} a_{11} & \cdots & a_{1n} \\ & \cdots & \\ a_{m1} & \cdots & a_{mn} \end{pmatrix}, \quad \mathbf{x} = \begin{pmatrix} x_1 \\ \vdots \\ x_n \end{pmatrix}, \quad \text{and } \mathbf{b} = \begin{pmatrix} b_1 \\ \vdots \\ b_m \end{pmatrix}.$$

Recall the discussion of least squares in Section 6.2. The equation $A\mathbf{x} = \mathbf{b}$ has a solution if and only if $\mathbf{b}$ is in the column space W of A. If $b \notin W$ then there is no $\mathbf{x} \in \mathfrak{M}_{n \times 1}$ such that $A\mathbf{x} = \mathbf{b}$, so we look for an $\mathbf{x} \in \mathfrak{M}_{n \times 1}$ that makes $A\mathbf{x}$ as close as possible to $\mathbf{b}$. Such an $\mathbf{x}$ is called a *least squares approximate solution* of the equation $A\mathbf{x} = \mathbf{b}$. As shown in Chapter 6, $\mathbf{x}$ is a least squares approximate solution of $A\mathbf{x} = \mathbf{b}$ if and only if $A\mathbf{x} = \mathrm{Proj}_W\mathbf{b}$.

Let $\mathbf{w} = \mathrm{Proj}_W\mathbf{b}$. Then $\mathbf{w}$ is the unique vector in W with the property that $\mathbf{w} + \mathbf{v} = \mathbf{b}$ for some vector $\mathbf{v}$ perpendicular to W. In other words, $\mathbf{w}$ is the unique vector in W such that $\mathbf{b} - \mathbf{w}$ is perpendicular to W. Said yet another way, $\mathbf{w}$ is the unique vector in W such that $\mathbf{a}_i \cdot (\mathbf{b} - \mathbf{w}) = 0$ for $i = 1, \ldots, n$, where $\mathbf{a}_1, \ldots, \mathbf{a}_n$ are the column vectors of A.

Thus $\mathbf{x}$ is a least squares approximate solution of $A\mathbf{x} = \mathbf{b}$ if and only if

$$A\mathbf{x} = \mathbf{w},$$

that is, if and only if

$$\mathbf{a}_i \cdot (\mathbf{b} - A\mathbf{x}) = 0$$

for $i = 1, \ldots, n$. But the left hand side of this equation is just the ith row of the matrix $A^t(\mathbf{b} - A\mathbf{x})$. Hence $\mathbf{x}$ is a least squares approximate solution of the equation $A\mathbf{x} = \mathbf{b}$ if and only if

$$A^t(\mathbf{b} - A\mathbf{x}) = 0,$$

that is, if and only if

$$A^tA\mathbf{x} = A^t\mathbf{b}.$$

If the square matrix A^tA is invertible, we can solve this equation explicitly for $\mathbf{x}$. We have proved the following theorem.

Theorem 5. *The least squares approximate solutions of the linear system $A\mathbf{x} = \mathbf{b}$ are the solutions of the equation*

$$A^tA\mathbf{x} = A^t\mathbf{b}.$$

If the matrix A^tA is invertible, then the least squares approximate solution is unique and is given by the formula

$$\mathbf{x} = (A^tA)^{-1}A^t\mathbf{b}.$$

Example 7.   Let us use the formula of Theorem 5 to find the least squares approximate solution of the system

$$x_1 + x_2 = 9$$
$$3x_1 - x_2 = 1$$
$$x_1 + 2x_2 = 10.$$

Here,

$$A = \begin{pmatrix} 1 & 1 \\ 3 & -1 \\ 1 & 2 \end{pmatrix} \quad \text{and} \quad \mathbf{b} = \begin{pmatrix} 9 \\ 1 \\ 10 \end{pmatrix}$$

so

$$A^tA = \begin{pmatrix} 1 & 3 & 1 \\ 1 & -1 & 2 \end{pmatrix} \begin{pmatrix} 1 & 1 \\ 3 & -1 \\ 1 & 2 \end{pmatrix} = \begin{pmatrix} 11 & 0 \\ 0 & 6 \end{pmatrix}$$

and

$$\mathbf{x} = (A^tA)^{-1}A^t\mathbf{b} = \begin{pmatrix} \frac{1}{11} & 0 \\ 0 & \frac{1}{6} \end{pmatrix} \begin{pmatrix} 1 & 3 & 1 \\ 1 & -1 & 2 \end{pmatrix} \begin{pmatrix} 9 \\ 1 \\ 10 \end{pmatrix}$$

$$= \begin{pmatrix} \frac{1}{11} & 0 \\ 0 & \frac{1}{6} \end{pmatrix} \begin{pmatrix} 22 \\ 28 \end{pmatrix} = \begin{pmatrix} 2 \\ \frac{14}{3} \end{pmatrix}.$$

Compare this solution with the solution to this same problem in Example 5 of Section 6.2. ∎

EXERCISES

1. Find the inverse of each of the following linear maps:
    (a) $\phi: V \to V$, $\phi(\mathbf{v}) = 2\mathbf{v}$
    (b) $\phi: V \to V$, $\phi(\mathbf{v}) = -\mathbf{v}$
    (c) $\phi: \mathbb{R}^2 \to \mathbb{R}^2$, $\phi(x_1, x_2) = (x_2, x_1)$
    (d) $\phi: \mathbb{R}^2 \to \mathbb{R}^2$, $\phi(x_1, x_2) = (2x_1, 3x_2)$
    (e) $\phi: \mathbb{R}^3 \to \mathcal{P}^3$, $\phi(a, b, c) = a + bx + cx^2$
    (f) $\phi: \mathcal{P}^2 \to \mathbb{R}^2$, $\phi(p(x)) = (p(0), p(1))$

2. Which of the following linear maps are invertible?
    (a) $D: \mathcal{P}^2 \to \mathcal{P}^2$ (b) $D + 1: \mathcal{P}^2 \to \mathcal{P}^2$
    (c) $D: \mathcal{P}^3 \to \mathcal{P}^2$ (d) $\phi: \mathcal{P}^2 \to \mathcal{P}^3$, $\phi(p(x)) = xp(x)$

3. Determine which of the following matrices are invertible and, for each that is, find its inverse.

(a) $\begin{pmatrix} 1 & 1 \\ 1 & -1 \end{pmatrix}$

(b) $\begin{pmatrix} 1 & 3 \\ -2 & 2 \end{pmatrix}$

(c) $\begin{pmatrix} 2 & 3 & -1 \\ 1 & 2 & 3 \\ -1 & -1 & 4 \end{pmatrix}$

(d) $\begin{pmatrix} 1 & -1 & 1 \\ -1 & 2 & -1 \\ 2 & -1 & 1 \end{pmatrix}$

(e) $\begin{pmatrix} 1 & 1 & 1 \\ 1 & 2 & 3 \\ 1 & 4 & 9 \end{pmatrix}$

(f) $\begin{pmatrix} 2 & 1 & 4 \\ 3 & 2 & 5 \\ 0 & -1 & 1 \end{pmatrix}$

4. Find the inverse of each of the given matrices.

(a) $\begin{pmatrix} 1 & a \\ 0 & 1 \end{pmatrix}$

(b) $\begin{pmatrix} a & 0 \\ 0 & b \end{pmatrix}$ where $ab \neq 0$

(c) $\begin{pmatrix} 1 & 0 & a \\ 0 & 1 & 0 \\ 0 & 0 & 1 \end{pmatrix}$

(d) $\begin{pmatrix} 1 & a & b \\ 0 & 1 & c \\ 0 & 0 & 1 \end{pmatrix}$

(e) $\begin{pmatrix} 1 & 0 & 0 & a \\ 0 & 1 & 0 & 0 \\ 0 & 0 & 1 & 0 \\ 0 & 0 & 0 & 1 \end{pmatrix}$

5. Find the inverse of

(a) $\begin{pmatrix} 1 & 2 & -1 & 3 \\ 0 & 1 & 4 & -2 \\ 0 & 0 & 1 & 5 \\ 0 & 0 & 0 & 1 \end{pmatrix}$

(b) $\begin{pmatrix} 1 & 2 & 3 & 4 \\ 1 & 3 & 2 & 6 \\ -2 & -3 & -6 & 0 \\ 1 & 4 & 1 & 7 \end{pmatrix}$

6. A square matrix A is invertible if and only if $\det A \neq 0$. Why?

7. Show that if A and $B \in \mathfrak{M}_{n \times n}$ are both invertible, then so is AB, and $(AB)^{-1} = B^{-1}A^{-1}$.

8. Use the method of Example 5 to find the inverse of each of the following linear maps.

(a) $\phi: \mathbb{R}^2 \to \mathbb{R}^2$, $\phi(x_1, x_2) = (x_1 + x_2, x_1)$

(b) $\phi: \mathbb{R}^2 \to \mathbb{R}^2$, $\phi(x_1, x_2) = (4x_1 + 3x_2, x_1 + x_2)$

(c) $\phi: \mathbb{R}^3 \to \mathbb{R}^3$, $\phi(x_1, x_2, x_3) = (\sqrt{3}x_2 + x_3, -x_2 + \sqrt{3}x_3, x_1)$

(d) $\phi: \mathbb{R}^n \to \mathbb{R}^n$, $\phi(x_1, \ldots, x_n) = (x_1, x_1 + x_2, \ldots, x_1 + x_2 + \cdots + x_n)$

9. Find A^{-1} and use it to solve $A\mathbf{x} = \mathbf{b}$, where $A = \begin{pmatrix} 3 & 4 \\ 17 & 23 \end{pmatrix}$ and $\mathbf{b} =$

(a) $\begin{pmatrix} 2 \\ 1 \end{pmatrix}$

(b) $\begin{pmatrix} -3 \\ 5 \end{pmatrix}$

(c) $\begin{pmatrix} \pi \\ 0 \end{pmatrix}$

(d) $\begin{pmatrix} 0 \\ 1 \end{pmatrix}$

10. Find A^{-1} and use it to solve $A\mathbf{x} = \mathbf{b}$, where $A = \begin{pmatrix} -3 & 1 & 1 \\ 2 & 6 & 5 \\ -5 & 7 & 6 \end{pmatrix}$ and $\mathbf{b} =$

(a) $\begin{pmatrix} 1 \\ -1 \\ 0 \end{pmatrix}$

(b) $\begin{pmatrix} 8 \\ 4 \\ 8 \end{pmatrix}$

(c) $\begin{pmatrix} -2 \\ 3 \\ 7 \end{pmatrix}$

(d) $\begin{pmatrix} 16 \\ -16 \\ 32 \end{pmatrix}$

11. Let σ be a permutation of the integers $\{1, \ldots, n\}$. Show that the matrix M_σ of σ is invertible and that $M_\sigma^{-1} = M_{\sigma^{-1}}$. [Hint: Use Exercise 14(a) of the previous section.]

12. Show that if $E \in \mathfrak{M}_{n \times n}$ is an elementary matrix (see Exercise 15 of the previous section), then E is invertible and E^{-1} is an elementary matrix.

13. Let $\phi: V \to W$ be a linear map and suppose $\psi: W \to V$ is a function such that $\psi \circ \phi = i_V$ and $\phi \circ \psi = i_W$. Show that ψ is a *linear* map.

14. Let $\phi: V \to W$ be a linear map. A linear map $\psi: W \to V$ is said to be a **left inverse** of ϕ if $\psi \circ \phi = i_V$. On the other hand, ψ is a **right inverse** of ϕ if $\phi \circ \psi = i_W$.

    (a) Show that if V and W are finite dimensional and ϕ has a left inverse, then $\dim V \leq \dim W$.

    (b) Show that if V and W are finite dimensional and ϕ has a right inverse, then $\dim V \geq \dim W$.

15. (a) Show that, for each $A \in \mathfrak{M}_{n \times n}$,

    $$\sum_{k=1}^{n} (-1)^{i+k} a_{ik} \det A_{jk} = \begin{cases} \det A & \text{if } i = j \\ 0 & \text{if } i \neq j \end{cases}$$

    where A_{ij} is the (i, j)-minor of A. [Hint: The left hand side is equal to the determinant of some matrix. If $i \neq j$, can you find two rows that are equal?]

    (b) Suppose $\det A \neq 0$. Show that

    $$(-1)^{i+j} \det A_{ji}/\det A$$

    is the (i, j)-entry of A^{-1}.

16. Use the formula of Exercise 15(b) to find the inverse of each of the following matrices:

    (a) $\begin{pmatrix} 2 & 3 \\ 1 & 2 \end{pmatrix}$ (b) $\begin{pmatrix} 2 & 0 \\ 1 & 3 \end{pmatrix}$ (c) $\begin{pmatrix} 1 & a \\ 0 & 1 \end{pmatrix}$

    (d) $\begin{pmatrix} 1 & a & b \\ 0 & 1 & c \\ 0 & 0 & 1 \end{pmatrix}$ (e) $\begin{pmatrix} 2 & 3 & -5 \\ 3 & -1 & 2 \\ 5 & 4 & -6 \end{pmatrix}$

17. Prove **Cramer's rule:** If $A \in \mathfrak{M}_{n \times n}$ and $\det A \neq 0$, then the unique solution of the equation $A\mathbf{x} = \mathbf{b}$ ($\mathbf{b} \in \mathfrak{M}_{n \times 1}$) is

    $$\mathbf{x} = \begin{pmatrix} x_1 \\ \vdots \\ x_n \end{pmatrix}$$

    where $x_j = \det A_j/\det A$ and A_j is the matrix obtained by replacing the jth column of A by $\mathbf{b}$. [Hint: Use Exercise 15(b) and the fact that $\mathbf{x} = A^{-1}\mathbf{b}$.]

18. Use Cramer's rule (Exercise 17) to solve each of the following linear systems:

    (a) $\begin{aligned} 3x_1 + 4x_2 &= 1 \\ 2x_1 + 5x_2 &= -1 \end{aligned}$ (c) $\begin{aligned} 2x_1 + x_2 + 5x_3 &= 0 \\ x_1 \qquad + 3x_3 &= 1 \\ -x_1 + 2x_2 \qquad &= -1 \end{aligned}$

    (b) $\begin{aligned} 7x_1 + 3x_2 &= 5 \\ -4x_1 + 2x_2 &= 1 \end{aligned}$ (d) $\begin{aligned} 2x_1 + 3x_2 - 5x_3 &= 2 \\ 3x_1 - x_2 + 2x_3 &= -1 \\ 5x_1 + 4x_2 - 6x_3 &= 3 \end{aligned}$

19. Find least squares approximate solutions to the following linear systems.

    (a) $\begin{aligned} 2x_1 + x_2 &= 1 \\ -x_1 + 3x_2 &= 1 \\ x_1 - x_2 &= 0 \end{aligned}$ (b) $\begin{aligned} 2x_1 - 3x_2 &= -1 \\ 3x_1 - 2x_2 &= 1 \\ x_1 + x_2 &= 1 \end{aligned}$

(c) $2x_1 + x_2 = 1$
$3x_1 + x_2 = 2$
$5x_1 + x_2 = 3$

(d) $3x_1 - 4x_2 = 6$
$-3x_1 + 4x_2 = -5$

(e) $x_1 + x_2 + x_3 = 1$
$x_1 \quad\quad + x_3 = 0$
$\quad x_2 + x_3 = 0$
$x_1 + x_2 \quad\quad = 0$

20. If $\mathbf{x}$ is an approximate solution to the linear system $A\mathbf{x} = \mathbf{b}$, then $E = \|A\mathbf{x} - \mathbf{b}\|$ measures how far $\mathbf{x}$ is from being a true solution. Compute E for the least squares approximate solutions to the systems of Exercise 19.

21. Show that the linear system $A\mathbf{x} = \mathbf{b}$ ($A \in \mathfrak{M}_{m \times n}$, $\mathbf{b} \in \mathfrak{M}_{m \times 1}$) has a *unique* least squares approximate solution if and only if the columns of A form a linearly independent set.

22. A square matrix A is **orthogonal** if its row vectors form an orthonormal set.

(a) Show that A is orthogonal if and only if $AA^t = I$.

(b) Show that A is orthogonal if and only if $A^{-1} = A^t$.

(c) Show that A is orthogonal if and only if A^t is orthogonal. (Thus the *rows* of A form an orthonormal set if and only if the *columns* of A form an orthonormal set.)

(d) Show that $A \in \mathfrak{M}_{n \times n}$ is orthogonal if and only if $\phi_A: \mathbb{R}^n \to \mathbb{R}^n$ has the property $\phi_A(\mathbf{u}) \cdot \phi_A(\mathbf{v}) = \mathbf{u} \cdot \mathbf{v}$ for all $\mathbf{u}, \mathbf{v} \in \mathbb{R}^n$. [Hint: Check this property first when $\mathbf{u} = \mathbf{e}_i$, $\mathbf{v} = \mathbf{e}_j$.]

(e) Show that $A \in \mathfrak{M}_{n \times n}$ is orthogonal if and only if $\phi_A: \mathbb{R}^n \to \mathbb{R}^n$ has the property $\|\phi_A(\mathbf{v})\| = \|\mathbf{v}\|$ for all $\mathbf{v} \in V$. [Hint: Use (d) and the fact that $\mathbf{u} \cdot \mathbf{v} = \frac{1}{4}(\|\mathbf{u} + \mathbf{v}\|^2 - \|\mathbf{u} - \mathbf{v}\|^2).$]

23. Prove the converse of Theorem 1: if V and W are finite dimensional vector spaces of the same dimension, then there exists an invertible linear map $\phi: V \to W$. [Hint: Let $\{\mathbf{v}_1, \ldots, \mathbf{v}_n\}$ be a basis for V and $\{\mathbf{w}_1, \ldots, \mathbf{w}_n\}$ be a basis for W. Define $\phi: V \to W$ to be the linear map such that $\phi(\mathbf{v}_i) = \mathbf{w}_i$ for $1 \leq i \leq n$. What is the inverse map?]

7.4 THE MATRIX OF A LINEAR MAP

We have seen that for every linear map $\phi: \mathbb{R}^n \to \mathbb{R}^m$ there is a matrix $A \in \mathfrak{M}_{m \times n}$ such that ϕ is equal to the linear map ϕ_A associated with A. If we write vectors in $\mathbb{R}^n$ and $\mathbb{R}^m$ vertically, this says that for every linear map $\phi: \mathbb{R}^n \to \mathbb{R}^m$ there is an $m \times n$ matrix A such that

$$\phi(\mathbf{x}) = A\mathbf{x}$$

for all $\mathbf{x} \in \mathbb{R}^n$. We shall now show that to any linear map $\phi: V \to W$, where V and W are *any* finite dimensional vector spaces, there is associated a matrix such that the effect of ϕ on vectors in V can be computed by matrix multiplication. The matrix associated with ϕ will depend on a choice of an ordered basis for V and one for W.

Recall that if $\mathbf{B} = (\mathbf{v}_1, \ldots, \mathbf{v}_n)$ is an ordered basis for V, then the **B**-coordinate n-tuple of $\mathbf{v} \in V$ is the n-tuple $(c_1, \ldots, c_n) \in \mathbb{R}^n$ such that

$$\mathbf{v} = c_1\mathbf{v}_1 + \cdots + c_n\mathbf{v}_n.$$

We define the **B**-*coordinate matrix* $M_{\mathbf{B}}(\mathbf{v})$ of $\mathbf{v}$ to be the **B**-coordinate n-tuple of $\mathbf{v}$ written vertically. Thus

$$M_{\mathbf{B}}(\mathbf{v}) = \begin{pmatrix} c_1 \\ \vdots \\ c_n \end{pmatrix}$$

where $\mathbf{v} = c_1\mathbf{v}_1 + \cdots + c_n\mathbf{v}_n$.

Given ordered bases $\mathbf{B} = (\mathbf{v}_1, \ldots, \mathbf{v}_n)$ for V and $\mathbf{B}' = (\mathbf{w}_1, \ldots, \mathbf{w}_m)$ for W and a linear map $\phi: V \to W$, we seek a matrix $A \in \mathfrak{M}_{m \times n}$ such that

$$M_{\mathbf{B}'}(\phi(\mathbf{v})) = AM_{\mathbf{B}}(\mathbf{v})$$

for all $\mathbf{v} \in V$. It is not difficult to see what the matrix A must be. Its jth column must be

$$A\begin{pmatrix} 0 \\ \vdots \\ 1 \\ \vdots \\ 0 \end{pmatrix} \overset{j\text{th row}}{=} AM_{\mathbf{B}}(\mathbf{v}_j) = M_{\mathbf{B}'}(\phi(\mathbf{v}_j)).$$

The $m \times n$ matrix, whose jth column ($j = 1, \ldots, n$) is the **B**'-coordinate matrix of $\phi(\mathbf{v}_j)$, is called the $(\mathbf{B}, \mathbf{B}')$-*coordinate matrix of* ϕ, or *the matrix of* ϕ *relative to the ordered bases* $\mathbf{B}, \mathbf{B}'$. This matrix is denoted $M^{\mathbf{B}}{}_{\mathbf{B}'}(\phi)$. Thus,

$$M^{\mathbf{B}}{}_{\mathbf{B}'}(\phi) = \begin{pmatrix} | & & | \\ M_{\mathbf{B}'}(\phi(\mathbf{v}_1)) & \cdots & M_{\mathbf{B}'}(\phi(\mathbf{v}_n)) \\ \downarrow & & \downarrow \end{pmatrix}$$

where $\mathbf{B} = (\mathbf{v}_1, \ldots, \mathbf{v}_n)$. That this matrix actually has the property required of it is the content of the following theorem.

Theorem 1. *Let V and W be finite dimensional vector spaces with ordered bases* $\mathbf{B}$ *and* $\mathbf{B}'$ *and let* $\phi: V \to W$ *be a linear map. Then, for each* $\mathbf{v} \in V$,

$$M_{\mathbf{B}'}(\phi(\mathbf{v})) = M^{\mathbf{B}}{}_{\mathbf{B}'}(\phi)\, M_{\mathbf{B}}(\mathbf{v}).$$

Proof. Let

$$M_{\mathbf{B}}(\mathbf{v}) = \begin{pmatrix} c_1 \\ \vdots \\ c_n \end{pmatrix}$$

Then

$$\mathbf{v} = c_1\mathbf{v}_1 + \cdots + c_n\mathbf{v}_n$$

so

$$\phi(\mathbf{v}) = c_1\phi(\mathbf{v}_1) + \cdots + c_n\phi(\mathbf{v}_n)$$

and

$$M_{\mathbf{B}'}(\phi(\mathbf{v})) = c_1 M_{\mathbf{B}'}(\phi(\mathbf{v}_1)) + \cdots + c_n M_{\mathbf{B}'}(\phi(\mathbf{v}_n))$$

$$= \left(M_{\mathbf{B}'}(\phi(\mathbf{v}_1)) \cdots M_{\mathbf{B}'}(\phi(\mathbf{v}_n)) \right)\begin{pmatrix} c_1 \\ \vdots \\ c_n \end{pmatrix}$$

$$= M^{\mathbf{B}}{}_{\mathbf{B}'}(\phi) M_{\mathbf{B}}(\mathbf{v}). \quad \blacksquare$$

Example 1. Let us find the $(\mathbf{B}, \mathbf{B}')$-coordinate matrix for $D\colon \mathcal{P}^4 \to \mathcal{P}^3$, where $\mathbf{B} = (1, x, x^2, x^3)$ and $\mathbf{B}' = (1, x, x^2)$. Since

$$D(1) = 0 = 0 \cdot 1 + 0x + 0x^2$$

$$D(x) = 1 = 1 \cdot 1 + 0x + 0x^2$$

$$D(x^2) = 2x = 0 \cdot 1 + 2x + 0x^2$$

$$D(x^3) = 3x^2 = 0 \cdot 1 + 0x + 3x^2$$

we find that

$$M^{\mathbf{B}}{}_{\mathbf{B}'}(D) = \left(M_{\mathbf{B}'}(D(1)) \quad M_{\mathbf{B}'}(D(x)) \quad M_{\mathbf{B}'}(D(x^2)) \quad M_{\mathbf{B}'}(D(x^3)) \right)$$

or

$$M^{\mathbf{B}}{}_{\mathbf{B}'}(D) = \begin{pmatrix} 0 & 1 & 0 & 0 \\ 0 & 0 & 2 & 0 \\ 0 & 0 & 0 & 3 \end{pmatrix}.$$

Notice that we can, if we wish, use this matrix together with Theorem 1 to compute the derivative of any degree 3 polynomial by matrix multiplication. If

$$p(x) = a_0 + a_1 x + a_2 x^2 + a_3 x^3$$

then

$$M_{\mathbf{B}}(p(x)) = \begin{pmatrix} a_0 \\ a_1 \\ a_2 \\ a_3 \end{pmatrix}$$

and

$$M_{\mathbf{B}'}(D(p(x))) = M^{\mathbf{B}}{}_{\mathbf{B}'}(D)\, M_{\mathbf{B}}(p(x)) = \begin{pmatrix} 0 & 1 & 0 & 0 \\ 0 & 0 & 2 & 0 \\ 0 & 0 & 0 & 3 \end{pmatrix} \begin{pmatrix} a_0 \\ a_1 \\ a_2 \\ a_3 \end{pmatrix} = \begin{pmatrix} a_1 \\ 2a_2 \\ 3a_3 \end{pmatrix}$$

so

$$D(p(x)) = a_1 + 2a_2 x + 3a_3 x^2. \quad \blacksquare$$

Example 2. Consider the linear map $\phi: \mathcal{P}^4 \to \mathcal{P}^4$, $\phi(p(x)) = p(x + 1)$. (You should verify that this map is, in fact, linear.) To find the matrix $M^{\mathbf{B}}{}_{\mathbf{B}'}(\phi)$, where $\mathbf{B} = \mathbf{B}' = (x^3, x^2, x, 1)$, we compute

$$\phi(x^3) = (x + 1)^3 = 1x^3 + 3x^2 + 3x + 1 \cdot 1$$
$$\phi(x^2) = (x + 1)^2 = 0x^3 + 1x^2 + 2x + 1 \cdot 1$$
$$\phi(x) = x + 1 = 0x^3 + 0x^2 + 1x + 1 \cdot 1$$
$$\phi(1) = 1 = 0x^3 + 0x^2 + 0x + 1 \cdot 1$$

and read off the entries:

$$M^{\mathbf{B}}{}_{\mathbf{B}'}(\phi) = \begin{pmatrix} 1 & 0 & 0 & 0 \\ 3 & 1 & 0 & 0 \\ 3 & 2 & 1 & 0 \\ 1 & 1 & 1 & 1 \end{pmatrix}$$

Using this matrix, we can quickly compute $p(x + 1)$ for any degree 3 polynomial $p(x)$. For example, if $p(x) = x^3 + 4x^2 - 3x + 2$, then

$$M_{\mathbf{B}'}(p(x + 1)) = M^{\mathbf{B}}{}_{\mathbf{B}'}(\phi)\, M_{\mathbf{B}}(p(x)) = \begin{pmatrix} 1 & 0 & 0 & 0 \\ 3 & 1 & 0 & 0 \\ 3 & 2 & 1 & 0 \\ 1 & 1 & 1 & 1 \end{pmatrix} \begin{pmatrix} 1 \\ 4 \\ -3 \\ 2 \end{pmatrix} = \begin{pmatrix} 1 \\ 7 \\ 8 \\ 4 \end{pmatrix}$$

so

$$p(x + 1) = x^3 + 7x^2 + 8x + 4. \quad \blacksquare$$

Theorem 2. *Let V and W be finite dimensional vector spaces with ordered bases* **B** *and* **B**', *let $\phi: V \to W$ and $\psi: V \to W$ be linear maps, and let $c \in \mathbb{R}$. Then*

(i) $$M^{\mathbf{B}}{}_{\mathbf{B}'}(\phi + \psi) = M^{\mathbf{B}}{}_{\mathbf{B}'}(\phi) + M^{\mathbf{B}}{}_{\mathbf{B}'}(\psi)$$

(ii) $$M^{\mathbf{B}}{}_{\mathbf{B}'}(c\phi) = c\, M^{\mathbf{B}}{}_{\mathbf{B}'}(\phi).$$

Moreover, if X is another finite dimensional vector space with ordered basis $\mathbf{B}''$ *and* $\zeta\colon W \to X$ *is a linear map, then*

(iii) $M^{\mathbf{B}}{}_{\mathbf{B}''}(\zeta \circ \phi) = M^{\mathbf{B}'}{}_{\mathbf{B}''}(\zeta)\, M^{\mathbf{B}}{}_{\mathbf{B}'}(\phi).$

Finally, if $\phi\colon V \to W$ *is invertible then*

(iv) $M^{\mathbf{B}'}{}_{\mathbf{B}}(\phi^{-1}) = (M^{\mathbf{B}}{}_{\mathbf{B}'}(\phi))^{-1}.$

Proof. (i) follows from the fact that the jth column of $M^{\mathbf{B}}{}_{\mathbf{B}'}(\phi + \psi)$ is equal to

$$M_{\mathbf{B}'}((\phi + \psi)(\mathbf{v}_j)) = M_{\mathbf{B}'}(\phi(\mathbf{v}_j) + \psi(\mathbf{v}_j))$$
$$= M_{\mathbf{B}'}(\phi(\mathbf{v}_j)) + M_{\mathbf{B}'}(\psi(\mathbf{v}_j))$$

where $\mathbf{v}_j$ is the jth basis vector in $\mathbf{B}$. The proof of (ii) is similar.

For (iii) observe that, for each $\mathbf{v} \in V$ we have

$$M^{\mathbf{B}}{}_{\mathbf{B}''}(\zeta \circ \phi)\, M_{\mathbf{B}}(\mathbf{v}) = M_{\mathbf{B}''}(\zeta \circ \phi(\mathbf{v}))$$
$$= M_{\mathbf{B}''}(\zeta(\phi(\mathbf{v})))$$
$$= M^{\mathbf{B}'}{}_{\mathbf{B}''}(\zeta)\, M_{\mathbf{B}'}(\phi(\mathbf{v}))$$
$$= M^{\mathbf{B}'}{}_{\mathbf{B}''}(\zeta)\, M^{\mathbf{B}}{}_{\mathbf{B}'}(\phi)\, M_{\mathbf{B}}(\mathbf{v}).$$

Taking $\mathbf{v} = \mathbf{v}_j$, so that $M_B(\mathbf{v}) = \begin{pmatrix} 0 \\ \vdots \\ 1 \\ \vdots \\ 0 \end{pmatrix}\leftarrow j\text{th row}$, this equation says that, for each

j, the jth column of $M^{\mathbf{B}}{}_{\mathbf{B}''}(\zeta \circ \phi)$ equals the jth column of the product $M^{\mathbf{B}'}{}_{\mathbf{B}''}(\zeta)\, M^{\mathbf{B}}{}_{\mathbf{B}'}(\phi)$. This proves (iii).

Finally, (iv) follows from (iii) since

$$M^{\mathbf{B}'}{}_{\mathbf{B}}(\phi^{-1})\, M^{\mathbf{B}}{}_{\mathbf{B}'}(\phi) = M^{\mathbf{B}}{}_{\mathbf{B}}(\phi^{-1} \circ \phi) = M^{\mathbf{B}}{}_{\mathbf{B}}(i_V) = I. \quad\blacksquare$$

Using Theorem 2, we can calculate how the matrix of a linear map changes when we change the ordered bases being used. We shall carry out the calculation here only in the important special case where $V = W$.

Suppose V is a finite dimensional vector space and that $\mathbf{B}_1$ and $\mathbf{B}_2$ are two ordered bases for V. Then for any linear map $\phi\colon V \to V$ we can calculate the matrices $M_1 = M^{\mathbf{B}_1}{}_{\mathbf{B}_1}(\phi)$ and $M_2 = M^{\mathbf{B}_2}{}_{\mathbf{B}_2}(\phi)$. We seek a formula relating these two matrices. Since $i_V \circ \phi = \phi = \phi \circ i_V$, we can apply Theorem 2(iii) to obtain

$$M^{\mathbf{B}_2}{}_{\mathbf{B}_1}(i_V)\, M^{\mathbf{B}_2}{}_{\mathbf{B}_2}(\phi) = M^{\mathbf{B}_2}{}_{\mathbf{B}_1}(\phi) = M^{\mathbf{B}_1}{}_{\mathbf{B}_1}(\phi)\, M^{\mathbf{B}_2}{}_{\mathbf{B}_1}(i_V).$$

Solving for $M^{\mathbf{B}_2}{}_{\mathbf{B}_2}(\phi)$ we find that

$$M^{\mathbf{B}_2}{}_{\mathbf{B}_2}(\phi) = (M^{\mathbf{B}_2}{}_{\mathbf{B}_1}(i_V))^{-1}\, M^{\mathbf{B}_1}{}_{\mathbf{B}_1}(\phi)\, M^{\mathbf{B}_2}{}_{\mathbf{B}_1}(i_V)$$

or

$$M_2 = P^{-1}M_1P$$

where $P = M^{B_2}{}_{B_1}(i_V)$ is the matrix whose columns are the B_1-coordinate vectors of the B_2-basis vectors.

We have just proved the following theorem.

Theorem 3. *Let V be a finite dimensional vector space, let $B_1 = (v_1, \ldots, v_n)$ and $B_2 = (w_1, \ldots, w_n)$ be ordered bases for V, and let $\phi: V \to V$ be a linear map. Then the matrix $M_2 = M^{B_2}{}_{B_2}(\phi)$ of ϕ relative to the basis B_2 is related to the matrix $M_1 = M^{B_1}{}_{B_1}(\phi)$ of ϕ relative to the basis B_1 by*

$$M_2 = P^{-1}M_1P$$

where

$$P = \left(M_{B_1}(w_1) \quad \cdots \quad M_{B_1}(w_n) \right).$$

Example 3. Let $\phi: \mathbb{R}^2 \to \mathbb{R}^2$ be the linear map $\phi(x_1, x_2) = (2x_1 + 3x_2, 12x_1 + 2x_2)$, let $B_1 = ((1, 0), (0, 1))$, and let $B_2 = ((1, 2), (-1, 2))$. Then

$$\phi(1, 0) = (2, 12) = 2(1, 0) + 12(0, 1)$$
$$\phi(0, 1) = (3, 2) = 3(1, 0) + 2(0, 1)$$

so $M_1 = \begin{pmatrix} 2 & 3 \\ 12 & 2 \end{pmatrix}$

and

$$\phi(1, 2) = (8, 16) = 8(1, 2) + 0(-1, 2)$$
$$\phi(-1, 2) = (4, -8) = 0(1, 2) + (-4)(-1, 2)$$

so $M_2 = \begin{pmatrix} 8 & 0 \\ 0 & -4 \end{pmatrix}$.

Furthermore,

$$(1, 2) = 1(1, 0) + 2(0, 1)$$
$$(-1, 2) = -1(1, 0) + 2(0, 1)$$

so $P = \begin{pmatrix} 1 & -1 \\ 2 & 2 \end{pmatrix}$.

To check the equation $M_2 = P^{-1}M_1P$ it is easiest to check the equivalent equation $PM_2 = M_1P$:

$$PM_2 = \begin{pmatrix} 1 & -1 \\ 2 & 2 \end{pmatrix}\begin{pmatrix} 8 & 0 \\ 0 & -4 \end{pmatrix} = \begin{pmatrix} 8 & 4 \\ 16 & -8 \end{pmatrix}$$

$$M_1P = \begin{pmatrix} 2 & 3 \\ 12 & 2 \end{pmatrix}\begin{pmatrix} 1 & -1 \\ 2 & 2 \end{pmatrix} = \begin{pmatrix} 8 & 4 \\ 16 & -8 \end{pmatrix}. \quad \blacksquare$$

EXERCISES

1. Let $\mathbf{B} = ((1, 1, 1), (1, 1, 0), (1, 0, 0))$ and $\mathbf{B}' = ((1, 0), (0, 1))$. Find $M^{\mathbf{B}}_{\mathbf{B}'}(\phi)$, where $\phi\colon \mathbb{R}^3 \to \mathbb{R}^2$ is the linear map
   (a) $\phi(x_1, x_2, x_3) = (x_1, x_2)$
   (b) $\phi(x_1, x_2, x_3) = (x_1 + x_2 + x_3, x_2 - x_3)$
   (c) $\phi(x_1, x_2, x_3) = (2x_1 + 3x_2 - x_3, -x_1 + x_2 + 5x_3)$

2. Let $\phi\colon \mathbb{R}^2 \to \mathbb{R}^2$ be the linear map $\phi(x_1, x_2) = (2x_1 + x_2, -x_1 + 3x_2)$. Find $M^{\mathbf{B}}_{\mathbf{B}'}(\phi)$ where
   (a) $\mathbf{B} = \mathbf{B}' = ((1, 0), (0, 1))$
   (b) $\mathbf{B} = ((1, 0), (0, 1))$, $\mathbf{B}' = ((2, -1), (1, 3))$
   (c) $\mathbf{B} = ((2, 1), (1, -2))$, $\mathbf{B}' = ((1, 1), (-1, 1))$
   (d) $\mathbf{B} = ((2, 1), (1, -2))$, $\mathbf{B}' = ((-1, 1), (1, 1))$

3. (a) Find the $(\mathbf{B}, \mathbf{B}')$-coordinate matrix of $D\colon \mathcal{P}^5 \to \mathcal{P}^4$, where $\mathbf{B} = (1, x, x^2, x^3, x^4)$ and $\mathbf{B}' = (1, x, x^2, x^3)$.
   (b) Use the result of part (a) to compute the derivative of $p(x) = x^4 + 2x^3 - 2x^2 + x + 1$.

4. Let $\phi\colon \mathcal{P}^4 \to \mathcal{P}^4$ be the linear map $\phi = D + 1$. Find the matrix $M^{\mathbf{B}}_{\mathbf{B}}(\phi)$, where
   (a) $\mathbf{B} = (1, x, x^2, x^3)$
   (b) $\mathbf{B} = (x^3, x^2, x, 1)$
   (c) $\mathbf{B} = (1, 1 + x, 1 + x + x^2, 1 + x + x^2 + x^3)$

5. Let $\phi\colon \mathcal{P}^3 \to \mathcal{P}^3$, $\phi(p(x)) = p(x + 2)$. Find $M^{\mathbf{B}}_{\mathbf{B}'}(\phi)$, where
   (a) $\mathbf{B} = \mathbf{B}' = (x^2, x, 1)$
   (b) $\mathbf{B} = (x^2, x, 1)$, $\mathbf{B}' = (1, x, x^2)$
   (c) $\mathbf{B} = (1, x, x^2)$, $\mathbf{B}' = (x^2, x, 1)$
   (d) $\mathbf{B} = (1, x, x^2)$, $\mathbf{B}' = (1, x + 2, (x + 2)^2)$

6. Let $V = \mathfrak{L}(e^t, e^{2t}, e^{3t})$ and let $\mathbf{B} = (e^t, e^{2t}, e^{3t})$. Find $M^{\mathbf{B}}_{\mathbf{B}}(\phi)$ where $\phi\colon V \to V$ is defined by
   (a) $\phi(y) = y'$
   (b) $\phi(y) = y' - y$
   (c) $\phi(y) = y'' - 3y' + 2y$
   (d) $\phi(y) = y''' - 6y'' + 11y' - 6y$

7. Let $\phi\colon \mathcal{P}^3 \to \mathcal{P}^2$ be the linear map whose matrix relative to the ordered bases $\mathbf{B} = (x^2, x, 1)$ for $\mathcal{P}^3$ and $\mathbf{B}' = (x, 1)$ for $\mathcal{P}^2$ is
$$M^{\mathbf{B}}_{\mathbf{B}'}(\phi) = \begin{pmatrix} 1 & -1 & 0 \\ 2 & 3 & 1 \end{pmatrix}.$$
   Find
   (a) $\phi(x + 1)$ (c) $\phi(3x^2 - 6x + 4)$
   (b) $\phi(x^2 - x + 2)$ (d) $\phi((x - 2)^2)$

8. Let $\phi\colon \mathcal{P}^4 \to \mathcal{P}^4$ be the linear map $\phi(p(x)) = p(x + 1)$.
   (a) Find a formula for $\phi \circ \phi(p(x))$.
   (b) Find $M^{\mathbf{B}}_{\mathbf{B}}(\phi \circ \phi)$, where $\mathbf{B} = (x^3, x^2, x, 1)$, in two different ways: (i) by direct computation, using the result of part (a), and (ii) by using Theorem 2(iii).

9. Let $V = \mathcal{L}(e^t, e^{-t})$, let $\mathbf{B}_1 = (e^t, e^{-t})$, and let $\mathbf{B}_2 = (\cosh t, \sinh t)$. Find $M_1 = M^{\mathbf{B}_1}{}_{\mathbf{B}_1}(\phi)$, $M_2 = M^{\mathbf{B}_2}{}_{\mathbf{B}_2}(\phi)$, and P such that $M_2 = P^{-1}M_1P$, where
   (a) $\phi(y) = y'$
   (b) $\phi(y) = y' - y$

10. Let $\phi: \mathbb{R}^2 \to \mathbb{R}^2$ be the linear map $\phi(x_1, x_2) = (2x_1, 3x_2)$.
    (a) Find $M_1 = M^{\mathbf{B}_1}{}_{\mathbf{B}_1}(\phi)$ where $\mathbf{B}_1 = ((1, 0), (0, 1))$
    (b) Find $M_2 = M^{\mathbf{B}_2}{}_{\mathbf{B}_2}(\phi)$ when $\mathbf{B}_2 = ((-1, 1), (1, 1))$
    (c) Find P such that $M_2 = P^{-1}M_1P$.

11. Repeat Exercise 10 for $\phi(x_1, x_2) = (x_1 + 3x_2, 3x_1 + x_2)$.

12. Let V and W be finite dimensional vector spaces with ordered bases $\mathbf{B}_1$ and $\mathbf{B}_1'$ respectively, and let $\phi: V \to W$ be a linear map.
    (a) Show that if $\mathbf{B}_2$ is obtained from $\mathbf{B}_1$ by interchanging two vectors, then $M^{\mathbf{B}_2}{}_{\mathbf{B}_1'}(\phi)$ can be obtained from $M^{\mathbf{B}_1}{}_{\mathbf{B}_1'}(\phi)$ by interchanging two columns.
    (b) Show that if $\mathbf{B}_2'$ is obtained from $\mathbf{B}_1'$ by interchanging two vectors, then $M^{\mathbf{B}_1}{}_{\mathbf{B}_2'}(\phi)$ can be obtained from $M^{\mathbf{B}_1}{}_{\mathbf{B}_1'}(\phi)$ by interchanging two rows.

13. Let V and W be finite dimensional vector spaces. Suppose $\mathbf{B}_1$ and $\mathbf{B}_2$ are two ordered bases for V and $\mathbf{B}_1'$ and $\mathbf{B}_2'$ are two ordered bases for W. Let $\phi: V \to W$ be a linear map. Show that
$$M^{\mathbf{B}_2}{}_{\mathbf{B}_2'}(\phi) = Q^{-1}M^{\mathbf{B}_1}{}_{\mathbf{B}_1'}(\phi)P$$
where $P = M^{\mathbf{B}_2}{}_{\mathbf{B}_1}(i_V)$ and $Q = M^{\mathbf{B}_2'}{}_{\mathbf{B}_1'}(i_W)$.

14. Let V be a finite dimensional inner product space. A linear map $\phi: V \to V$ is said to be **self-adjoint** if $\langle \phi(\mathbf{v}), \mathbf{w} \rangle = \langle \mathbf{v}, \phi(\mathbf{w}) \rangle$ for all $\mathbf{v}, \mathbf{w} \in V$. Show that $\phi: V \to V$ is self-adjoint if and only if the matrix for ϕ with respect to each ordered orthonormal basis for V is symmetric. ($A \in \mathfrak{M}_{n \times n}$ is **symmetric** if $A = A^t$.)

15. Let V be a finite dimensional vector space and let $\phi: V \to V$ be a linear map. Show that if M_1 is the matrix of ϕ relative to one ordered basis for V and if M_2 is the matrix of ϕ relative to another ordered basis for V, then $\det M_2 = \det M_1$. [Hint: Use Exercise 16(b) of Section 7.2.]

Remark. It follows from this exercise that we can define the **determinant of a linear map** $\phi: V \to V$, where V is any finite dimensional vector space, by $\det \phi = \det M$, where M is the matrix for ϕ relative to any ordered basis for V. The number $\det M$ will not depend on which matrix representation of ϕ is used.

7.5 KERNEL AND IMAGE

Associated with each linear map $\phi: V \to W$ there are two important subspaces. The first of these is the set of all vectors in V that are mapped by ϕ to the zero vector in W. This subspace of V is called the **kernel** of ϕ, and is denoted $\ker \phi$. Thus,

$$\ker \phi = \{\mathbf{v} \in V \,|\, \phi(\mathbf{v}) = \mathbf{0}\}.$$

Let us check that ker ϕ is, in fact, a subspace of V. From Theorem 4 of Section 7.1, we know that $\phi(\mathbf{0}) = \mathbf{0}$, hence $\mathbf{0} \in \ker \phi$. If $\mathbf{u}$ and $\mathbf{v}$ are vectors in ker ϕ, then $\phi(\mathbf{u}) = \mathbf{0}$ and $\phi(\mathbf{v}) = \mathbf{0}$, hence $\phi(\mathbf{u} + \mathbf{v}) = \phi(\mathbf{u}) + \phi(\mathbf{v}) = \mathbf{0} + \mathbf{0} = \mathbf{0}$, so $\mathbf{u} + \mathbf{v} \in \ker \phi$. And finally, if $c \in \mathbb{R}$ and $\mathbf{v} \in \ker \phi$, then $\phi(\mathbf{v}) = \mathbf{0}$, hence $\phi(c\mathbf{v}) = c\phi(\mathbf{v}) = c\mathbf{0} = \mathbf{0}$, and so $c\mathbf{v} \in \ker \phi$. Therefore ker ϕ *is* a subspace of V.

Example 1.   Let $D\colon \mathfrak{D} \to \mathfrak{F}(\mathbb{R})$ denote differentiation. Then $\ker D = \{f \,|\, f' = 0\}$ is the subspace of $\mathfrak{D}$ consisting of all functions whose derivative is the zero function. Thus ker D is the subspace of $\mathfrak{D}$ consisting of all constant functions.   ∎

Example 2.   Let $\phi\colon \mathbb{R}^2 \to \mathbb{R}^2$ be defined by $\phi(x_1, x_2) = (-x_1, x_2)$. Thus ϕ is reflection in the x_2-axis. It is clear geometrically that $\mathbf{0} = (0, 0)$ is the only vector whose reflection in the x_2-axis is $\mathbf{0}$, hence $\ker \phi = \{\mathbf{0}\}$. We can see this algebraically, as well, by noting that $\phi(x_1, x_2) = (-x_1, x_2) = (0, 0)$ if and only if $-x_1 = 0$ and $x_2 = 0$; that is, if and only if $(x_1, x_2) = (0, 0)$.   ∎

Example 3.   Let $\mathbf{a} \in \mathbb{R}^n$, $\mathbf{a} \neq \mathbf{0}$, and define $\phi_\mathbf{a}\colon \mathbb{R}^n \to \mathbb{R}$ by $\phi_\mathbf{a}(\mathbf{v}) = \mathbf{a} \cdot \mathbf{v}$ (see Example 7, Section 7.1). Then

$$\ker \phi_\mathbf{a} = \{\mathbf{v} \,|\, \mathbf{a} \cdot \mathbf{v} = 0\}$$

is the subspace of $\mathbb{R}^n$ consisting of all vectors whose dot product with $\mathbf{a}$ is 0. Thus ker $\phi_\mathbf{a}$ is the set of all vectors perpendicular to $\mathbf{a}$; that is, ker $\phi_\mathbf{a}$ is the hyperplane in $\mathbb{R}^n$ through $\mathbf{0}$ perpendicular to $\mathbf{a}$ (see Figure 7.11).   ∎

Example 4.   Let $\phi\colon \mathbb{R}^3 \to \mathbb{R}^2$ be the linear map

$$\phi(x_1, x_2, x_3) = (2x_1 + 3x_2 - x_3, -x_1 + 4x_2 + 6x_3).$$

Then

$$\ker \phi = \{(x_1, x_2, x_3) \,|\, 2x_1 + 3x_2 - x_3 = 0 \text{ and } -x_1 + 4x_2 + 6x_3 = 0\}.$$

So ker ϕ is the solution space of the homogeneous linear system

$$2x_1 + 3x_2 - x_3 = 0$$
$$-x_1 + 4x_2 + 6x_3 = 0.$$

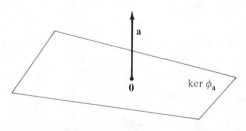

FIGURE 7.11

Row reduction yields

$$\begin{pmatrix} 2 & 3 & -1 & 0 \\ -1 & 4 & 6 & 0 \end{pmatrix} \rightarrow \begin{pmatrix} 1 & 0 & -2 & 0 \\ 0 & 1 & 1 & 0 \end{pmatrix}.$$

Hence the general solution of this system is

$$\mathbf{x} = t(2, -1, 1)$$

and

$$\ker \phi = \mathcal{L}((2, -1, 1))$$

is the line in $\mathbb{R}^3$ through $\mathbf{0}$ in the direction of $(2, -1, 1)$. ■

Example 5. Let

$$A = \begin{pmatrix} a_{11} & \cdots & a_{1n} \\ & \cdots & \\ a_{m1} & \cdots & a_{mn} \end{pmatrix} \in \mathfrak{M}_{m \times n}$$

and let $\phi_A : \mathbb{R}^n \rightarrow \mathbb{R}^m$ be the linear map associated with A (see Example 8, Section 7.1). Then

$$\ker \phi_A = \{\mathbf{v} \mid \phi_A(\mathbf{v})\} = \mathbf{0}\}$$
$$= \{\mathbf{v} \mid \mathbf{a}_1 \cdot \mathbf{v} = 0, \ldots, \mathbf{a}_m \cdot \mathbf{v} = 0\},$$

where $\mathbf{a}_1, \ldots, \mathbf{a}_m$ are the row vectors of A. Thus the kernel of ϕ_A is just the solution space of the homogeneous linear system

$$a_{11}x_1 + \cdots + a_{1n}x_n = 0$$

$$\cdots$$

$$a_{m1}x_1 + \cdots + a_{mn}x_n = 0.$$

The solution technique described in Chapter 1 leads directly to a basis for $\ker \phi_A$. Thus, for example, if

$$A = \begin{pmatrix} 1 & -1 & 1 & 1 & -2 \\ -2 & 2 & -1 & 0 & 1 \\ 1 & -1 & 2 & 3 & -5 \end{pmatrix}$$

then

$$\{(1, 1, 0, 0, 0), (1, 0, -2, 1, 0), (-1, 0, 3, 0, 1)\}$$

is a basis for $\ker \phi_A$ (see Example 1 in Section 2.4). ■

Example 6. Let $\mathfrak{D}^{\infty}$ denote the vector space of all infinitely differentiable functions on $\mathbb{R}$ and let $L: \mathfrak{D}^{\infty} \to \mathfrak{D}^{\infty}$ be a constant coefficient linear differential operator:

$$L = a_n D^n + \cdots + a_1 D + a_0$$

where $a_0, a_1, \ldots, a_n \in \mathbb{R}$. Then L is a linear map. Its kernel,

$$\ker L = \{y \mid Ly = 0\},$$

is the solution space of the homogeneous linear differential equation $Ly = 0$. The solution technique described in Chapter 3 leads directly to a basis for ker L. Thus, for example, if $L = D^4 - 16$ then

$$\{e^{2x}, e^{-2x}, \cos 2x, \sin 2x\}$$

is a basis for ker L (see Example 6 in Section 3.3). ∎

The second subspace associated with a linear map $\phi: V \to W$ is a subspace of W called the image of ϕ. The **image** of ϕ, denoted im ϕ, is defined by

$$\text{im } \phi = \{\mathbf{w} \in W \mid \mathbf{w} = \phi(\mathbf{v}) \text{ for some } \mathbf{v} \in V\}$$

Thus, im ϕ is the set of vectors in W that are "hit" by ϕ (see Figure 7.12). If im $\phi = W$, then we say that ϕ maps V **onto** W, or simply that ϕ is **onto**.

Let us verify that im ϕ is, in fact, a subspace of W. Since $\phi(\mathbf{0}) = \mathbf{0}$, we see that $\mathbf{0} \in \text{im } \phi$. If $\mathbf{w}_1$ and $\mathbf{w}_2 \in \text{im } \phi$, then there exist $\mathbf{v}_1$ and $\mathbf{v}_2 \in V$ with $\phi(\mathbf{v}_1) = \mathbf{w}_1$ and $\phi(\mathbf{v}_2) = \mathbf{w}_2$; hence $\mathbf{w}_1 + \mathbf{w}_2 = \phi(\mathbf{v}_1) + \phi(\mathbf{v}_2) = \phi(\mathbf{v}_1 + \mathbf{v}_2)$, so $\mathbf{w}_1 + \mathbf{w}_2 \in \text{im } \phi$. Finally, if $c \in \mathbb{R}$ and $\mathbf{w} \in \text{im } \phi$, then $\mathbf{w} = \phi(\mathbf{v})$ for some $\mathbf{v} \in V$; hence $c\mathbf{w} = c\phi(\mathbf{v}) = \phi(c\mathbf{v})$, so $c\mathbf{w} \in \text{im } \phi$. Thus im ϕ is a subspace of V.

Example 7. Let $\phi: \mathbb{R}^3 \to \mathbb{R}^3$ be the linear map defined by

$$\phi(x_1, x_2, x_3) = (x_1 - x_3, x_1 - 2x_2 + x_3, x_2 - x_3).$$

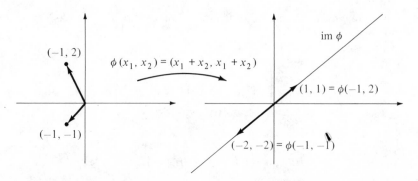

FIGURE 7.12

The image of ϕ is the set of all $\mathbf{y} = (y_1, y_2, y_3) \in \mathbb{R}^3$ such that $\phi(x_1, x_2, x_3) = \mathbf{y}$ for some $(x_1, x_2, x_3) \in \mathbb{R}^3$. Thus $\mathbf{y} \in \mathrm{im}\ \phi$ if and only if the equation

$$(x_1 - x_3, x_1 - 2x_2 + x_3, x_2 - x_3) = (y_1, y_2, y_3)$$

has a solution. But this vector equation is equivalent to the linear system

$$x_1 \qquad\qquad - x_3 = y_1$$
$$x_1 - 2x_2 + x_3 = y_2$$
$$x_2 - x_3 = y_3.$$

Row reduction of the matrix of this system yields

$$\begin{pmatrix} 1 & 0 & -1 & y_1 \\ 1 & -2 & 1 & y_2 \\ 0 & 1 & -1 & y_3 \end{pmatrix} \rightarrow \begin{pmatrix} 1 & 0 & -1 & y_1 \\ 0 & -2 & 2 & -y_1 + y_2 \\ 0 & 1 & -1 & y_3 \end{pmatrix}$$

$$\rightarrow \begin{pmatrix} 1 & 0 & -1 & y_1 \\ 0 & 1 & -1 & y_3 \\ 0 & -2 & 2 & -y_1 + y_2 \end{pmatrix} \rightarrow \begin{pmatrix} 1 & 0 & -1 & y_1 \\ 0 & 1 & -1 & y_3 \\ 0 & 0 & 0 & -y_1 + y_2 + 2y_3 \end{pmatrix}.$$

Hence im ϕ, which is the set of all (y_1, y_2, y_3) for which this system has a solution, is the set of all (y_1, y_2, y_3) such that $-y_1 + y_2 + 2y_3 = 0$. In other words, im ϕ is the plane in $\mathbb{R}^3$ through $\mathbf{0}$ perpendicular to $(-1, 1, 2)$. ∎

Example 8. Let $A = \begin{pmatrix} a_{11} & \cdots & a_{1n} \\ & \cdots & \\ a_{m1} & \cdots & a_{mn} \end{pmatrix} \in \mathfrak{M}_{m \times n}$ and let $\phi_A \colon \mathbb{R}^n \to \mathbb{R}^m$ be the linear map associated with A. The image of ϕ_A is the set of all $(y_1, \ldots, y_m)$ in $\mathbb{R}^m$ such that $(y_1, \ldots, y_m) = \phi_A(x_1, \ldots, x_n)$ for some $(x_1, \ldots, x_n) \in \mathbb{R}^n$. But the vector equation $\phi_A(x_1, \ldots, x_n) = (y_1, \ldots, y_m)$ is equivalent to the system of linear equations

$$a_{11}x_1 + \qquad \cdots \qquad + a_{1n}x_n = y_1$$
$$\cdots$$
$$a_{m1}x_1 + \qquad \cdots \qquad + a_{mn}x_n = y_m.$$

This system has a solution if and only if $\begin{pmatrix} y_1 \\ \vdots \\ y_m \end{pmatrix}$ is in the column space of A (Theorem 1 of Section 2.3). We conclude, then, that *the image of ϕ_A is equal to column space of A.* ∎

Often one wishes to describe the image of a linear map by exhibiting a basis for it. For linear maps $\phi: \mathbb{R}^n \to \mathbb{R}^m$ there is an easy algorithm for doing this, based on the facts that (i) each such ϕ is of the form $\phi = \phi_A$ for some $A \in \mathfrak{M}_{m \times n}$ and (ii) the image of ϕ_A is the column space of A. If we write the column vectors of A horizontally rather than vertically, we see that the image of ϕ_A is the row space of the transposed matrix

$$A^t = \begin{pmatrix} a_{11} & a_{21} & \cdots & a_{m1} \\ a_{12} & a_{22} & \cdots & a_{m2} \\ & & \cdots & \\ a_{1n} & a_{2n} & \cdots & a_{mn} \end{pmatrix}$$

whose rows are the columns of A. The row space of A^t has a basis consisting of the nonzero row vectors in the row echelon matrix of A^t (Example 5 of Section 4.2). Hence *the nonzero row vectors in the row echelon matrix of A^t form a basis for the image of ϕ_A.*

Example 9. For $\phi(x_1, x_2, x_3) = (x_1 - x_3, x_1 - 2x_2 + x_3, x_2 - x_3)$, as in Example 7, we have $\phi = \phi_A$ where

$$A = \begin{pmatrix} 1 & 0 & -1 \\ 1 & -2 & 1 \\ 0 & 1 & -1 \end{pmatrix}$$

and

$$A^t = \begin{pmatrix} 1 & 1 & 0 \\ 0 & -2 & 1 \\ -1 & 1 & -1 \end{pmatrix} \xrightarrow{\text{row reduction}} \begin{pmatrix} 1 & 0 & \frac{1}{2} \\ 0 & 1 & -\frac{1}{2} \\ 0 & 0 & 0 \end{pmatrix}$$

so $\{(1, 0, \frac{1}{2}), (0, 1, -\frac{1}{2})\}$ is a basis for im ϕ. ■

The main result of this section is the following theorem, which relates the dimensions of the spaces V, ker ϕ, and im ϕ.

Theorem 1. (*The Dimension Theorem*) *If V is a finite dimensional vector space and $\phi: V \to W$ is a linear map, then*

$$\dim \ker \phi + \dim \operatorname{im} \phi = \dim V.$$

Proof. Let $\{v_1, \ldots, v_k\}$ be a basis for ker ϕ. Since ker $\phi \subseteq V$, we know that this basis for ker ϕ can be extended to a basis $\{v_1, \ldots, v_k, v_{k+1}, \ldots, v_n\}$ for V. We shall show that $\{\phi(v_{k+1}), \ldots, \phi(v_n)\}$ is a basis for im ϕ. The formula of the theorem will then follow immediately, since

$$\dim \ker \phi + \dim \operatorname{im} \phi = k + (n - k) = n = \dim V.$$

To show that $\{\phi(v_{k+1}), \ldots, \phi(v_n)\}$ is a basis for im ϕ, we must show that this set is linearly independent and spans im ϕ. To establish linear independence, assume that

$$c_{k+1}\phi(v_{k+1}) + \cdots + c_n\phi(v_n) = 0,$$

where $c_{k+1}, \ldots, c_n \in \mathbb{R}$. Then

$$\phi(c_{k+1}v_{k+1} + \cdots + c_n v_n) = 0$$

and so

$$c_{k+1}v_{k+1} + \cdots + c_n v_n \in \ker \phi.$$

Since $\{v_1, \ldots, v_k\}$ is a basis for ker ϕ, there exist $c_1, \ldots, c_k \in \mathbb{R}$ such that

$$c_{k+1}v_{k+1} + \cdots + c_n v_n = c_1 v_1 + \cdots + c_k v_k$$

or

$$(-c_1)v_1 + \cdots + (-c_k)v_k + c_{k+1}v_{k+1} + \cdots + c_n v_n = 0.$$

But $\{v_1, \ldots, v_n\}$ is a *basis* for V; hence this equation implies that

$$-c_1 = -c_2 = \cdots = -c_k = c_{k+1} = \cdots = c_n = 0,$$

and, in particular, that

$$c_{k+1} = \cdots = c_n = 0,$$

so $\{\phi(v_{k+1}), \ldots, \phi(v_n)\}$ is linearly independent.

Finally, we must show that every vector $w \in \operatorname{im} \phi$ is a linear combination of $\phi(v_{k+1}), \ldots, \phi(v_n)$. We have $w = \phi(v)$ for some $v \in V$. But since $\{v_1, \ldots, v_n\}$ is a basis for V,

$$v = c_1 v_1 + \cdots + c_k v_k + c_{k+1}v_{k+1} + \cdots c_n v_n$$

for some $c_1, \ldots, c_n \in \mathbb{R}$. If we apply ϕ to both sides of this equation we get

$$w = \phi(v) = \phi(c_1 v_1 + \cdots + c_n v_n)$$
$$= c_1\phi(v_1) + \cdots + c_k\phi(v_k) + c_{k+1}\phi(v_{k+1}) + \cdots + c_n\phi(v_n)$$
$$= c_{k+1}\phi(v_{k+1}) + \cdots + c_n\phi(v_n) \quad (\text{since } v_i \in \ker \phi \text{ for } 1 \leq i \leq k).$$

Thus w is a linear combination of $\phi(v_{k+1}), \ldots, \phi(v_n)$, as required. ∎

Example 10.   Let $\phi\colon \mathbb{R}^3 \to \mathbb{R}^2$ be the linear map

$$\phi(x_1, x_2, x_3) = (2x_1 + 3x_2 - x_3, -x_1 + 4x_2 + 6x_3).$$

Since $\dim \mathbb{R}^3 = 3$ and since we know from Example 4 that $\dim \ker \phi = \dim \mathcal{L}((2, -1, 1)) = 1$, we see that

$$\dim \operatorname{im} \phi = \dim \mathbb{R}^3 - \dim \ker \phi = 3 - 1 = 2.$$

Since $\operatorname{im} \phi$ is a 2-dimensional subspace of $\mathbb{R}^2$, it must be equal to $\mathbb{R}^2$. Hence we can conclude that ϕ is onto. This, in turn, means that the linear system

$$2x_1 + 3x_2 - x_3 = b_1$$
$$-x_1 + 4x_2 + 6x_3 = b_2$$

is consistent for all $(b_1, b_2) \in \mathbb{R}^2$. ∎

Example 11.   Let

$$A = \begin{pmatrix} a_{11} & \cdots & a_{1n} \\ & \cdots & \\ a_{m1} & \cdots & a_{mn} \end{pmatrix} \in \mathfrak{M}_{m \times n}$$

and let $\phi_A\colon \mathbb{R}^n \to \mathbb{R}^m$ be the linear map associated with A. Then the kernel of ϕ_A is the solution space of the homogeneous linear system

$$a_{11}x_1 + \cdots + a_{1n}x_n = 0$$
$$\cdots$$
$$a_{m1}x_1 + \cdots + a_{mn}x_n = 0.$$

Its dimension is $n - r$ where r is the rank of the matrix A (see Example 4 of Section 4.4). By the Dimension Theorem,

$$\dim \operatorname{im} \phi_A = n - \dim \ker \phi_A = n - (n - r) = r.$$

But $\operatorname{im} \phi_A$ is just the column space of A, so we can conclude that *the dimension of the column space of a matrix A is equal to the rank of A*. Since the rank of A is just the dimension of the row space of A, we can also conclude that *the row space of A and the column space of A have the same dimension*. ∎

The image of a linear map $\phi\colon V \to W$ determines whether or not ϕ is onto: ϕ is onto if and only if $\operatorname{im} \phi = W$. Similarly, the kernel of a linear map can be used to determine whether or not the map is one-to-one. A linear map $\phi\colon V \to W$ is called ***one-to-one*** if $\phi(\mathbf{u}) \neq \phi(\mathbf{v})$ whenever $\mathbf{u} \neq \mathbf{v}$ ($\mathbf{u}, \mathbf{v} \in V$). In other words, a linear map is one-to-one if it never sends two distinct vectors in V to the same vector in W. Said yet another way, ϕ is one-to-one if $\phi(\mathbf{u}) = \phi(\mathbf{v})$ only when $\mathbf{u} = \mathbf{v}$.

Theorem 2. *A linear map* $\phi: V \to W$ *is one-to-one if and only if* $\ker \phi = \{\mathbf{0}\}$.

Proof. For each $\mathbf{v} \in \ker \phi$ we have $\phi(\mathbf{v}) = \mathbf{0} = \phi(\mathbf{0})$. If ϕ is one-to-one it follows that $\mathbf{v} = \mathbf{0}$, hence $\ker \phi = \{\mathbf{0}\}$.

Conversely, for each $\mathbf{u}, \mathbf{v} \in V$ such that $\phi(\mathbf{u}) = \phi(\mathbf{v})$ we have $\phi(\mathbf{u} - \mathbf{v}) = \phi(\mathbf{u}) - \phi(\mathbf{v}) = \mathbf{0}$, hence $\mathbf{u} - \mathbf{v} \in \ker \phi$. If $\ker \phi = \{\mathbf{0}\}$ it follows that $\mathbf{u} - \mathbf{v} = \mathbf{0}$, that is, $\mathbf{u} = \mathbf{v}$, and hence ϕ is one-to-one. ∎

Example 12. Referring to Examples 1 through 5, we observe that

  (i) $D: \mathfrak{D} \to \mathfrak{F}(\mathbb{R})$ is not one-to-one
  (ii) In $\mathbb{R}^2$, reflection in the x_2-axis is one-to-one
  (iii) $\phi_{\mathbf{a}}: \mathbb{R}^n \to \mathbb{R}(\mathbf{a} \neq \mathbf{0})$ is one-to-one if and only if there are *no* nonzero vectors $\mathbf{v} \in \mathbb{R}^n$ perpendicular to $\mathbf{a}$. This happens if and only if $n = 1$
  (iv) $\phi: \mathbb{R}^3 \to \mathbb{R}^2$ defined by $\phi(x_1, x_2, x_3) = (2x_1 + 3x_2 - x_3, -x_1 + 4x_2 + 6x_3)$ is not one-to-one
  (v) $\phi_A: \mathbb{R}^n \to \mathbb{R}^m$ is one-to-one if and only if the homogeneous linear system with coefficient matrix A has the unique solution $(x_1, \ldots, x_n) = (0, \ldots, 0)$. This happens if and only if A has rank n. ∎

Next we investigate the connection between the invertibility of a linear map $\phi: V \to W$ and the subspaces $\ker \phi$ and $\operatorname{im} \phi$.

Theorem 3. *A linear map* $\phi: V \to W$ *is invertible if and only if it is one-to-one and onto.*

Proof. Suppose $\phi: V \to W$ is invertible. Then there exists a linear map $\psi: W \to V$ such that

$$\psi \circ \phi = i_V \quad \text{and} \quad \phi \circ \psi = i_W.$$

Now if $\mathbf{u}$ and $\mathbf{v}$ are vectors in V such that $\phi(\mathbf{u}) = \phi(\mathbf{v})$, then $\mathbf{u} = \psi(\phi(\mathbf{u})) = \psi(\phi(\mathbf{v})) = \mathbf{v}$ so ϕ is one-to-one. Furthermore, for each $\mathbf{w} \in W$ we have $\mathbf{w} = \phi(\psi(\mathbf{w}))$, so ϕ is onto. Thus if ϕ is invertible, then it is both one-to-one and onto.

Conversely, suppose $\phi: V \to W$ is both one-to-one and onto. Then for each $\mathbf{w} \in W$ there is one and only one $\mathbf{v} \in V$ such that $\phi(\mathbf{v}) = \mathbf{w}$. Define $\psi: W \to V$ by $\phi(\mathbf{w}) = \mathbf{v}$ where $\mathbf{v}$ is the unique vector such that $\phi(\mathbf{v}) = \mathbf{w}$. Thus

$$\psi(\mathbf{w}) = \mathbf{v} \text{ if and only if } \phi(\mathbf{v}) = \mathbf{w}$$

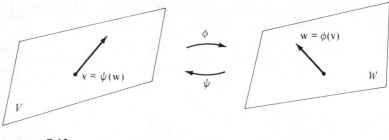

FIGURE 7.13

(see Figure 7.13). It follows immediately that for each $\mathbf{v} \in V$ we have $\psi(\phi(\mathbf{v})) = \mathbf{v}$, and for each $\mathbf{w} \in W$ we have $\phi(\psi(\mathbf{w})) = \mathbf{w}$. Thus $\psi \circ \phi = i_V$ and $\phi \circ \psi = i_W$ as required. But we must still check that the map ψ is a *linear* map. This is easy, using the linearity of ϕ.

(i) For $\mathbf{w}_1$ and $\mathbf{w}_2 \in W$ we have

$$\phi(\psi(\mathbf{w}_1 + \mathbf{w}_2)) = \mathbf{w}_1 + \mathbf{w}_2$$
$$= \phi(\psi(\mathbf{w}_1)) + \phi(\psi(\mathbf{w}_2))$$
$$= \phi(\psi(\mathbf{w}_1) + \psi(\mathbf{w}_2)).$$

Since ϕ is one-to-one this implies that $\psi(\mathbf{w}_1 + \mathbf{w}_2) = \psi(\mathbf{w}_1) + \psi(\mathbf{w}_2)$.

(ii) Similarly, for $\mathbf{w} \in W$ and $c \in \mathbb{R}$,

$$\phi(\psi(c\mathbf{w})) = c\mathbf{w} = c\phi(\psi(\mathbf{w})) = \phi(c\psi(\mathbf{w}))$$

so $\psi(c\mathbf{w}) = c\psi(\mathbf{w})$, since ϕ is one-to-one.

Thus if ϕ is one-to-one and onto, then it is invertible. ∎

According to Theorem 3, we can determine whether a linear map $\phi: V \to W$ is invertible by checking two conditions:

(i) that ϕ is one-to-one (or, equivalently, that $\ker \phi = \{\mathbf{0}\}$), and

(ii) that ϕ is onto (or, equivalently, that $\text{im } \phi = W$). When $\dim V = \dim W$, it suffices to check only one of these conditions.

Theorem 4. *Let V and W be finite dimensional vector spaces of the same dimension and let $\phi: V \to W$ be a linear map. Then ϕ is one-to-one if and only if ϕ is onto.*

Proof. Assume ϕ is one-to-one. Then $\ker \phi = \{\mathbf{0}\}$ so, by the Dimension Theorem,

$$\dim \text{im } \phi = \dim V - \dim \ker \phi = \dim V = \dim W.$$

By Theorem 5 of Section 4.4, this implies that im $\phi = W$, so ϕ is onto.
Conversely, assume ϕ is onto. Then, by the Dimension Theorem,

$$\dim \ker \phi = \dim V - \dim \operatorname{im} \phi = \dim V - \dim W = 0$$

so $\ker \phi = \{\mathbf{0}\}$ and ϕ is one-to-one. ∎

Theorem 4 is especially useful when applied to linear maps from a finite
dimensional vector space to itself, for then $W = V$ so certainly $\dim W = \dim V$.
But the finite dimensionality of the vector space is essential. When V is infinite
dimensional, it is *not* true that $\phi: V \to V$ is one-to-one if and only if ϕ is onto, as
the following example shows.

Example 13. Let $\phi: \mathbb{R}^\infty \to \mathbb{R}^\infty$ be defined by

$$\phi(x_1, x_2, x_3, \ldots) = (0, x_1, x_2, x_3, \ldots).$$

This linear map is one-to-one (its kernel is $\{\mathbf{0}\}$) but it is not onto $((1, 0, 0, \ldots) \notin$
im ϕ). ∎

Example 14. Consider the linear map $\phi: \mathcal{P}^3 \to \mathcal{P}^3$ given by $\phi = D - 1$. Since

$$\phi(a_0 + a_1 x + a_2 x^2) = (a_1 - a_0) + (2a_2 - a_1)x - a_2 x^2$$

we see that $a_0 + a_1 x + a_2 x^2 \in \ker \phi$ if and only if $a_1 - a_0 = 0$, $2a_2 - a_1 = 0$,
and $a_2 = 0$; that is, if and only if $a_0 = a_1 = a_2 = 0$. Hence $\ker \phi = \{\mathbf{0}\}$ and ϕ is
one-to-one. By Theorem 4, ϕ is also onto, hence invertible.
   In contrast, note that the linear map $D: \mathcal{P}^3 \to \mathcal{P}^3$ is not invertible: its kernel
is $\mathcal{P}^1$ and its image is $\mathcal{P}^2$. ∎

EXERCISES

1. Describe the kernel of the linear map $D^2: \mathfrak{D}^\infty \to \mathfrak{D}^\infty$.

2. Let $\phi: \mathbb{R}^3 \to \mathbb{R}^3$ be the linear map
$$\phi(x_1, x_2, x_3) = (x_1 - x_2, x_2 - x_3, x_3 - x_1).$$
   (a) Show that the kernel of ϕ is a line and find an equation for it.
   (b) Show that the image of ϕ is a plane and find an equation for it.

3. Find a basis for the kernel of each of the following linear maps $L: \mathfrak{D}^\infty \to \mathfrak{D}^\infty$.
   (a) $L = D^2 + 1$ (c) $L = (D - 1)^3$
   (b) $L = D^2 - 1$ (d) $L = D^6 - 3D^4 - 4D^2$

4. Find a basis for the kernel of the linear map ϕ_A where
   (a) $A = \begin{pmatrix} 1 & 1 \\ 1 & -1 \end{pmatrix}$

(b) $A = \begin{pmatrix} 1 & 1 \\ -1 & -1 \end{pmatrix}$

(c) $A = \begin{pmatrix} 0 & 0 & 1 \\ 0 & 1 & 0 \\ 1 & 0 & 0 \end{pmatrix}$

(d) $A = \begin{pmatrix} 1 & -1 & 2 \\ 0 & 1 & 1 \\ 2 & -1 & 5 \end{pmatrix}$

(e) $A = \begin{pmatrix} 2 & -1 & 3 \\ 1 & 0 & -1 \\ 3 & -1 & 1 \\ -1 & 1 & -1 \end{pmatrix}$

(f) $A = \begin{pmatrix} 2 & 1 & 3 & -1 \\ -1 & 0 & -1 & 1 \\ 3 & -1 & 1 & -1 \end{pmatrix}$

5.   Find a basis for the image of each of the linear maps ϕ_A of Exercise 4.

6.   Use the Dimension Theorem and your solutions to Exercise 4 to compute the dimension of the image of ϕ_A for each of the linear maps ϕ_A of Exercise 4. Check that your answers are consistent with your solutions to Exercise 5.

7.   Describe the kernel of each of the following linear maps and determine which of these maps are one-to-one.

   (a) $\phi(x_1, x_2) = (x_1, x_1 - x_2, x_2)$, $\phi: \mathbb{R}^2 \to \mathbb{R}^3$
   (b) $\phi(\mathbf{x}) = 2\mathbf{x}$, $\phi: \mathbb{R}^n \to \mathbb{R}^n$
   (c) $\phi(x_1, x_2, x_3) = (x_1 - x_2, x_2 - x_3, x_3 - x_1)$, $\phi: \mathbb{R}^3 \to \mathbb{R}^3$
   (d) $\phi(x_1, x_2, x_3) = (x_1 + x_2 + x_3, x_1 + x_2 + x_3)$,$\phi: \mathbb{R}^3 \to \mathbb{R}^2$
   (e) $\phi(x_1, x_2, x_3, \ldots) = (x_2, x_3, \ldots)$, $\phi: \mathbb{R}^\infty \to \mathbb{R}^\infty$

8.   Describe the image of each of the linear maps in Exercise 7 and determine which of these maps are onto.

9.   Show that the linear map $D - a: \mathcal{P}^3 \to \mathcal{P}^3$ is invertible if and only if $a \neq 0$.

10.   Let $V = \mathcal{L}(e^x, e^{2x}, e^{3x})$, let $a \in \mathbb{R}$, and let $\phi = D - a$.
   (a)   Show that $\phi(y) \in V$ whenever $y \in V$.
   (b)   For what values of a is $\phi: V \to V$ one-to-one?
   (c)   For what values of a is $\phi: V \to V$ onto?

11.   Let $\phi: \mathcal{P}^n \to \mathbb{R}$ be the map $\phi(f) = f(0)$.
   (a)   Show that ϕ is a linear map.
   (b)   Find a basis for the kernel of ϕ.
   (c)   Describe the image of ϕ.

12.   Let $a \in \mathbb{R}$ and define $\phi: \mathcal{P}^n \to \mathbb{R}$ by $\phi(f) = f(a)$.
   (a)   Show that ϕ is a linear map.
   (b)   Describe ker ϕ.
   (c)   Describe im ϕ.

13.   Let $\phi: \mathcal{P} \to \mathcal{P}$ be the linear map $\phi(p(x)) = \int_0^x p(t)\, dt$.

   (a)   Describe the kernel of ϕ. Is ϕ one-to-one?
   (b)   Describe the image of ϕ. Is ϕ onto?

14. Let $\phi: \mathcal{P} \to \mathcal{P}$ be the map $\phi(p(x)) = p(x - 1)$.
    (a) Show that ϕ is a linear map.
    (b) Show that ϕ is one-to-one.
    (c) Show that ϕ is onto.
    (d) Describe ϕ^{-1}.

15. Let $\mathbf{a}_1, \ldots, \mathbf{a}_{n-1} \in \mathbb{R}^n$ and let $\phi: \mathbb{R}^n \to \mathbb{R}$ be the linear map defined by

$$\phi(\mathbf{x}) = \det \begin{pmatrix} \mathbf{a}_1 \\ \cdots \\ \mathbf{a}_{n-1} \\ \mathbf{x} \end{pmatrix}.$$

    (a) Show that if $\{\mathbf{a}_1, \ldots, \mathbf{a}_{n-1}\}$ is linearly dependent, then ϕ is the zero map.
    (b) Show that if $\{\mathbf{a}_1, \ldots, \mathbf{a}_{n-1}\}$ is linearly independent, then $\ker \phi = \mathcal{L}(\mathbf{a}_1, \ldots, \mathbf{a}_{n-1})$.

16. Let $A \in \mathfrak{M}_{m \times n}$. Show that the linear map $\phi_A: \mathbb{R}^n \to \mathbb{R}^m$ is onto if and only if A has rank m.

17. Let V and W be finite dimensional vector spaces and let $\phi: V \to W$ be a linear map.
    (a) Show that if $\dim V > \dim W$, then ϕ cannot be one-to-one.
    (b) Show that if $\dim V < \dim W$, then ϕ cannot be onto.

18. Let $\phi: V \to W$ be a linear map.
    (a) Show that if ϕ has a left inverse (see Exercise 14, Section 7.3), then ϕ is one-to-one.
    (b) Show that if ϕ has a right inverse, then ϕ is onto.

Eigenvectors

EIGHT

8.1 EIGENVECTORS AND EIGENVALUES

In this chapter we shall study linear maps $\phi\colon V \to V$ that map a vector space V into itself. The simplest such linear maps are the maps ϕ_λ that multiply each vector $\mathbf{v} \in V$ by some real number λ (see Figure 8.1). If $\lambda > 1$, then ϕ_λ is expansion by the factor λ. If $0 < \lambda < 1$, then ϕ_λ is contraction by the factor λ. If $\lambda = -1$, then ϕ_λ is reflection in $\mathbf{0}$. If $\lambda < 0$ and $\lambda \neq -1$, then ϕ_λ can be viewed as the composition

$$\phi_\lambda = \phi_{-1} \circ \phi_{|\lambda|}$$

of an expansion or a contraction followed by reflection in $\mathbf{0}$.

Most linear maps $\phi\colon V \to V$ are not so simple. Frequently, however, it is possible to find certain vectors on which a given linear map acts like ϕ_λ for some λ. It is then easy to visualize what the linear map ϕ is doing, geometrically, to these special vectors. These vectors are called eigenvectors of the linear map ϕ.

Let V be a vector space and let $\phi\colon V \to V$ be a linear map. An ***eigenvector*** of ϕ is a nonzero vector $\mathbf{v} \in V$ such that $\phi(\mathbf{v}) = \lambda\mathbf{v}$ for some $\lambda \in \mathbb{R}$. The real number λ such that $\phi(\mathbf{v}) = \lambda\mathbf{v}$ is called the ***eigenvalue*** of ϕ corresponding to the eigenvector $\mathbf{v}$.

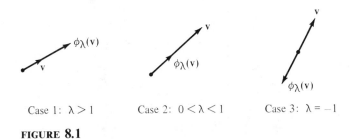

Case 1: $\lambda > 1$ Case 2: $0 < \lambda < 1$ Case 3: $\lambda = -1$

FIGURE 8.1

243

Example 1. Let $\phi\colon \mathbb{R}^2 \to \mathbb{R}^2$ be the linear map $\phi(x_1, x_2) = (2x_1, -3x_2)$. Then $\phi(e_1) = 2e_1$ and $\phi(e_2) = -3e_2$. Hence e_1 is an eigenvector of ϕ with eigenvalue 2, and e_2 is an eigenvector of ϕ with eigenvalue -3 (see Figure 8.2). So ϕ stretches e_1 by a factor of 2, and it acts on e_2 by stretching by a factor of 3 *and* reversing direction.

Now let us see if there are any other eigenvectors of ϕ. If $v = (x_1, x_2)$ is any eigenvector of ϕ, with eigenvalue λ, then $\phi(x_1, x_2) = \lambda(x_1, x_2)$, hence $(2x_1, -3x_2) = (\lambda x_1, \lambda x_2)$. Thus (x_1, x_2) must be a solution of the linear system

$$2x_1 = \lambda x_1$$
$$-3x_2 = \lambda x_2.$$

Since $v \neq 0$, either $x_1 \neq 0$ or $x_2 \neq 0$. If $x_1 \neq 0$, then $\lambda = 2$. If $x_2 \neq 0$, then $\lambda = -3$. Hence 2 and -3 are the only eigenvalues of ϕ. Moreover, $\lambda = 2$ implies that $x_2 = 0$ so $v = (x_1, 0)$ for some x_1, and $\lambda = -3$ implies that $x_1 = 0$ so $v = (0, x_2)$ for some x_2. We conclude, then, that the only eigenvectors of ϕ are the nonzero multiples of e_1 (with eigenvalue 2) and the nonzero multiples of e_2 (with eigenvalue -3). ∎

Remark. Let $\phi\colon V \to V$ be a linear map. For each real number λ, the set V_λ of all vectors $v \in V$ such that $\phi(v) = \lambda v$ is a subspace of V. Indeed, the condition $\phi(v) = \lambda v$ can be rewritten as $\phi(v) - \lambda v = 0$ or as $(\phi - \lambda i)(v) = 0$, where $i\colon V \to V$ is the identity map. Hence $V_\lambda = \ker(\phi - \lambda i)$. Usually, for $\lambda \in \mathbb{R}$, we find $V_\lambda = \{0\}$. But if λ is an eigenvalue of ϕ then $V_\lambda \neq \{0\}$. V_λ is called the λ-*eigenspace* of ϕ. Notice that V_λ consists of the eigenvectors of ϕ with eigenvalue λ together with the zero vector.

Example 2. Let $\phi\colon \mathbb{R}^2 \to \mathbb{R}^2$ be defined by $\phi(x_1, x_2) = (x_1 + 2x_2, 2x_1 + x_2)$. Then $v = (x_1, x_2)$ is an eigenvector of ϕ if there is a real number λ for which

$$x_1 + 2x_2 = \lambda x_1$$
$$2x_1 + x_2 = \lambda x_2.$$

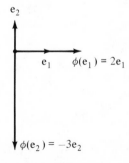

FIGURE 8.2

In other words, (x_1, x_2) is an eigenvector of ϕ if and only if (x_1, x_2) is a nonzero solution of the homogeneous system

(*)
$$
\begin{aligned}
(1 - \lambda)x_1 + \quad\quad 2x_2 &= 0 \\
2x_1 + (1 - \lambda)x_2 &= 0.
\end{aligned}
$$

This system has a nonzero solution if and only if its coefficient matrix has rank <2; that is, if and only if the determinant of its coefficient matrix is zero. Since

$$
\det\begin{pmatrix} 1 - \lambda & 2 \\ 2 & 1 - \lambda \end{pmatrix} = \lambda^2 - 2\lambda - 3,
$$

the eigenvalues of ϕ must be the roots of the polynomial equation

$$
\lambda^2 - 2\lambda - 3 = 0,
$$

namely -1 and 3.

Taking $\lambda = -1$ in (*), we see that the eigenvectors of ϕ with eigenvalue -1 are the nonzero solutions of the homogeneous system

$$
\begin{aligned}
2x_1 + 2x_2 &= 0 \\
2x_1 + 2x_2 &= 0,
\end{aligned}
$$

namely, the nonzero multiples of $(-1, 1)$. Similarly, taking $\lambda = 3$ in (*), we find that the eigenvectors of ϕ with eigenvalue 3 are the nonzero solutions of the homogeneous system

$$
\begin{aligned}
-2x_1 + 2x_2 &= 0 \\
2x_1 - 2x_2 &= 0,
\end{aligned}
$$

namely, the nonzero multiples of $(1, 1)$.

Notice that ϕ reflects through $\mathbf{0}$ each vector in the eigenspace $V_{-1} = \mathcal{L}((-1, 1))$, and ϕ stretches by a factor of 3 each vector in the eigenspace $V_3 = \mathcal{L}((1, 1))$ (see Figure 8.3). ∎

Example 3. Let $\mathcal{D}^\infty$ be the vector space of all infinitely differentiable functions with domain $\mathbb{R}$. Let us find the eigenvectors and eigenvalues of $D: \mathcal{D}^\infty \to \mathcal{D}^\infty$. A nonzero function $f \in \mathcal{D}^\infty$ is an eigenvector of D with eigenvalue λ if and only if

$$
Df = \lambda f;
$$

that is, if and only if f is a solution of the linear differential equation

$$
y' - \lambda y = 0.
$$

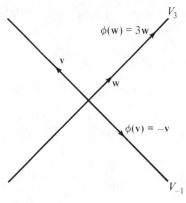

FIGURE 8.3

Since this equation has general solution $y = ce^{\lambda x}$, we see that every real number λ is an eigenvalue of D. For each $\lambda \in \mathbb{R}$, the λ-eigenspace of D is

$$V_\lambda = \mathcal{L}(e^{\lambda x}). \qquad \blacksquare$$

Now let A be a square matrix, $A \in \mathfrak{M}_{n \times n}$ for some n. A nonzero vector $\mathbf{v} \in \mathbb{R}^n$ is called an **eigenvector** of A with **eigenvalue** λ if $\phi_A(\mathbf{v}) = \lambda\mathbf{v}$, where $\phi_A \colon \mathbb{R}^n \to \mathbb{R}^n$ is the linear map associated with A. Thus the eigenvectors and eigenvalues of A are the same as the eigenvectors and eigenvalues of ϕ_A. If we write vectors in $\mathbb{R}^n$ vertically, then ϕ_A is just multiplication by A, so $\mathbf{v}$ is an eigenvector of A with eigenvalue ϕ if and only if

$$A\mathbf{v} = \lambda\mathbf{v},$$

or equivalently, if and only if

$$(A - \lambda I)\mathbf{v} = \mathbf{0}.$$

Thus the eigenvectors of A with eigenvalue λ are the nonzero solutions of the homogeneous linear system whose coefficient matrix is $A - \lambda I$. This system will have a nonzero solution if and only if $A - \lambda I$ has rank $< n$; that is, if and only if $\det(A - \lambda I) = 0$. Thus the eigenvalues of A are the zeros of the polynomial $\det(A - \lambda I)$.

The polynomial $p(\lambda) = \det(A - \lambda I)$ is called the **characteristic polynomial** of A. Note that the degree of the characteristic polynomial of $A \in \mathfrak{M}_{n \times n}$ is equal to n.

Example 4. Let

$$A = \begin{pmatrix} 2 & 1 & 2 \\ 1 & 2 & 2 \\ 1 & 1 & 3 \end{pmatrix}.$$

The eigenvalues of A are the zeros of the characteristic polynomial

$$\det(A - \lambda I) = \det \begin{pmatrix} 2 - \lambda & 1 & 2 \\ 1 & 2 - \lambda & 2 \\ 1 & 1 & 3 - \lambda \end{pmatrix}$$

$$= (2 - \lambda) \det \begin{pmatrix} 2 - \lambda & 2 \\ 1 & 3 - \lambda \end{pmatrix} - 1 \det \begin{pmatrix} 1 & 2 \\ 1 & 3 - \lambda \end{pmatrix}$$

$$+ 2 \det \begin{pmatrix} 1 & 2 - \lambda \\ 1 & 1 \end{pmatrix}$$

$$= -\lambda^3 + 7\lambda^2 - 11\lambda + 5$$

$$= -(\lambda - 1)(\lambda - 1)(\lambda - 5).$$

Thus the eigenvalues of A are 1 and 5.
   When $\lambda = 1$,

$$A - \lambda I = \begin{pmatrix} 1 & 1 & 2 \\ 1 & 1 & 2 \\ 1 & 1 & 2 \end{pmatrix}$$

so the eigenvectors of A with eigenvalue 1 are the nonzero solutions of the system

$$x_1 + x_2 + 2x_3 = 0$$
$$x_1 + x_2 + 2x_3 = 0$$
$$x_1 + x_2 + 2x_3 = 0,$$

that is, all nonzero vectors of the form $c_1(-1, 1, 0) + c_2(-2, 0, 1)$.
   When $\lambda = 5$,

$$A - \lambda I = \begin{pmatrix} -3 & 1 & 2 \\ 1 & -3 & 2 \\ 1 & 1 & -2 \end{pmatrix}$$

so the eigenvectors of A with eigenvalue 5 are the nonzero solutions of the system

$$-3x_1 + x_2 + 2x_3 = 0$$
$$x_1 - 3x_2 + 2x_3 = 0$$
$$x_1 + x_2 - 2x_3 = 0.$$

Since

$$\begin{pmatrix} -3 & 1 & 2 \\ 1 & -3 & 2 \\ 1 & 1 & -2 \end{pmatrix} \text{ reduces to } \begin{pmatrix} 1 & 0 & -1 \\ 0 & 1 & -1 \\ 0 & 0 & 0 \end{pmatrix},$$

the eigenvectors of A with eigenvalue 5 are the nonzero multiples of $(1, 1, 1)$.

Notice that the results of the above computation give us a good geometric understanding of the linear map $\phi_A: \mathbb{R}^3 \to \mathbb{R}^3$ (see Figure 8.4). On the eigenspace $V_1 = \mathcal{L}((-1, 1, 0), (-2, 0, 1))$, ϕ_A is the identity map ($\phi_A(\mathbf{v}) = \mathbf{v}$ for $\mathbf{v} \in V_1$). On the eigenspace $V_5 = \mathcal{L}((1, 1, 1))$, ϕ_A stretches each vector by a factor of 5 ($\phi_A(\mathbf{v}) = 5\mathbf{v}$ for $\mathbf{v} \in V_5$). ■

Since each linear map $\phi: \mathbb{R}^n \to \mathbb{R}^n$ is ϕ_A for some matrix $A \in \mathfrak{M}_{n \times n}$, it is clear that the eigenvalues of $\phi: \mathbb{R}^n \to \mathbb{R}^n$ can always be found as the zeros of a polynomial [the polynomial $\det(A - \lambda I)$, where A is the matrix of ϕ], and the eigenspace V_λ of ϕ corresponding to an eigenvalue λ can be found as the solution space of a homogeneous linear system [the system $(A - \lambda I)\mathbf{x} = \mathbf{0}$.] This procedure is also valid for linear maps $\phi: V \to V$ where V is any finite dimensional vector space.

Theorem 1. *Let $\phi: V \to V$ be a linear map from a finite dimensional vector space V into itself, let $\mathbf{B}$ be any ordered basis for V, and let $A = M^{\mathbf{B}}{}_{\mathbf{B}}(\phi)$ be the matrix of ϕ with respect to $\mathbf{B}$. Then*

(i)   *a real number λ is an eigenvalue of ϕ if and only if it is an eigenvalue of A, and*

(ii)   *a vector $\mathbf{v} \in V$ is an eigenvector of ϕ with eigenvalue λ if and only if its coordinate matrix $M_\mathbf{B}(\mathbf{v})$ is an eigenvector of A with the same eigenvalue.*

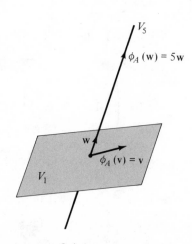

FIGURE 8.4

Proof.   The vector equation

$$\phi(\mathbf{v}) = \lambda \mathbf{v}$$

is equivalent to the matrix equation

$$M^{\mathbf{B}}{}_{\mathbf{B}}(\phi)\, M_{\mathbf{B}}(\mathbf{v}) = M_{\mathbf{B}}(\lambda \mathbf{v}).$$

Since $M^{\mathbf{B}}{}_{\mathbf{B}}(\phi) = A$ and $M_{\mathbf{B}}(\lambda \mathbf{v}) = \lambda M_{\mathbf{B}}(\mathbf{v})$, this last equation is equivalent to the equation

$$A\, M_{\mathbf{B}}(\mathbf{v}) = \lambda M_{\mathbf{B}}(\mathbf{v}). \quad \blacksquare$$

Example 5.   Let $D: \mathcal{P}^4 \to \mathcal{P}^4$ denote differentiation and let $\mathbf{B} = (1, x, x^2, x^3)$. Then

$$A = M^{\mathbf{B}}{}_{\mathbf{B}}(D) = \begin{pmatrix} 0 & 1 & 0 & 0 \\ 0 & 0 & 2 & 0 \\ 0 & 0 & 0 & 3 \\ 0 & 0 & 0 & 0 \end{pmatrix}$$

and

$$\det(A - \lambda I) = \det \begin{pmatrix} -\lambda & 1 & 0 & 0 \\ 0 & -\lambda & 2 & 0 \\ 0 & 0 & -\lambda & 3 \\ 0 & 0 & 0 & -\lambda \end{pmatrix} = \lambda^4.$$

Hence $\lambda = 0$ is the only eigenvalue of $D: \mathcal{P}^4 \to \mathcal{P}^4$.

The eigenvectors of A with eigenvalue 0 are the nonzero multiples of $(1, 0, 0, 0)$. Hence the 0-eigenspace of $D: \mathcal{P}^4 \to \mathcal{P}^4$ is

$$V_0 = \{p(x) \in \mathcal{P}^4 | M_{\mathbf{B}}(p(x)) = (c, 0, 0, 0) \text{ for some } c \in \mathbb{R}\};$$

this is the space of constant polynomials. $\quad \blacksquare$

Since any multiple of an eigenvector is also an eigenvector with the same eigenvalue, it is clear that eigenvectors with distinct eigenvalues cannot be multiples of one another. In other words, each pair of eigenvectors with distinct eigenvalues is linearly independent. More generally, we can prove the following.

Theorem 2.   *Let $\phi: V \to V$ be a linear map and suppose $\mathbf{v}_1, \ldots, \mathbf{v}_k$ are eigenvectors of ϕ with distinct eigenvalues $\lambda_1, \ldots, \lambda_k$. Then $\{\mathbf{v}_1, \ldots, \mathbf{v}_k\}$ is linearly independent.*

Proof.   We shall prove independence by showing that dependence would imply that two of the eigenvalues were equal.

Suppose, then, that $\{v_1, \ldots, v_k\}$ is dependent. Let m be the largest integer for which $\{v_1, \ldots, v_m\}$ is independent. Thus $m < k$, $\{v_1, \ldots, v_m\}$ is independent, and

$$v_{m+1} = a_1 v_1 + \cdots + a_m v_m$$

for some $a_1, \ldots, a_m \in \mathbb{R}$. If we apply ϕ to both sides of this equation, we get

$$\phi(v_{m+1}) = a_1 \phi(v_1) + \cdots + a_m \phi(v_m)$$

or

$$\lambda_{m+1} v_{m+1} = a_1 \lambda_1 v_1 + \cdots + a_m \lambda_m v_m.$$

On the other hand,

$$\lambda_{m+1} v_{m+1} = \lambda_{m+1}(a_1 v_1 + \cdots + a_m v_m) = a_1 \lambda_{m+1} v_1 + \cdots + a_m \lambda_{m+1} v_m.$$

Subtracting, we get

$$0 = a_1(\lambda_{m+1} - \lambda_1)v_1 + \cdots + a_m(\lambda_{m+1} - \lambda_m)v_m$$

or, since $\{v_1, \ldots, v_m\}$ is independent,

$$a_1(\lambda_{m+1} - \lambda_1) = \cdots = a_m(\lambda_{m+1} - \lambda_m) = 0.$$

But a_i must be different from zero for at least one i (since $v_{m+1} \neq 0$). For this i we must have $\lambda_i = \lambda_{m+1}$. ∎

Corollary.   *Let V be a vector space of dimension n and let $\phi: V \to V$ be a linear map with n distinct eigenvalues. Then V has a basis consisting of eigenvectors of ϕ.*

Remark.   It is very desirable to be able to find an ordered basis **B** consisting of eigenvectors of ϕ, because the matrix $M^B{}_B(\phi)$ of ϕ with respect to such a basis will be especially simple: it will be a diagonal matrix (all entries off the main diagonal will be zero) whose diagonal entries will be the eigenvalues of ϕ. According to the corollary, this will always be possible when the characteristic polynomial $\det(A - \lambda I)$ has n distinct real zeros, where $n = \dim V$ and A is the matrix of ϕ with respect to any ordered basis for V.

Example 6.   Let

$$A = \begin{pmatrix} 0 & -2 & 2 \\ -3 & 1 & 3 \\ -1 & 1 & 3 \end{pmatrix}.$$

Let us find an ordered basis for $\mathbb{R}^3$ with respect to which the matrix of the linear map $\phi_A\colon \mathbb{R}^3 \to \mathbb{R}^3$ is diagonal.

First we compute the characteristic polynomial:

$$\det(A - \lambda I) = \det\begin{pmatrix} -\lambda & -2 & 2 \\ -3 & 1-\lambda & 3 \\ -1 & 1 & 3-\lambda \end{pmatrix}$$

$$= -\lambda[(1-\lambda)(3-\lambda) - 3] - (-2)[-3(3-\lambda) + 3] + 2[-3 + (1-\lambda)]$$

$$= -\lambda(-4\lambda + \lambda^2) + 2(3\lambda - 6) + 2(-\lambda - 2)$$

$$= -\lambda^3 + 4\lambda^2 + 4\lambda - 16 = -(\lambda - 2)(\lambda + 2)(\lambda - 4).$$

Hence ϕ_A has eigenvalues $2, -2, 4$.

Next we solve the three homogeneous systems whose coefficient matrices are $A - \lambda I, \lambda = 2, -2, 4$.

$$A - 2I = \begin{pmatrix} -2 & -2 & 2 \\ -3 & -1 & 3 \\ -1 & 1 & 1 \end{pmatrix} \rightarrow \begin{pmatrix} 1 & 0 & -1 \\ 0 & 1 & 0 \\ 0 & 0 & 0 \end{pmatrix}$$

$$A - (-2)I = \begin{pmatrix} 2 & -2 & 2 \\ -3 & 3 & 3 \\ -1 & 1 & 5 \end{pmatrix} \rightarrow \begin{pmatrix} 1 & -1 & 0 \\ 0 & 0 & 1 \\ 0 & 0 & 0 \end{pmatrix}$$

$$A - 4I = \begin{pmatrix} -4 & -2 & 2 \\ -3 & -3 & 3 \\ -1 & 1 & -1 \end{pmatrix} \rightarrow \begin{pmatrix} 1 & 0 & 0 \\ 0 & 1 & -1 \\ 0 & 0 & 0 \end{pmatrix}.$$

Hence eigenvectors corresponding to $\lambda = 2, -2, 4$ are, respectively,

$$(1, 0, 1), \ (1, 1, 0), \ (0, 1, 1).$$

Since the eigenvalues are distinct, these vectors form an independent set, hence a basis for $\mathbb{R}^3$. If $\mathbf{B} = ((1, 0, 1), (1, 1, 0), (0, 1, 1))$ then

$$M^{\mathbf{B}}{}_{\mathbf{B}}(\phi_A) = \begin{pmatrix} 2 & 0 & 0 \\ 0 & -2 & 0 \\ 0 & 0 & 4 \end{pmatrix}. \ \blacksquare$$

Example 7. Let $\phi_A\colon \mathbb{R}^3 \to \mathbb{R}^3$ be the linear map associated with the matrix

$$A = \begin{pmatrix} 2 & 1 & 2 \\ 1 & 2 & 2 \\ 1 & 1 & 3 \end{pmatrix}.$$

We have already seen, in Example 4, that ϕ_A has eigenvalues 1 and 5 and corresponding eigenspaces

$$V_1 = \mathcal{L}((-1, 1, 0), (-2, 0, 1))$$

and

$$V_5 = \mathcal{L}((1, 1, 1)).$$

This linear map does not have three distinct eigenvalues. Nevertheless, the set $\{(-1, 1, 0), (-2, 0, 1), (1, 1, 1)\}$ of eigenvectors is linearly independent, hence a basis for $\mathbb{R}^3$. The matrix for ϕ_A with respect to the ordered basis $\mathbf{B} = ((-1, 1, 0), (-2, 0, 1), (1, 1, 1))$ is

$$M^{\mathbf{B}}{}_{\mathbf{B}}(\phi_A) = \begin{pmatrix} 1 & 0 & 0 \\ 0 & 1 & 0 \\ 0 & 0 & 5 \end{pmatrix}. \quad \blacksquare$$

Example 8. Let $D: \mathcal{P}^4 \to \mathcal{P}^4$, as in Example 5. This linear map does not have an eigenvector basis, since the only eigenvectors are the constant polynomials. Thus there is no ordered basis for $\mathcal{P}^4$ with respect to which the matrix for D is a diagonal matrix. $\blacksquare$

EXERCISES

1. Find the eigenvalues and corresponding eigenvectors of the following matrices.

(a) $\begin{pmatrix} 2 & 1 \\ -1 & 2 \end{pmatrix}$ (d) $\begin{pmatrix} 33 & -10 \\ 105 & -32 \end{pmatrix}$ (g) $\begin{pmatrix} 13 & -4 \\ 42 & -13 \end{pmatrix}$

(b) $\begin{pmatrix} 5 & -2 \\ -2 & 5 \end{pmatrix}$ (e) $\begin{pmatrix} -11 & -14 \\ 7 & 10 \end{pmatrix}$ (h) $\begin{pmatrix} 0 & -1 \\ 1 & 0 \end{pmatrix}$

(c) $\begin{pmatrix} 1 & -1 \\ 1 & 1 \end{pmatrix}$ (f) $\begin{pmatrix} 0 & -2 \\ -2 & 0 \end{pmatrix}$

2. Find the eigenvalues and corresponding eigenvectors of

(a) $\begin{pmatrix} 1 & -1 & 1 \\ 1 & 1 & 2 \\ 2 & 0 & 3 \end{pmatrix}$ (c) $\begin{pmatrix} 31 & -100 & 70 \\ 18 & -59 & 42 \\ 12 & -40 & 29 \end{pmatrix}$

(b) $\begin{pmatrix} 1 & 1 & 1 \\ 1 & 0 & -2 \\ 1 & -1 & 1 \end{pmatrix}$ (d) $\begin{pmatrix} 7 & -9 & -15 \\ 0 & 4 & 0 \\ 3 & -9 & -11 \end{pmatrix}$

(e) $\begin{pmatrix} 1 & 0 & 1 \\ 0 & 1 & 0 \\ 1 & 0 & -1 \end{pmatrix}$

3. Find the eigenvalues and corresponding eigenvectors of each of the following linear maps.

   (a) $\phi: \mathbb{R}^2 \to \mathbb{R}^2$, $\phi(x_1, x_2) = (-2x_1 + 2x_2, 2x_1 + 3x_2)$

   (b) $\phi: \mathfrak{D}^\infty \to \mathfrak{D}^\infty$, $\phi = D^2$

   (c) $\phi: \mathfrak{D}^\infty \to \mathfrak{D}^\infty$, $\phi = D^2 + 1$

   (d) $\phi: \mathfrak{D}^\infty \to \mathfrak{D}^\infty$, $\phi = D^2 + D$

   (e) $\phi: \mathcal{C}(\mathbb{R}) \to \mathcal{C}(\mathbb{R})$, $\phi(f) = \int_0^x f$

4. Find the eigenvalues and corresponding eigenvectors of

   (a) $\phi: \mathcal{P}^2 \to \mathcal{P}^2$, $\phi(p(x)) = p(-x)$

   (b) $\phi: \mathcal{P} \to \mathcal{P}$, $\phi(p(x)) = p(-x)$

   (c) $\phi: \mathcal{P} \to \mathcal{P}$, $\phi(p(x)) = xp'(x)$

   (d) $\phi: \mathfrak{D}^\infty \to \mathfrak{D}^\infty$, $\phi = xD$

5. Let $r_\theta: \mathbb{R}^2 \to \mathbb{R}^2$ be rotation counterclockwise through the angle θ.

   (a) Show that if θ is not an integer multiple of π, then r_θ has no eigenvectors.

   (b) What are the eigenvalues and corresponding eigenvectors of r_θ if θ is an integer multiple of π?

6. (a) Show that if $\phi: V \to V$ is a linear map and $\mathbf{v} \in V$ is an eigenvector of ϕ with eigenvalue λ, then $\mathbf{v}$ is also an eigenvector of $\phi^2 = \phi \circ \phi$. What is the corresponding eigenvalue?

   (b) Show that $\mathbf{v} \in V$ may be an eigenvector of ϕ^2 even though it is not an eigenvector of ϕ. [Hint: Consider the linear map $D: \mathfrak{D}^\infty \to \mathfrak{D}^\infty$.]

7. Show that if $\phi: V \to V$ is an invertible linear map, then $\mathbf{v} \in V$ is an eigenvector of ϕ if and only if it is an eigenvector of ϕ^{-1}. What is the relationship between the eigenvalues?

8. Show that if the dimension of V is odd, then each linear map $\phi: V \to V$ must have at least one eigenvector. [Hint: What is the degree of the characteristic polynomial of $M^{\mathbf{B}}{}_{\mathbf{B}}(\phi)$, where $\mathbf{B}$ is an ordered basis for V?]

9. A linear map $\phi: \mathbb{R}^n \to \mathbb{R}^n$ is said to be *self-adjoint* if $\phi(\mathbf{v}) \cdot \mathbf{w} = \mathbf{v} \cdot \phi(\mathbf{w})$ for all $\mathbf{v}, \mathbf{w} \in V$.

   (a) Show that if $\phi: \mathbb{R}^n \to \mathbb{R}^n$ is self-adjoint and $\mathbf{v}, \mathbf{w} \in V$ are eigenvectors of ϕ with distinct eigenvalues, then $\mathbf{v}$ is perpendicular to $\mathbf{w}$. [Hint: Evaluate $\phi(\mathbf{v}) \cdot \mathbf{w}$ in two different ways.] Conclude that, if dim $V = n$ and if $\phi: V \to V$ is self-adjoint and has n distinct eigenvalues, then there is an orthonormal basis for V consisting of eigenvectors of ϕ.

   (b) Show that the results of part (a) are valid in any inner product space V, where $\phi: V \to V$ is self-adjoint if $\langle \phi(\mathbf{v}), \mathbf{w} \rangle = \langle \mathbf{v}, \phi(\mathbf{w}) \rangle$ for all $\mathbf{v}, \mathbf{w} \in V$.

8.2 DIAGONALIZATION OF MATRICES

A linear map $\phi: V \to V$, where V is a finite dimensional vector space, is said to be *diagonable* if there is an ordered basis $\mathbf{B}$ for V such that the matrix $M^{\mathbf{B}}{}_{\mathbf{B}}(\phi)$ of ϕ with respect to $\mathbf{B}$ is a diagonal matrix. We say that such a basis $\mathbf{B}$ *diagonalizes* ϕ. From the previous section, we know that $\mathbf{B}$ diagonalizes ϕ if and only if each vector in **B** is an eigenvector of ϕ. Furthermore, the diagonal elements of $M^\mathbf{B}_\mathbf{B}(\phi)$, where **B** is a diagonalizing basis, are the corresponding eigenvalues of ϕ.

A square matrix A is said to be *diagonable* if the associated linear map $\phi_A \colon \mathbb{R}^n \to \mathbb{R}^n$ is diagonable.

Theorem 1. $A \in \mathfrak{M}_{n \times n}$ *is diagonable if and only if there is an invertible matrix* $P \in \mathfrak{M}_{n \times n}$ *such that* $P^{-1}AP$ *is a diagonal matrix.*

Proof. Let $\mathbf{B}_1 = (\mathbf{e}_1, \ldots, \mathbf{e}_n)$ be the standard ordered basis for $\mathbb{R}^n$ and let $\mathbf{B}_2 = (\mathbf{v}_1, \ldots, \mathbf{v}_n)$ be any other ordered basis for $\mathbb{R}^n$. Then, by Theorem 3 of Section 7.4, we have, for each $A \in \mathfrak{M}_{n \times n}$,

$$M^{\mathbf{B}_2}_{\mathbf{B}_2}(\phi_A) = P^{-1} M^{\mathbf{B}_1}_{\mathbf{B}_1}(\phi_A) P$$

where P is the matrix whose columns are the vectors $M_{\mathbf{B}_1}(\mathbf{v}_j), j = 1, \ldots, n$. Since $\mathbf{B}_1$ is the standard ordered basis for $\mathbb{R}^n$, we find $M^{\mathbf{B}_1}_{\mathbf{B}_1}(\phi_A) = A$ and $M_{\mathbf{B}_1}(\mathbf{v}_j) = \mathbf{v}_j$ for each j. Hence

$$M^{\mathbf{B}_2}_{\mathbf{B}_2}(\phi_A) = P^{-1}AP$$

where P is the matrix whose columns are the vectors $\mathbf{v}_j, j = 1, \ldots, n$.

Now, if A is diagonable then there exists an ordered basis $\mathbf{B}_2$ such that $M^{\mathbf{B}_2}_{\mathbf{B}_2}(\phi_A)$ is a diagonal matrix, so $M^{\mathbf{B}_2}_{\mathbf{B}_2}(\phi) = P^{-1}AP$ is diagonal, where P is the matrix whose columns are the vectors in $\mathbf{B}_2$. Conversely, if $P^{-1}AP$ is diagonal for some P, then we may take $\mathbf{B}_2$ to be the ordered basis for $\mathbb{R}^n$ consisting of the column vectors of P. Since $M^{\mathbf{B}_2}_{\mathbf{B}_2}(\phi_A) = P^{-1}AP$ is diagonal, we see that $\mathbf{B}_2$ diagonalizes ϕ_A, so A is diagonable. ∎

We say that an invertible matrix P *diagonalizes* a matrix A if $P^{-1}AP$ is a diagonal matrix. Given a diagonable matrix $A \in \mathfrak{M}_{n \times n}$, we can construct a diagonalizing matrix P as follows: (i) find a basis for $\mathbb{R}^n$ consisting of eigenvectors of A and (ii) use these vectors as the columns of P. The diagonal entries in $P^{-1}AP$ will be the corresponding eigenvalues of A.

Example 1. Let us find an invertible matrix P and a diagonal matrix D such that $P^{-1}AP = D$, where

$$A = \begin{pmatrix} -17 & 30 \\ -10 & 18 \end{pmatrix}.$$

The characteristic polynomial of A is

$$\det(A - \lambda I) = \det \begin{pmatrix} -17 - \lambda & 30 \\ -10 & 18 - \lambda \end{pmatrix}$$

$$= (-17 - \lambda)(18 - \lambda) + 300 = \lambda^2 - \lambda - 6$$

$$= (\lambda - 3)(\lambda + 2).$$

Hence the eigenvalues of A are 3 and -2. We need to find eigenvectors corresponding to these eigenvalues. Row reduction yields, for $\lambda = 3$,

$$\begin{pmatrix} -20 & 30 \\ -10 & 15 \end{pmatrix} \rightarrow \begin{pmatrix} 1 & -\frac{3}{2} \\ 0 & 0 \end{pmatrix},$$

and, for $\lambda = -2$,

$$\begin{pmatrix} -15 & 30 \\ -10 & 20 \end{pmatrix} \rightarrow \begin{pmatrix} 1 & -2 \\ 0 & 0 \end{pmatrix}.$$

Hence $(\frac{3}{2}, 1)$ is an eigenvector with eigenvalue 3 and $(2, 1)$ is an eigenvector with eigenvalue 2. Since $\mathbf{B} = ((\frac{3}{2}, 1), (2, 1))$ is a basis for $\mathbb{R}^2$ consisting of eigenvectors of A, the matrix

$$P = \begin{pmatrix} \frac{3}{2} & 2 \\ 1 & 1 \end{pmatrix}$$

diagonalizes A. Since the diagonal entries in $D = P^{-1}AP$ are the corresponding eigenvalues, we conclude that $D = \begin{pmatrix} 3 & 0 \\ 0 & -2 \end{pmatrix}$.

As a check, notice that

$$PD = \begin{pmatrix} \frac{3}{2} & 2 \\ 1 & 1 \end{pmatrix} \begin{pmatrix} 3 & 0 \\ 0 & -2 \end{pmatrix} = \begin{pmatrix} \frac{9}{2} & -4 \\ 3 & -2 \end{pmatrix}$$

and that

$$AP = \begin{pmatrix} -17 & 30 \\ -10 & 18 \end{pmatrix} \begin{pmatrix} \frac{3}{2} & 2 \\ 1 & 1 \end{pmatrix} = \begin{pmatrix} \frac{9}{2} & -4 \\ 3 & -2 \end{pmatrix};$$

hence $PD = AP$, or $D = P^{-1}AP$. ∎

Example 2. Let

$$A = \begin{pmatrix} 1 & 3 & -3 \\ 15 & -5 & 21 \\ 3 & -3 & 7 \end{pmatrix}.$$

Then

$$\det(A - \lambda I) = \det \begin{pmatrix} 1 - \lambda & 3 & -3 \\ 15 & -5 - \lambda & 21 \\ 3 & -3 & 7 - \lambda \end{pmatrix} = (1 - \lambda)(\lambda - 4)(\lambda + 2)$$

so the eigenvalues of A are $1, 4, -2$. Row reduction of the appropriate matrices shows that the corresponding eigenvectors are, respectively,

$$(-1, 1, 1), (-2, -1, 1) \text{ and } (-1, 2, 1).$$

Hence

$$P^{-1}AP = D$$

where

$$P = \begin{pmatrix} -1 & -2 & -1 \\ 1 & -1 & 2 \\ 1 & 1 & 1 \end{pmatrix} \text{ and } D = \begin{pmatrix} 1 & 0 & 0 \\ 0 & 4 & 0 \\ 0 & 0 & -2 \end{pmatrix}. \blacksquare$$

EXERCISES

1. Which of the following matrices are diagonable?

   (a) $\begin{pmatrix} 2 & 3 \\ 3 & 2 \end{pmatrix}$ (c) $\begin{pmatrix} 1 & 1 \\ 0 & 1 \end{pmatrix}$

   (b) $\begin{pmatrix} 0 & 1 \\ 0 & 0 \end{pmatrix}$ (d) $\begin{pmatrix} 2 & 1 & 1 \\ 1 & 2 & 1 \\ 2 & 2 & 3 \end{pmatrix}$

2. For each of the matrices A in Exercise 1 that is diagonable, find a diagonalizing matrix P and a diagonal matrix D such that $P^{-1}AP = D$.

3. Show that each $A \in \mathfrak{M}_{2 \times 2}$ of the form

$$A = \begin{pmatrix} a & b \\ b & c \end{pmatrix}$$

   is diagonable. [Hint: Show that, unless $b = 0$, the characteristic polynomial of A has two distinct real zeros.]

4. Diagonalize each of the following matrices by finding an invertible matrix P and a diagonal matrix D such that $P^{-1}AP = D$.

   (a) $A = \begin{pmatrix} -3 & 4 \\ -2 & 3 \end{pmatrix}$ (c) $A = \begin{pmatrix} -8 & -9 \\ 18 & 19 \end{pmatrix}$ (e) $A = \begin{pmatrix} -11 & -21 \\ 10 & 18 \end{pmatrix}$

   (b) $A = \begin{pmatrix} 47 & -50 \\ 40 & -43 \end{pmatrix}$ (d) $A = \begin{pmatrix} 1 & 1 \\ 1 & -1 \end{pmatrix}$

5. Diagonalize each of the following matrices by finding an invertible matrix P and a diagonal matrix D such that $P^{-1}AP = D$.

   (a) $A = \begin{pmatrix} 5 & 1 & -2 \\ 6 & 4 & -4 \\ 9 & 3 & -4 \end{pmatrix}$ (c) $A = \begin{pmatrix} 1 & -1 & 1 \\ -3 & 5 & -1 \\ -7 & 11 & -3 \end{pmatrix}$

   (b) $A = \begin{pmatrix} 11 & 6 & 6 \\ 29 & 16 & 16 \\ -48 & -24 & -25 \end{pmatrix}$ (d) $A = \begin{pmatrix} 8 & 11 & -27 \\ 9 & 18 & -39 \\ 5 & 9 & -20 \end{pmatrix}$

6. Let $A \in \mathfrak{M}_{n \times n}$ and let $P \in \mathfrak{M}_{n \times n}$ be invertible.
    (a) Show that $(P^{-1}AP)^2 = P^{-1}A^2P$.
    (b) Show that $(P^{-1}AP)^k = P^{-1}A^kP$ for each positive integer k.

7. Exercise 6(b) suggests a method for computing powers of diagonable matrices. If $D = P^{-1}AP$ is diagonal, then $A = PDP^{-1}$ so $A^k = PD^kP^{-1}$. Use this method to find A^5 where $A =$

    (a) $\begin{pmatrix} 1 & 2 \\ 2 & 1 \end{pmatrix}$ (c) $\begin{pmatrix} 13 & -4 \\ 42 & -13 \end{pmatrix}$

    (b) $\begin{pmatrix} -17 & 30 \\ -10 & 18 \end{pmatrix}$ (d) $\begin{pmatrix} 33 & -10 \\ 105 & -32 \end{pmatrix}$

8.3 LINEAR SYSTEMS OF DIFFERENTIAL EQUATIONS

An important application of the diagonalization of matrices is in solving systems of first order linear differential equations with constant coefficients. These are systems of the form:

$$\frac{dx_1}{dt} = a_{11}x_1 + a_{12}x_2 + \cdots + a_{1n}x_n$$

$$\frac{dx_2}{dt} = a_{21}x_1 + a_{22}x_2 + \cdots + a_{2n}x_n$$

$$\cdots$$

$$\frac{dx_n}{dt} = a_{n1}x_1 + a_{n2}x_2 + \cdots + a_{nn}x_n,$$

where $x_1, x_2, \ldots, x_n$ are differentiable functions of t and $a_{ij} \in \mathbb{R}$ for $1 \leq i, j \leq n$. A *solution* of this system is an n-tuple $(x_1, \ldots, x_n)$ of functions that satisfies these n equations simultaneously.

Using matrices we can write this system as

$$\begin{pmatrix} \dfrac{dx_1}{dt} \\ \vdots \\ \dfrac{dx_n}{dt} \end{pmatrix} = \begin{pmatrix} a_{11} & \cdots & a_{1n} \\ & \cdots & \\ a_{n1} & \cdots & a_{nn} \end{pmatrix} \begin{pmatrix} x_1 \\ \vdots \\ x_n \end{pmatrix}$$

or, more compactly, as

$$\frac{d\mathbf{x}}{dt} = A\mathbf{x},$$

where

$$\mathbf{x} = \begin{pmatrix} x_1 \\ \vdots \\ x_n \end{pmatrix}, \quad A = \begin{pmatrix} a_{11} & \cdots & a_{1n} \\ & \cdots & \\ a_{n1} & \cdots & a_{nn} \end{pmatrix},$$

and

$$\frac{d\mathbf{x}}{dt} = \begin{pmatrix} \dfrac{dx_1}{dt} \\ \vdots \\ \dfrac{dx_n}{dt} \end{pmatrix}.$$

Notice first that if A is a diagonal matrix, say

$$A = \begin{pmatrix} \lambda_1 & & \\ & \ddots & \\ & & \lambda_n \end{pmatrix},$$

then our system reduces to

$$\frac{dx_1}{dt} = \lambda_1 x_1$$

$$\frac{dx_2}{dt} = \lambda_2 x_2$$

$$\cdots$$

$$\frac{dx_n}{dt} = \lambda_n x_n$$

with general solution

$$\begin{aligned} x_1 &= c_1 e^{\lambda_1 t} \\ x_2 &= c_2 e^{\lambda_2 t} \\ &\cdots \\ x_n &= c_n e^{\lambda_n t} \end{aligned} \quad \text{or} \quad \mathbf{x} = \begin{pmatrix} c_1 e^{\lambda_1 t} \\ \vdots \\ c_n e^{\lambda_n t} \end{pmatrix}$$

where $c_1, \ldots, c_n \in \mathbb{R}$ are arbitrary constants.

Now suppose that we wish to solve $\dfrac{d\mathbf{x}}{dt} = A\mathbf{x}$, where A is not *diagonal* but is *diagonable*. Let $P \in \mathfrak{M}_{n \times n}$ diagonalize A, so that $P^{-1}AP = D$, or $A = PDP^{-1}$, where D is a diagonal matrix. Then the equation $\dfrac{d\mathbf{x}}{dt} = A\mathbf{x}$ becomes

$$\frac{d\mathbf{x}}{dt} = PDP^{-1}\mathbf{x}$$

or

$$P^{-1} \frac{d\mathbf{x}}{dt} = D(P^{-1}\mathbf{x}).$$

Let $\mathbf{u} = P^{-1}\mathbf{x}$. Then $\dfrac{d\mathbf{u}}{dt} = P^{-1} \dfrac{d\mathbf{x}}{dt}$, so our equation becomes

$$\frac{d\mathbf{u}}{dt} = D\mathbf{u}.$$

Since D is diagonal, the general solution of this equation is

$$\mathbf{u} = \begin{pmatrix} c_1 e^{\lambda_1 t} \\ \vdots \\ c_n e^{\lambda_n t} \end{pmatrix}$$

where $\lambda_1, \ldots, \lambda_n$ are the diagonal entries in D; that is, $\lambda_1, \ldots, \lambda_n$ are the eigenvalues of A. Thus the general solution of the equation $\dfrac{d\mathbf{x}}{dt} = A\mathbf{x}$ is $\mathbf{x} = P\mathbf{u}$ where P diagonalizes A and $\mathbf{u}$ is as above. In other words,

$$\mathbf{x} = \begin{pmatrix} | & & | \\ \mathbf{v}_1 & \cdots & \mathbf{v}_n \\ | & & | \end{pmatrix} \begin{pmatrix} c_1 e^{\lambda_1 t} \\ \vdots \\ c_n e^{\lambda_n t} \end{pmatrix}$$

where $\{\mathbf{v}_1, \ldots, \mathbf{v}_n\}$ is a basis for $\mathbb{R}^n$ consisting of eigenvectors of A and $\lambda_1, \ldots, \lambda_n$ are the corresponding eigenvalues. If we carry out the matrix multiplication, we see that $\mathbf{x}$ can be rewritten as

$$\mathbf{x} = c_1 e^{\lambda_1 t} \mathbf{v}_1 + \cdots + c_n e^{\lambda_n t} \mathbf{v}_n.$$

We have proved the following theorem.

Theorem 1. *Suppose $A \in \mathfrak{M}_{n \times n}$ has n linearly independent eigenvectors $\mathbf{v}_1, \ldots, \mathbf{v}_n$ with corresponding eigenvalues $\lambda_1, \ldots, \lambda_n$. Then the general solution of the system*

$$\frac{d\mathbf{x}}{dt} = A\mathbf{x}$$

is

$$\mathbf{x} = c_1 e^{\lambda_1 t} \mathbf{v}_1 + \cdots + c_n e^{\lambda_n t} \mathbf{v}_n$$

where $c_1, \ldots, c_n$ are arbitrary constants.

Example 1. To solve the system

$$\frac{dx_1}{dt} = -17x_1 + 30x_2$$

$$\frac{dx_2}{dt} = -10x_1 + 18x_2,$$

recall that, in Example 1 of the previous section, we found that the vectors $(\frac{3}{2}, 1)$ and $(2, 1)$ are eigenvectors of the matrix

$$A = \begin{pmatrix} -17 & 30 \\ -10 & 18 \end{pmatrix}$$

with corresponding eigenvalues 3 and -2. The general solution is therefore

$$\begin{pmatrix} x_1 \\ x_2 \end{pmatrix} = c_1 e^{3t} \begin{pmatrix} \frac{3}{2} \\ 1 \end{pmatrix} + c_2 e^{-2t} \begin{pmatrix} 2 \\ 1 \end{pmatrix}$$

or

$$x_1 = \tfrac{3}{2} c_1 e^{3t} + 2 c_2 e^{-2t}$$
$$x_2 = c_1 e^{3t} + c_2 e^{-2t}. \quad \blacksquare$$

Example 2. We can solve the system

$$\frac{dx_1}{dt} = x_1 + 3x_2 - 3x_3$$

$$\frac{dx_2}{dt} = 15x_1 - 5x_2 + 21x_3$$

$$\frac{dx_3}{dt} = 3x_1 - 3x_2 + 7x_3$$

using the results of our computations in Example 2 of the previous section. We found there that $(-1, 1, 1)$, $(-2, -1, 1)$, and $(-1, 2, 1)$ are eigenvectors of

$$A = \begin{pmatrix} 1 & 3 & -3 \\ 15 & -5 & 21 \\ 3 & -3 & 7 \end{pmatrix}$$

with corresponding eigenvalues 1, 4, and -2. Hence the general solution is

$$\begin{pmatrix} x_1 \\ x_2 \\ x_3 \end{pmatrix} = c_1 e^{t} \begin{pmatrix} -1 \\ 1 \\ 1 \end{pmatrix} + c_2 e^{4t} \begin{pmatrix} -2 \\ -1 \\ 1 \end{pmatrix} + c_3 e^{-2t} \begin{pmatrix} -1 \\ 2 \\ 1 \end{pmatrix}$$

or

$$
\begin{aligned}
x_1 &= -c_1e^t - 2c_2e^{4t} - c_3e^{-2t} \\
x_2 &= c_1e^t - c_2e^{4t} + 2c_3e^{-2t} \\
x_3 &= c_1e^t + c_2e^{4t} + c_3e^{-2t}. \quad \blacksquare
\end{aligned}
$$

Theorem 1 tells us how to solve $\dfrac{d\mathbf{x}}{dt} = A\mathbf{x}$ when A is a diagonable matrix. We shall now derive a formula for the general solution of this equation that is valid whether or not A is diagonable. The formula depends on the exponential of a matrix, which we shall now define.

Recall that the exponential of any real number x can be defined by the power series

$$
e^x = 1 + x + \frac{x^2}{2!} + \frac{x^3}{3!} + \cdots.
$$

Since we can compute powers of matrices as well as powers of real numbers, we can also define the exponential of any square matrix A by the formula

$$
e^A = I + A + \frac{1}{2!}A^2 + \frac{1}{3!}A^3 + \cdots.
$$

It can be shown that the series converges for all $A \in \mathfrak{M}_{n \times n}$ in the sense that if

$$
S_n = I + A + \frac{1}{2!}A^2 + \cdots + \frac{1}{n!}A^n
$$

then $\lim\limits_{n \to \infty} s_{ij,n}$ exists for all i and j, where $s_{ij,n}$ is the (i,j)-entry in S_n. The limiting value, $\lim\limits_{n \to \infty} s_{ij,n}$, is, by definition, the (i,j)-entry of e^A.

Example 3. Let

$$
A = \begin{pmatrix} a & 0 \\ 0 & b \end{pmatrix}.
$$

Then

$$
\begin{aligned}
e^A &= I + A + \frac{1}{2!}A^2 + \frac{1}{3!}A^3 + \cdots \\[2mm]
&= \begin{pmatrix} 1 & 0 \\ 0 & 1 \end{pmatrix} + \begin{pmatrix} a & 0 \\ 0 & b \end{pmatrix} + \frac{1}{2!}\begin{pmatrix} a^2 & 0 \\ 0 & b^2 \end{pmatrix} + \frac{1}{3!}\begin{pmatrix} a^3 & 0 \\ 0 & b^3 \end{pmatrix} + \cdots \\[2mm]
&= \begin{pmatrix} 1 + a + \dfrac{a^2}{2!} + \dfrac{a^3}{3!} + \cdots & 0 \\[3mm] 0 & 1 + b + \dfrac{b^2}{2!} + \dfrac{b^3}{3!} + \cdots \end{pmatrix} \\[2mm]
&= \begin{pmatrix} e^a & 0 \\ 0 & e^b \end{pmatrix}.
\end{aligned}
$$

Similarly we can verify that, for any diagonal matrix

$$D = \begin{pmatrix} \lambda_1 & & \\ & \ddots & \\ & & \lambda_n \end{pmatrix}, \quad e^D = \begin{pmatrix} e^{\lambda_1} & & \\ & \ddots & \\ & & e^{\lambda_n} \end{pmatrix}. \quad \blacksquare$$

Example 4. Let us compute e^{tA}, where

$$A = \begin{pmatrix} 0 & -1 \\ 1 & 0 \end{pmatrix}.$$

Since

$$A^2 = \begin{pmatrix} -1 & 0 \\ 0 & -1 \end{pmatrix} = -I,$$

it is easy to compute all the powers of A. We find that

$$A^3 = A^2 A = (-I)A = -A$$
$$A^4 = A^3 A = (-A)A = I$$
$$A^5 = A^4 A = IA = A$$
$$A^6 = A^5 A = A^2 = -I$$

and so on. Hence

$$e^{tA} = I + tA + \frac{1}{2!}(tA)^2 + \frac{1}{3!}(tA)^3 + \frac{1}{4!}(tA)^4 + \frac{1}{5!}(tA)^5 + \frac{1}{6!}(tA)^6 + \cdots$$

$$= I + tA + \frac{1}{2!}t^2 A^2 + \frac{1}{3!}t^3 A^3 + \frac{1}{4!}t^4 A^4 + \frac{1}{5!}t^5 A^5 + \frac{1}{6!}t^6 A^6 + \cdots$$

$$= I + tA - \frac{1}{2!}t^2 I - \frac{1}{3!}t^3 A + \frac{1}{4!}t^4 I + \frac{1}{5!}t^5 A - \frac{1}{6!}t^6 I - \cdots$$

$$= \left(1 - \frac{t^2}{2!} + \frac{t^4}{4!} - \frac{t^6}{6!} + \cdots\right)I + \left(t - \frac{t^3}{3!} + \frac{t^5}{5!} - \cdots\right)A$$

$$= (\cos t)I + (\sin t)A.$$

Thus, for $A = \begin{pmatrix} 0 & -1 \\ 1 & 0 \end{pmatrix}$, $e^{tA} = \begin{pmatrix} \cos t & -\sin t \\ \sin t & \cos t \end{pmatrix}$. $\blacksquare$

Given a matrix A whose entries a_{ij} are functions of a real variable t, we define the derivative $\dfrac{dA}{dt}$ to be the matrix whose entries are $\dfrac{da_{ij}}{dt}$.

Theorem 2. Let $A \in \mathfrak{M}_{n \times n}$. *Then*

(i) $\dfrac{d}{dt}e^{tA} = Ae^{tA}$, *and*

(ii) $e^{-tA} = (e^{tA})^{-1}$.

Proof. First notice that the familiar rules for differentiation of sums and products apply to matrix functions exactly as they do to scalar functions. (To see this, simply write out the definitions of matrix sum and matrix product, and differentiate each entry.) Hence

$$\frac{d}{dt}(e^{tA}) = \frac{d}{dt}(I + tA + \frac{1}{2!}t^2A^2 + \frac{1}{3!}t^3A^3 + \cdots)$$

$$= A + tA^2 + \frac{1}{2!}t^2A^3 + \cdots$$

$$= A(I + tA + \frac{1}{2!}(tA)^2 + \cdots)$$

$$= Ae^{tA},$$

proving (i).

To prove (ii), notice that

$$\frac{d}{dt}(e^{tA}e^{-tA}) = (Ae^{tA})e^{-tA} + e^{tA}(-Ae^{-tA})$$

$$= Ae^{tA}e^{-tA} - Ae^{tA}e^{-tA} = 0.$$

Hence $e^{tA}e^{-tA} = C$ for some constant matrix C. Taking $t = 0$ we see that $C = I$, so $e^{tA}e^{-tA} = I$, and hence $e^{-tA} = (e^{tA})^{-1}$. ■

Now let us return to the equation

$$\frac{d\mathbf{x}}{dt} = A\mathbf{x}$$

where $A \in \mathfrak{M}_{n \times n}$. This equation is very similar to the first order linear differential equation

$$\frac{dx}{dt} = ax,$$

where $a \in \mathbb{R}$, which has general solution $x = ce^{at}$, or $x = e^{ta}c$, where c is an arbitrary constant. So we might guess that the general solution of

$$\frac{d\mathbf{x}}{dt} = A\mathbf{x}$$

is $\mathbf{x} = e^{tA}\mathbf{c}$, where $\mathbf{c} \in \mathbb{R}^n$ is an arbitrary vector constant. (We have written $\mathbf{c}$ to the right of e^{tA} because we want $\mathbf{x}$ to be a column vector!) It is easy to check that this is the correct answer.

Theorem 3. *Let $A \in \mathfrak{M}_{n \times n}$. Then the general solution of the matrix differential equation*

$$\frac{d\mathbf{x}}{dt} = A\mathbf{x}$$

is $\mathbf{x} = e^{tA}\mathbf{c}$, where $\mathbf{c} \in \mathbb{R}^n$ is an arbitrary vector constant.

Proof

$$\frac{d\mathbf{x}}{dt} = A\mathbf{x} \Leftrightarrow \frac{d\mathbf{x}}{dt} - A\mathbf{x} = \mathbf{0}$$

$$\Leftrightarrow e^{-tA}\left(\frac{d\mathbf{x}}{dt} - A\mathbf{x}\right) = \mathbf{0}$$

$$\Leftrightarrow e^{-tA}\frac{d\mathbf{x}}{dt} - Ae^{-tA}\mathbf{x} = \mathbf{0}$$

$$\Leftrightarrow \frac{d}{dt}(e^{-tA}\mathbf{x}) = \mathbf{0}$$

$$\Leftrightarrow e^{-tA}\mathbf{x} = \mathbf{c} \text{ for some } \mathbf{c} \in \mathbb{R}^n$$

$$\Leftrightarrow \mathbf{x} = e^{tA}\mathbf{c} \text{ for some } \mathbf{c} \in \mathbb{R}^n. \quad \blacksquare$$

Example 5. Let us find the general solution of the system

$$\frac{dx_1}{dt} = -x_2$$

$$\frac{dx_2}{dt} = x_1.$$

In matrix form, this system becomes

$$\frac{d\mathbf{x}}{dt} = A\mathbf{x}$$

where $A = \begin{pmatrix} 0 & -1 \\ 1 & 0 \end{pmatrix}$. Its general solution is

$$\mathbf{x} = e^{tA}\mathbf{c}$$

or, since $e^{tA} = \begin{pmatrix} \cos t & -\sin t \\ \sin t & \cos t \end{pmatrix}$ (see Example 4),

$$\begin{pmatrix} x_1 \\ x_2 \end{pmatrix} = \begin{pmatrix} \cos t & -\sin t \\ \sin t & \cos t \end{pmatrix}\begin{pmatrix} c_1 \\ c_2 \end{pmatrix}.$$

We can, if we wish, write this solution as

$$x_1 = c_1 \cos t - c_2 \sin t$$
$$x_2 = c_1 \sin t + c_2 \cos t. \quad \blacksquare$$

Remark. The formula of Theorem 3 is so simple and so appealing that you might wonder why we bothered to prove Theorem 1. The reason is that the matrix e^{tA} is often difficult to compute, and we need to compute e^{tA} in order to get from Theorem 3 explicit formulas for the unknown functions $x_1, \ldots, x_n$. It is possible to prove Theorem 1 from Theorem 3 (see Exercise 10), but the proof we gave for Theorem 1 is more direct.

EXERCISES

1. Use the results of the computations in Examples 4 and 6 in Section 8.1 to help you solve the following systems.

    (a) $\dfrac{dx_1}{dt} = 2x_1 + x_2 + 2x_3$

    $\dfrac{dx_2}{dt} = x_1 + 2x_2 + 2x_3$

    $\dfrac{dx_3}{dt} = x_1 + x_2 + 3x_3$

    (b) $\dfrac{dx_1}{dt} = -2x_2 + 2x_3$

    $\dfrac{dx_2}{dt} = -3x_1 + x_2 + 3x_3$

    $\dfrac{dx_3}{dt} = -x_1 + x_2 + 3x_3$

2. For each of the matrices A in Exercise 1 of the previous section that is diagonable, find the general solution of the system $\dfrac{d\mathbf{x}}{dt} = A\mathbf{x}$.

3. For each of the matrices A in Exercise 1 of the previous section that is *not* diagonable, find the general solution of $\dfrac{d\mathbf{x}}{dt} = A\mathbf{x}$. [Hint: Look for an easy way!]

4. For each matrix A in Exercise 4 of the previous section, find the general solution of the equation $\dfrac{d\mathbf{x}}{dt} = A\mathbf{x}$.

5. For each matrix A in Exercise 5 of the previous section, find the general solution of the equation $\dfrac{d\mathbf{x}}{dt} = A\mathbf{x}$.

6. Compute the matrix e^A, where

   (a) $A = \begin{pmatrix} 2 & 0 \\ 0 & -3 \end{pmatrix}$ (d) $A = \begin{pmatrix} 0 & 1 \\ 0 & 0 \end{pmatrix}$

   (b) $A = \begin{pmatrix} 0 & -\pi \\ \pi & 0 \end{pmatrix}$ (e) $A = \begin{pmatrix} 1 & 1 \\ 0 & 1 \end{pmatrix}$.

   (c) $A = \begin{pmatrix} 0 & \pi/2 \\ -\pi/2 & 0 \end{pmatrix}$

7. Compute e^A, where

   (a) $A = \begin{pmatrix} -1 & 0 & 0 \\ 0 & 1 & 0 \\ 0 & 0 & 0 \end{pmatrix}$ (b) $A = \begin{pmatrix} 0 & 1 & 0 \\ 0 & 0 & 1 \\ 0 & 0 & 0 \end{pmatrix}$

   (c) $A = \begin{pmatrix} 0 & 1 & 1 \\ 0 & 0 & 1 \\ 0 & 0 & 0 \end{pmatrix}$ (d) $A = \begin{pmatrix} 0 & a & b \\ 0 & 0 & c \\ 0 & 0 & 0 \end{pmatrix}$

   (e) $A = \begin{pmatrix} a & 0 & 0 \\ 0 & 0 & -b \\ 0 & b & 0 \end{pmatrix}$.

8. Since matrix addition is commutative but matrix multiplication is not commutative, we cannot expect that the equation $e^A e^B = e^{A+B}$ will always hold. Show that, in fact, $e^A e^B \neq e^{A+B}$ when

$$A = \begin{pmatrix} 0 & -t \\ 0 & 0 \end{pmatrix} \quad \text{and} \quad B = \begin{pmatrix} 0 & 0 \\ t & 0 \end{pmatrix}, t \neq 0.$$

9. Let $A \in \mathfrak{M}_{n \times n}$ and let $P \in \mathfrak{M}_{n \times n}$ be invertible. Show that $P^{-1} e^A P = e^{P^{-1}AP}$.

10. Let $A \in \mathfrak{M}_{n \times n}$ be diagonable. Use the result of Exercise 9 together with Theorem 3 to give an independent proof of Theorem 1. [Hint: If $D = P^{-1}AP$ then $A = PDP^{-1}$. Also, note that if $\mathbf{c}$ is an arbitrary vector constant then so is $P^{-1}\mathbf{c}$.]

11. Consider the inhomogeneous system of first order linear differential equations

$$\frac{d\mathbf{x}}{dt} = A\mathbf{x} + \mathbf{b}$$

where $A \in \mathfrak{M}_{n \times n}$ is invertible and $\mathbf{b} \in \mathbb{R}^n$. Show, by imitating the proof of Theorem 3, that the general solution of this system is

$$\mathbf{x} = e^{tA}\mathbf{c} - A^{-1}\mathbf{b}$$

where $\mathbf{c} \in \mathbb{R}^n$ is arbitrary.

12. Consider the nth order linear differential equation

   (*) $y^{(n)} + a_{n-1}y^{(n-1)} + \cdots + a_1 y' + a_0 y = 0$

   where $a_0, \ldots, a_{n-1} \in \mathbb{R}$.

   (a) Show that y is a solution of (*) if and only if $y = u_1$, where u_1 is the first entry in a solution $\mathbf{u} = \begin{pmatrix} u_1 \\ \vdots \\ u_n \end{pmatrix}$ of the system

$$\frac{d\mathbf{u}}{dt} = A\mathbf{u}, \quad A = \begin{pmatrix} 0 & 1 & 0 & \cdots & 0 \\ 0 & 0 & 1 & \cdots & 0 \\ & & \cdots & & \\ 0 & 0 & 0 & \cdots & 1 \\ -a_0 & -a_1 & -a_2 & \cdots & -a_{n-1} \end{pmatrix}.$$

(b) Calculate the characteristic polynomial of the matrix A in part (a) and show that it is equal to the auxiliary polynomial of the differential equation ($*$).

(c) Use the result of part (b) together with Theorem 3 to show that if the auxiliary polynomial of ($*$) has n distinct real zeros $\lambda_1, \ldots, \lambda_n$, then the solution space of ($*$) is spanned by $\{e^{\lambda_1 t}, \ldots, e^{\lambda_n t}\}$.

8.4 LINEAR DIFFERENCE EQUATIONS

Sequences $\mathbf{x} = (x_1, x_2, x_3, \ldots)$ of numbers arise in all branches of mathematics. Sometimes they are specified by a formula for the nth term. For example, the geometric sequence $(1, \frac{1}{2}, \frac{1}{4}, \frac{1}{8}, \ldots)$ can be described by the formula $x_n = (\frac{1}{2})^{n-1}$, $n = 1, 2, 3, \ldots$. In other cases, a sequence may be specified by a relationship between the nth term and one or more of the earlier terms. For example, the sequence $\mathbf{x}$ defined by the equation

$$x_n - 2x_{n-1} - x_{n-2} = 0, \quad n > 2$$

and satisfying the initial conditions $x_1 = 0$ and $x_2 = 1$ can be calculated to as many terms as desired by solving for x_n. For each $n > 2$ we have

$$x_n = 2x_{n-1} + x_{n-2}$$

so

$$x_3 = 2x_2 + x_1 = 2$$
$$x_4 = 2x_3 + x_2 = 5$$
$$x_5 = 2x_4 + x_3 = 12$$

and so on.

The equation $x_n - 2x_{n-1} - x_{n-2} = 0$, $n > 2$, is an example of a linear difference equation. A **linear difference equation** is an equation of the form

$$a_0 x_n + a_1 x_{n-1} + \cdots + a_k x_{n-k} = b_{n-k}, \quad n > k$$

relating, for all $n > k$, the nth term x_n of a sequence $\mathbf{x} = (x_1, x_2, x_3, \ldots)$ to the k previous terms. The integer k is called the **order** of the difference equation. The coefficients $a_0, \ldots, a_k$ are real numbers, with $a_0 \neq 0$ and $a_k \neq 0$, and $\mathbf{b} =$

$(b_1, b_2, b_3, \ldots)$ is an element of $\mathbb{R}^\infty$. A *solution* of this equation is any $\mathbf{x} \in \mathbb{R}^\infty$ that satisfies it.

It may appear, at first glance, that the equation

$$a_0 x_n + a_1 x_{n-1} + \cdots + a_k x_{n-k} = b_{n-k}, \quad n > k,$$

is not *one equation* but, rather, a *sequence of equations*. Technically, that is correct. However, we can rewrite this sequence of *scalar* equations as the single *vector* equation

$$(a_0 S^k + a_1 S^{k-1} + \cdots + a_k)(\mathbf{x}) = \mathbf{b}$$

where $S: \mathbb{R}^\infty \to \mathbb{R}^\infty$ is the *shift operator* defined by

$$S(x_1, x_2, x_3, \ldots) = (x_2, x_3, x_4, \ldots).$$

It is really the vector equation that is being discussed when we talk about *the* difference equation

$$a_0 x_n + a_1 x_{n-1} + \cdots + a_k x_{n-k} = b_{n-k}, \quad n > k.$$

For an explanation of why these equations are called difference equations, see Exercise 5.

The theory of linear difference equations is completely analogous to the theory of linear differential equations. You can easily check, for example, that if $\mathbf{x}_P \in \mathbb{R}^\infty$ is any particular solution of the equation

$$a_0 x_n + a_1 x_{n-1} + \cdots + a_k x_{n-k} = b_{n-k}, \quad n > k,$$

then every solution of this equation is of the form $\mathbf{x} = \mathbf{x}_P + \mathbf{x}_H$ where $\mathbf{x}_H \in \mathbb{R}^\infty$ satisfies the *associated homogeneous equation*

$$a_0 x_n + a_1 x_{n-1} + \cdots + a_k x_{n-k} = 0, \quad n > k$$

(see Exercise 6). You can also check that the set of solutions of the homogeneous equation is a vector space (a subspace of $\mathbb{R}^\infty$) and that the dimension of this solution space is equal to the order of the equation (see Exercise 7).

We shall confine our attention here to the general second order homogeneous linear difference equation

$$a x_n + b x_{n-1} + c x_{n-2} = 0, \quad n > 2,$$

or

$$x_n = -\frac{b}{a} x_{n-1} - \frac{c}{a} x_{n-2}, \quad n > 2,$$

where $a \neq 0$ and $c \neq 0$. To solve this equation, we let $y_n = x_{n-1}(n > 1)$ so that the equation is transformed into the pair of equations

$$x_n = -\frac{b}{a}x_{n-1} - \frac{c}{a}y_{n-1}$$

$$y_n = x_{n-1}, \qquad\qquad n > 2$$

or, equivalently, into the matrix equation

$$\begin{pmatrix} x_n \\ y_n \end{pmatrix} = \begin{pmatrix} -b/a & -c/a \\ 1 & 0 \end{pmatrix} \begin{pmatrix} x_{n-1} \\ y_{n-1} \end{pmatrix}, \quad n > 2.$$

Setting $A = \begin{pmatrix} -b/a & -c/a \\ 1 & 0 \end{pmatrix}$ we see that

$$\begin{pmatrix} x_n \\ y_n \end{pmatrix} = A \begin{pmatrix} x_{n-1} \\ y_{n-1} \end{pmatrix} = A^2 \begin{pmatrix} x_{n-2} \\ y_{n-2} \end{pmatrix} = \cdots = A^{n-2} \begin{pmatrix} x_2 \\ y_2 \end{pmatrix}$$

for all $n > 2$. Since $y_n = x_{n-1}$, it follows that

$$\begin{pmatrix} x_n \\ x_{n-1} \end{pmatrix} = A^{n-2} \begin{pmatrix} x_2 \\ x_1 \end{pmatrix}$$

for all $n > 2$. This formula is valid also when $n = 2$, since $A^0 = I$, leading us to the following theorem.

Theorem 1. *The general solution of the difference equation*

$$ax_n + bx_{n-1} + cx_{n-2} = 0, \quad n > 2,$$

can be calculated from the formula

$$\begin{pmatrix} x_n \\ x_{n-1} \end{pmatrix} = A^{n-2} \begin{pmatrix} x_2 \\ x_1 \end{pmatrix}, \quad n \geq 2,$$

where $A = \begin{pmatrix} -b/a & -c/a \\ 1 & 0 \end{pmatrix}$ *and where* x_1 *and* $x_2 \in \mathbb{R}$ *are arbitrary.*

In order to apply Theorem 1 to calculate an explicit formula for the nth term x_n of a sequence **x** satisfying the difference equation

$$ax_n + bx_{n-1} + cx_{n-2} = 0, \quad n > 2,$$

we must be able to compute the matrix power A^{n-2}, where

$$A = \begin{pmatrix} -b/a & -c/a \\ 1 & 0 \end{pmatrix}.$$

When A is diagonable, we can do this by diagonalizing A, as follows.
Since

$$\det(A - \lambda I) = \lambda^2 + \frac{b}{a}\lambda + \frac{c}{a} = \frac{1}{a}(a\lambda^2 + b\lambda + c),$$

the eigenvalues of A are the zeros of the polynomial $a\lambda^2 + b\lambda + c$. This polynomial is the **auxiliary polynomial** of the difference equation $ax_n + bx_{n-1} + cx_{n-2} = 0$, $n > 2$. When $b^2 - 4ac > 0$, the zeros of this polynomial, $\lambda_1 = (-b + \sqrt{b^2 - 4ac})/2a$ and $\lambda_2 = (-b - \sqrt{b^2 - 4ac})/2a$, are real and distinct. Hence, in this case, A is diagonable. It is easy to see that $(\lambda_1, 1)$ and $(\lambda_2, 1)$ are eigenvectors corresponding to the eigenvalues λ_1 and λ_2. Hence, if we let

$$P = \begin{pmatrix} \lambda_1 & \lambda_2 \\ 1 & 1 \end{pmatrix}, \text{ then } P^{-1}AP = D = \begin{pmatrix} \lambda_1 & 0 \\ 0 & \lambda_2 \end{pmatrix}.$$

It follows that $A = PDP^{-1}$ and

$$A^{n-2} = (PDP^{-1})(PDP^{-1})\cdots(PDP^{-1}) = PD^{n-2}P^{-1}$$

so, by Theorem 1,

$$\begin{pmatrix} x_n \\ x_{n-1} \end{pmatrix} = A^{n-2}\begin{pmatrix} x_2 \\ x_1 \end{pmatrix} = PD^{n-2}P^{-1}\begin{pmatrix} x_2 \\ x_1 \end{pmatrix}$$

$$= \begin{pmatrix} \lambda_1 & \lambda_2 \\ 1 & 1 \end{pmatrix}\begin{pmatrix} \lambda_1^{n-2} & 0 \\ 0 & \lambda_2^{n-2} \end{pmatrix}\begin{pmatrix} c_1 \\ c_2 \end{pmatrix}, \text{ for } n \geq 2,$$

where

$$\begin{pmatrix} c_1 \\ c_2 \end{pmatrix} = P^{-1}\begin{pmatrix} x_2 \\ x_1 \end{pmatrix}.$$

Note that (c_1, c_2) is an arbitrary vector in $\mathbb{R}^2$, since (x_2, x_1) is arbitrary and P^{-1} is invertible.

If we carry out the matrix multiplication we find that

$$x_n = c_1\lambda_1^{n-1} + c_2\lambda_2^{n-1}$$

$$x_{n-1} = c_1\lambda_1^{n-2} + c_2\lambda_2^{n-2}, \quad n \geq 2,$$

where c_1 and c_2 are arbitrary constants. In other words,

$$\mathbf{x} = c_1(1, \lambda_1, \lambda_1^2, \ldots) + c_2(1, \lambda_2, \lambda_2^2, \ldots).$$

In summary, we have shown the following:

If $b^2 - 4ac > 0$, then the solution space of the difference equation $ax_n + bx_{n-1} + cx_{n-2} = 0$, $n > 2$, is the subspace of $\mathbb{R}^\infty$ spanned by $\{(1, \lambda_1, \lambda_1^2, \ldots), (1, \lambda_2, \lambda_2^2, \ldots)\}$, where λ_1 and λ_2 are the zeros of the auxiliary polynomial $a\lambda^2 + b\lambda + c$.

Example 1. Consider the difference equation

$$6x_n - 5x_{n-1} + x_{n-2} = 0, \quad n > 2.$$

The auxiliary polynomial is

$$6\lambda^2 - 5\lambda + 1 = (2\lambda - 1)(3\lambda - 1).$$

Its zeros are $\lambda_1 = \frac{1}{2}$ and $\lambda_2 = \frac{1}{3}$. Hence the general solution of this difference equation is

$$\mathbf{x} = c_1(1, \tfrac{1}{2}, \tfrac{1}{4}, \tfrac{1}{8}, \ldots) + c_2(1, \tfrac{1}{3}, \tfrac{1}{9}, \tfrac{1}{27}, \ldots). \quad \blacksquare$$

Example 2. Let us find all sequences with the property that each term beyond the second in the sequence is the average of the preceding two terms. This condition may be expressed as $x_n = \frac{1}{2}(x_{n-1} + x_{n-2})$, $n > 2$, or as

$$2x_n - x_{n-1} - x_{n-2} = 0, \quad n > 2.$$

Since the auxiliary polynomial $2\lambda^2 - \lambda - 1$ has zeros $\lambda_1 = 1$ and $\lambda_2 = -\frac{1}{2}$, the general solution is

$$\mathbf{x} = c_1(1, 1, 1, \ldots) + c_2(1, -\tfrac{1}{2}, \tfrac{1}{4}, -\tfrac{1}{8}, \ldots). \quad \blacksquare$$

Example 3. The *Fibonacci sequence* is the sequence $\mathbf{x}$ with $x_1 = x_2 = 1$ and with the property that each term beyond the second is the sum of the preceding two. Thus

$$x_1 = 1, \, x_2 = 1, \, x_3 = 2, \, x_4 = 3, \, x_5 = 5, \, x_6 = 8,$$

and so on. Let us find a formula for the nth term x_n of the Fibonacci sequence.

The property defining the Fibonacci sequence may be expressed as $x_n = x_{n-1} + x_{n-2}$, $n > 2$, or as

$$x_n - x_{n-1} - x_{n-2} = 0, \quad n > 2.$$

The auxiliary polynomial $\lambda^2 - \lambda - 1$ has zeros $(1 \pm \sqrt{5})/2$, and hence this difference equation has general solution

$$\mathbf{x} = c_1\left(1, \frac{1 + \sqrt{5}}{2}, \left(\frac{1 + \sqrt{5}}{2}\right)^2, \dots\right) + c_2\left(1, \frac{1 - \sqrt{5}}{2}, \left(\frac{1 - \sqrt{5}}{2}\right)^2, \dots\right),$$

or

$$x_n = c_1\left(\frac{1 + \sqrt{5}}{2}\right)^{n-1} + c_2\left(\frac{1 - \sqrt{5}}{2}\right)^{n-1}, \quad n \geq 1.$$

The particular solution with $x_1 = x_2 = 1$ must satisfy

$$1 = x_1 = c_1 + c_2$$

$$1 = x_2 = c_1\left(\frac{1 + \sqrt{5}}{2}\right) + c_2\left(\frac{1 - \sqrt{5}}{2}\right).$$

Hence, for the Fibonacci sequence, we must have

$$c_1 = \frac{1 + \sqrt{5}}{2\sqrt{5}} \quad \text{and} \quad c_2 = -\frac{1 - \sqrt{5}}{2\sqrt{5}}.$$

Therefore, the nth term of the sequence is

$$x_n = \frac{1}{\sqrt{5}}\left[\left(\frac{1 + \sqrt{5}}{2}\right)^n - \left(\frac{1 - \sqrt{5}}{2}\right)^n\right], \quad n \geq 1.$$

Notice that $|(1 - \sqrt{5})/2| \approx 0.62 < 1$ so, for large n, x_n is approximately equal to $\left(\frac{1 + \sqrt{5}}{2}\right)^n / \sqrt{5}$. ∎

So far we have seen how to obtain an explicit solution of the difference equation $ax_n + bx_{n-1} + cx_{n-2} = 0$, $n > 2$, only when $b^2 - 4ac > 0$. In this case we were able to express A in the form $A = PDP^{-1}$, where D is a diagonal matrix, and hence we were able to compute the powers of A and apply the formula of Theorem 1. In the remaining two cases, when $b^2 + 4ac = 0$ and when $b^2 - 4ac < 0$, we cannot diagonalize A. However, we can express A in the form $A = PBP^{-1}$, where B is not diagonal but nevertheless is sufficiently simple that its powers are easy to compute.

Consider first the case where $b^2 - 4ac = 0$. Then the auxiliary polynomial $a\lambda^2 + b\lambda + c$ has a repeated zero $\lambda_1 = -b/2a$. We can express the entries of A in terms of λ_1. Indeed, since

$$a\lambda^2 + b\lambda + c = a(\lambda - \lambda_1)^2 = a\lambda^2 - 2a\lambda_1\lambda + a\lambda_1^2,$$

we see that $b = -2a\lambda_1$ and $c = a\lambda_1^2$. Hence

$$A = \begin{pmatrix} -b/a & -c/a \\ 1 & 0 \end{pmatrix} = \begin{pmatrix} 2\lambda_1 & -\lambda_1^2 \\ 1 & 0 \end{pmatrix}.$$

If we let

$$P = \begin{pmatrix} \lambda_1 & 1 \\ 1 & 0 \end{pmatrix} \quad \text{and} \quad B = \begin{pmatrix} \lambda_1 & 1 \\ 0 & \lambda_1 \end{pmatrix}$$

then it is easy to check that $A = PBP^{-1}$ (see Exercise 10a). The matrix B is called the "Jordan canonical form" of A. Its powers are easy to compute. We find that

$$B^2 = \begin{pmatrix} \lambda_1^2 & 2\lambda_1 \\ 0 & \lambda_1^2 \end{pmatrix}, \quad B^3 = \begin{pmatrix} \lambda_1^3 & 3\lambda_1^2 \\ 0 & \lambda_1^3 \end{pmatrix}, \quad B^4 = \begin{pmatrix} \lambda_1^4 & 4\lambda_1^3 \\ 0 & \lambda_1^4 \end{pmatrix},$$

and so on. In general,

$$B^k = \begin{pmatrix} \lambda_1^k & k\lambda_1^{k-1} \\ 0 & \lambda_1^k \end{pmatrix} \quad \text{for } k \geq 1.$$

Hence, using Theorem 1, we see that the general solution of the difference equation $ax_n + bx_{n-1} + cx_{n-2} = 0$, $n > 2$, when $b^2 - 4ac = 0$, is given by

$$\begin{pmatrix} x_n \\ x_{n-1} \end{pmatrix} = A^{n-2}\begin{pmatrix} x_2 \\ x_1 \end{pmatrix} = PB^{n-2}P^{-1}\begin{pmatrix} x_2 \\ x_1 \end{pmatrix}$$

$$= \begin{pmatrix} \lambda_1 & 1 \\ 1 & 0 \end{pmatrix}\begin{pmatrix} \lambda_1^{n-2} & (n-2)\lambda_1^{n-3} \\ 0 & \lambda_1^{n-2} \end{pmatrix}\begin{pmatrix} c_1 \\ c_2 \end{pmatrix}, \quad n \geq 2,$$

where

$$\begin{pmatrix} c_1 \\ c_2 \end{pmatrix} = P^{-1}\begin{pmatrix} x_1 \\ x_2 \end{pmatrix}.$$

Carrying out the matrix multiplication, we find that

$$\boxed{x_n = c_1\lambda_1^{n-1} + c_2(n-1)\lambda_1^{n-2}, \quad n \geq 1.}$$

The case when $b^2 - 4ac < 0$ can be handled in the same way. In this case the zeros of the auxiliary polynomial are complex numbers, which can be expressed in polar form as $r(\cos \theta \pm i \sin \theta)$ for some $r > 0$ and some θ not a multiple of π. Since

$$a\lambda^2 + b\lambda + c = a(\lambda - r(\cos \theta + i \sin \theta))(\lambda - r(\cos \theta - i \sin \theta))$$
$$= a\lambda^2 - (2ar \cos \theta)\lambda + ar^2,$$

we see that $b = -2ar \cos \theta$ and $c = ar^2$. Hence

$$A = \begin{pmatrix} -b/a & -c/a \\ 1 & 0 \end{pmatrix} = \begin{pmatrix} 2r \cos\theta & -r^2 \\ 1 & 0 \end{pmatrix}.$$

If we set

$$P = \begin{pmatrix} r \cos \theta & r \sin \theta \\ 1 & 0 \end{pmatrix} \quad \text{and} \quad B = \begin{pmatrix} r \cos \theta & r \sin \theta \\ -r \sin \theta & r \cos \theta \end{pmatrix}$$

then it is easy to check that $A = PBP^{-1}$ (see Exercise 10b). The powers of B are once again easy to compute. We find that

$$B^2 = \begin{pmatrix} r^2 \cos 2\theta & r^2 \sin 2\theta \\ -r^2 \sin 2\theta & r^2 \cos 2\theta \end{pmatrix}, \quad B^3 = \begin{pmatrix} r^3 \cos 3\theta & r^3 \sin 3\theta \\ -r^3 \sin 3\theta & r^3 \sin 3\theta \end{pmatrix}$$

and so on. In general,

$$B^k = \begin{pmatrix} r^k \cos k\theta & r^k \sin k\theta \\ -r^k \sin k\theta & r^k \cos k\theta \end{pmatrix} \quad \text{for } k \geq 1.$$

Hence, using Theorem 1, we see that the general solution of the difference equation $ax_n + bx_{n-1} + cx_{n-2} = 0$, $n > 2$, when $b^2 - 4ac < 0$, is given by

$$\begin{pmatrix} x_n \\ x_{n-1} \end{pmatrix} = A^{n-2} \begin{pmatrix} x_2 \\ x_1 \end{pmatrix} = PB^{n-2}P^{-1} \begin{pmatrix} x_2 \\ x_1 \end{pmatrix}$$

$$= \begin{pmatrix} r \cos \theta & r \sin \theta \\ 1 & 0 \end{pmatrix} \begin{pmatrix} r^{n-2} \cos(n-2)\theta & r^{n-2} \sin(n-2)\theta \\ -r^{n-2} \sin(n-2)\theta & r^{n-2} \cos(n-2)\theta \end{pmatrix} \begin{pmatrix} c_1 \\ c_2 \end{pmatrix}, \quad n \geq 2,$$

where

$$\begin{pmatrix} c_1 \\ c_2 \end{pmatrix} = P^{-1} \begin{pmatrix} x_1 \\ x_2 \end{pmatrix}.$$

Carrying out the matrix multiplication, we find that

$$x_n = c_1 r^{n-1} \cos(n-1)\theta + c_2 r^{n-1} \sin(n-1)\theta, \quad n \geq 1.$$

Here is a summary of the results of this section:

To solve a homogeneous second order linear difference equation $ax_n + bx_{n-1} + cx_{n-2} = 0$, $n > 2$, proceed as follows:

(1) Find the zeros of the auxiliary polynomial $a\lambda^2 + b\lambda + c$, and call these zeros λ_1 and λ_2.

(2) If λ_1 and λ_2 are real and unequal, then

$$x_n = c_1 \lambda_1^{n-1} + c_2 \lambda_2^{n-1}$$

for all $n \geq 1$, where c_1 and c_2 are arbitrary constants.

(3) If $\lambda_1 = \lambda_2$ then

$$x_n = c_1 \lambda_1^{n-1} + c_2 (n-1) \lambda_1^{n-2}$$

for all $n \geq 1$, where c_1 and c_2 are arbitrary constants.

(4) If λ_1 and λ_2 are the complex numbers $r(\cos\theta \pm i\sin\theta)$ ($r \neq 0$, $\theta \neq k\pi$ for any integer k) then

$$x_n = c_1 r^{n-1} \cos(n-1)\theta + c_2 r^{n-1} \sin(n-1)\theta$$

for all $n \geq 1$, where c_1 and c_2 are arbitrary constants.

Example 4. Let us find all sequences with the property that each term beyond the first in the sequence is the average of the term immediately preceding it and the term immediately following it. This condition can be expressed as $x_{n-1} = \frac{1}{2}(x_n + x_{n-2})$, $n > 2$, or as $x_n - 2x_{n-1} + x_{n-2} = 0$, $n > 2$. Since the auxiliary polynomial $\lambda^2 - 2\lambda + 1 = (\lambda - 1)^2$ has the repeated zero $\lambda_1 = \lambda_2 = 1$, the general solution is given by

$$x_n = c_1 1^{n-1} + c_2 (n-1) 1^{n-2} = c_1 + c_2(n-1), \quad n \geq 1.$$

In other words, the general solution is

$$\mathbf{x} = c_1(1, 1, 1, 1, \ldots) + c_2(0, 1, 2, 3, \ldots). \quad \blacksquare$$

Example 5. The sequence that begins with $x_1 = 0$ and $x_2 = 1$ and that satisfies the difference equation

$$x_n - 2x_{n-1} + 2x_{n-2} = 0, \quad n > 2,$$

starts out $x = (0, 1, 2, 2, 0, -4, -8, \ldots)$. Let us find a formula for the nth term of this sequence. Since the auxiliary polynomial $\lambda^2 - 2\lambda + 2$ has the complex zeros

$$\frac{2 \pm \sqrt{4 - 8}}{2} = 1 \pm i = \sqrt{2}\left(\cos\frac{\pi}{4} \pm i\sin\frac{\pi}{4}\right),$$

the general solution of this difference equation is given by

$$x_n = c_1(\sqrt{2})^{n-1}\cos(n-1)\frac{\pi}{4} + c_2(\sqrt{2})^{n-1}\sin(n-1)\frac{\pi}{4}, \quad n \geq 1.$$

The particular solution with $x_1 = 0$ and $x_2 = 1$ must satisfy

$$0 = x_1 = c_1$$

$$1 = x_2 = c_1\sqrt{2}\cos\frac{\pi}{4} + c_2\sqrt{2}\sin\frac{\pi}{4} = c_1 + c_2$$

Hence $c_1 = 0$ and $c_2 = 1$. Therefore

$$x_n = (\sqrt{2})^{n-1}\sin(n-1)\frac{\pi}{4}$$

for all $n \geq 1$. ■

EXERCISES

1. Find the general solution of each of the following difference equations:
   (a) $5x_n - 6x_{n-1} + x_{n-2} = 0, n > 2$
   (b) $x_n - 6x_{n-1} + 5x_{n-2} = 0, n > 2$
   (c) $x_n + 6x_{n-1} + 8x_{n-2} = 0, n > 2$
   (d) $x_n - 2x_{n-1} - 2x_{n-2} = 0, n > 2$
   (e) $x_n + 4x_{n-1} + 4x_{n-2} = 0, n > 2$
   (f) $4x_n + 4x_{n-1} + x_{n-2} = 0, n > 2$
   (g) $x_n + 6x_{n-1} + 9x_{n-2} = 0, n > 2$
   (h) $x_n + x_{n-1} + x_{n-2} = 0, n > 2$
   (i) $x_n + 2x_{n-1} + 2x_{n-2} = 0, n > 2$
   (j) $x_n + x_{n-2} = 0, n > 2$

2. Find a formula for the nth term of the sequence satisfying $x_1 = 0$, $x_2 = 1$, and $x_n - 2x_{n-1} - x_{n-2} = 0$, $n > 2$. (This is the sequence described in the first paragraph of this section.)

3. Write down the first five terms of each of the following sequences, and then find a formula for the nth term x_n.
    (a) the sequence with $x_1 = 0$, $x_2 = 1$, and $x_n = x_{n-1} - x_{n-2}$ for $n > 2$
    (b) the sequence with $x_1 = -1$, $x_2 = 1$, and $x_n = -2x_{n-2}$ for $n > 2$
    (c) the sequence with $x_1 = 0$, $x_2 = 1$, and $x_n = 2(x_{n-1} - x_{n-2})$ for $n > 2$
    (d) the sequence with $x_1 = 0$, $x_2 = 1$, and $x_n = 4(x_{n-1} - x_{n-2})$ for $n > 2$
    (e) the sequence with $x_1 = 0$, $x_2 = 1$, and $x_n = 5(x_{n-1} - x_{n-2})$ for $n > 2$

4. (a) Show that the general solution of the first order linear difference equation
    $$ax_n + bx_{n-1} = 0, \quad n > 1,$$
    is $\mathbf{x} = c(1, \lambda, \lambda^2, \lambda^3, \ldots)$ where $\lambda = -b/a$ and $c \in \mathbb{R}$ is arbitrary.
    (b) If an amount P_0 is deposited in a savings account at an interest rate of 6% compounded annually, then the amount P_n at the end of the nth year satisfies $P_n = 1.06 \, P_{n-1}$, $n \geq 1$. Use the result of (a) to find a formula for P_n in terms of P_0 and n.

5. Let $\Delta: \mathbb{R}^\infty \to \mathbb{R}^\infty$ be defined by
    $$\Delta(x_1, x_2, x_3, \ldots) = (x_2 - x_1, x_3 - x_2, x_4 - x_3, \ldots).$$
    (a) Show that Δ is a linear map.
    (b) Compute $\Delta^2 \mathbf{x}$ where $\mathbf{x} = (x_1, x_2, x_3, \ldots)$.
    (c) Show that each second order linear difference equation
    $$ax_n + bx_{n-1} + cx_{n-2} = b_{n-2}, \quad n > 2,$$
    can be rewritten in the form
    $$(p \Delta^2 + q \Delta + r)\mathbf{x} = \mathbf{b}$$
    where p, q, and $r \in \mathbb{R}$, $\mathbf{x} = (x_1, x_2, x_3, \ldots)$, and $\mathbf{b} = (b_1, b_2, b_3, \ldots)$.

    Remark. The sequence $\Delta\mathbf{x}$ is called the *first difference* of $\mathbf{x}$, and $\Delta^2\mathbf{x}$ is the *second difference* of $\mathbf{x}$. The linear map $p \Delta^2 + q \Delta + r$ is a second order *difference operator.* The equivalence established in (c) is the reason that the equation $ax_n + bx_{n-1} + cx_{n-2} = b_{n-2}$, $n > 2$, is called a *difference equation.*

6. Verify that if $\mathbf{x}$ and $\mathbf{x}_P$ are two solutions of the difference equation
    $$a_0x_n + a_1x_{n-1} + \cdots + a_kx_{n-k} = b_{n-k}, \quad n > k,$$
    then $\mathbf{x}_H = \mathbf{x} - \mathbf{x}_P$ is a solution of the associated homogeneous equation
    $$a_0x_n + a_1x_{n-1} + \cdots + a_kx_{n-k} = 0, \quad n > k.$$

7. (a) Show that the set of solutions of the homogeneous linear difference equation
    $$a_0x_n + a_1x_{n-1} + \cdots + a_kx_{n-k} = 0, \quad n > k,$$
    is a subspace of $\mathbb{R}^\infty$.
    (b) Show that, for each i, $1 \leq i \leq k$, there is a unique solution $\mathbf{y}_i = (y_{i1}, y_{i2}, y_{i3}, \ldots)$ of the equation in part (a) satisfying the conditions $y_{ii} = 1$ and $y_{ij} = 0$ for $j \neq i$, $1 \leq j \leq k$. Verify that the set $\{\mathbf{y}_1, \ldots, \mathbf{y}_k\}$ is a basis for the solution space of this equation, and conclude that the dimension of the solution space is k.

8. (a) Show that the general solution of the third order homogeneous linear
         difference equation

$$ax_n + bx_{n-1} + cx_{n-2} + dx_{n-3} = 0, \quad n > 3$$

can be calculated from the formula

$$\begin{pmatrix} x_n \\ x_{n-1} \\ x_{n-2} \end{pmatrix} = A^{n-3} \begin{pmatrix} x_3 \\ x_2 \\ x_1 \end{pmatrix}, \quad n \geq 3$$

where x_1, x_2, and x_3 are arbitrary, and where

$$A = \begin{pmatrix} -b/a & -c/a & -d/a \\ 1 & 0 & 0 \\ 0 & 1 & 0 \end{pmatrix}.$$

[Hint: Let $y_n = x_{n-1}$ and $z_n = x_{n-2}$.]

    (b) Generalize part (a) to describe the general solution of the kth order ho-
         mogeneous linear difference equation

$$a_0 x_n + a_1 x_{n-1} + \cdots + a_k x_{n-k} = 0, \quad n > k.$$

9. (a) Use the result of Exercise 8(a) to show that if the zeros $\lambda_1, \lambda_2, \lambda_3$ of the
         auxiliary polynomial $a\lambda^3 + b\lambda^2 + c\lambda + d$ are real and distinct, then the
         solution space of the equation $ax_n + bx_{n-1} + cx_{n-2} + dx_{n-3} = 0, n > 3$,
         is spanned by $\{(1, \lambda_1, \lambda_1^2, \ldots), (1, \lambda_2, \lambda_2^2, \ldots), (1, \lambda_3, \lambda_3^2, \ldots)\}$.

    (b) Generalizing part (a), describe a basis for the solution space of the differ-
         ence equation

$$a_0 x_n + a_1 x_{n-1} + \cdots + a_k x_{n-k} = 0, \quad n > k,$$

in   the   case   where   the   zeros   of   the   auxiliary   polynomial
$a_0 \lambda^n + a_1 \lambda^{n-1} + \cdots + a_k$ are real and distinct.

10. (a) Let $A = \begin{pmatrix} 2\lambda_1 & -\lambda_1^2 \\ 1 & 0 \end{pmatrix}$, $P = \begin{pmatrix} \lambda_1 & 1 \\ 1 & 0 \end{pmatrix}$, and $B = \begin{pmatrix} \lambda_1 & 1 \\ 0 & \lambda_1 \end{pmatrix}$ where $\lambda_1 \in \mathbb{R}$.
         Show that $AP = PB$, and hence that $A = PBP^{-1}$.

    (b) Let $A = \begin{pmatrix} 2r \cos \theta & -r^2 \\ 1 & 0 \end{pmatrix}$, $P = \begin{pmatrix} r \cos \theta & r \sin \theta \\ 1 & 0 \end{pmatrix}$, and

$B = \begin{pmatrix} r \cos \theta & r \sin \theta \\ -r \sin \theta & r \cos \theta \end{pmatrix}$. Show that $AP = PB$, and hence that
$A = PBP^{-1}$.

Additional Topics in
Differential Equations

9.1 VARIABLE COEFFICIENT EQUATIONS

In Chapter 3 you learned how to solve differential equations with constant coefficients. In this chapter we shall discuss equations with variable coefficients. We will first extend the integrating factor method to solve first order equations with variable coefficients, and then we will discuss the problem of solving higher order equations with nonconstant coefficients.

We will be concerned with differential equations of the following type. Let $a_n(x), a_{n-1}(x), \ldots, a_1(x), a_0(x)$ and $f(x)$ be functions that are defined and continuous on some interval $I = I_{(a,b)}$. Assume that $a_n(x) \neq 0$ for all $x \in I$. The equation

$$(1) \qquad a_n(x)y^{(n)} + a_{n-1}(x)y^{(n-1)} + \cdots + a_0(x)y = f(x)$$

is the **general nth order linear differential equation on** I. A **solution** of (1) is an n-times differentiable function $y \in \mathfrak{F}(I)$ that satisfies the equation.

First, we observe (without proof, but see Exercise 25) that the results of Section 3.1 on the structure of the solution sets still hold, with some minor modifications. Recall that $\mathfrak{F}(I)$ is the vector space of functions with domain I, and $\mathcal{C}(I)$ is the subspace of $\mathfrak{F}(I)$ consisting of continuous functions on I.

Theorem. Let $a_n(x), \ldots, a_0(x) \in \mathcal{C}(I)$, with $a_n(x) \neq 0$ for all $x \in I$. The set of solutions of the homogeneous linear differential equation

$$(A) \qquad a_n(x)y^{(n)} + \cdots + a_1(x)y' + a_0(x)y = 0$$

is a subspace of $\mathfrak{F}(I)$. Moreover, if $f \in \mathcal{C}(I)$ and y_P is a solution of

$$(B) \qquad a_n(x)y^{(n)} + \cdots + a_1(x)y' + a_0(x)y = f(x),$$

then the solution set of (B) is precisely the set of all functions in $\mathfrak{F}(I)$ of the form $y = y_P + y_H$, where y_H is a solution of the homogeneous equation (A). In particular, if $\{y_1, \ldots, y_k\}$ is a spanning set for the solution space of (A), then the general solution of (B) can be written in the form

$$y = y_P + c_1 y_1 + \cdots + c_k y_k.$$

279

Consider now the first order linear differential equation

$$a(x)y' + b(x)y = f(x)$$

where $a(x), b(x), f(x) \in \mathcal{C}(I)$ for some interval I and $a(x) \neq 0$ for all $x \in I$. Recall the method of Section 3.1, which is valid when a and b are constants: first divide by a, then multiply by the integrating factor e^{kx}, where $k = b/a$; the left hand side then becomes $(e^{kx}y)'$. A similar technique works when a and b are functions. First, rewrite the equation in the form

$$y' + p(x)y = \frac{1}{a(x)}f(x), \quad \text{where } p(x) = b(x)/a(x).$$

We seek an **integrating factor** for this equation; that is, we want to find a function $u(x)$ with the property that

$$u(x)(y' + p(x)y) = (u(x)y)'.$$

From the product rule for differentiation we know that

$$(u(x)y)' = u(x)y' + u'(x)y,$$

so the function sought must satisfy the equation

$$u(x)p(x)y = u'(x)y.$$

This equation will be satisfied if $u'(x)/u(x) = p(x)$, or $\ln u(x) = \int p(x)\, dx$, or

$$u(x) = e^{\int p(x)\, dx}.$$

The solution procedure is then clear:

To solve a first order linear differential equation

$$a(x)y' + b(x)y = f(x),$$

proceed as follows:

(i)   After dividing by the coefficient $a(x)$, multiply both sides of the equation by the integrating factor $e^{\int p(x)\, dx}$, where $p(x) = b(x)/a(x)$; the left hand side then becomes $(e^{\int p(x)\, dx}y)'$.

(ii) Integrate both sides and solve for y.

Example 1. Solve $y' + 2xy = x$ on $\mathbb{R}$. After multiplying by the integrating factor $e^{\int 2x\, dx} = e^{x^2}$, the equation becomes

$$(e^{x^2}y)' = xe^{x^2}.$$

Integrate to get $e^{x^2}y = \tfrac{1}{2}e^{x^2} + c$. Solve for y to get the general solution

$$y = \tfrac{1}{2} + ce^{-x^2}.$$

Notice that $y_p = \tfrac{1}{2}$ is a particular solution of the equation $y' + 2xy = x$ and $\{e^{-x^2}\}$ spans the solution space of the homogeneous equation $y' + 2xy = 0$. ■

Example 2. To solve $x^2y' + xy = \ln x$ on $I_{(0,\infty)}$, first divide by x^2 to obtain $y' + \dfrac{1}{x}y = \dfrac{1}{x^2}\ln x$. An integrating factor is $e^{\int (1/x)dx} = e^{\ln x} = x$. Multiply by x and the equation becomes

$$(xy)' = \frac{1}{x}\ln x.$$

Integrate to get $xy = \tfrac{1}{2}(\ln x)^2 + c$, or

$$y = \frac{1}{2x}(\ln x)^2 + c \cdot \frac{1}{x}.$$

Here $\left\{\dfrac{1}{x}\right\}$ spans the solution space of the associated homogeneous equation and $\dfrac{1}{2x}(\ln x)^2$ is a particular solution of the original equation. ■

Certain higher order linear differential equations with nonconstant coefficients can be solved by repeated applications of the technique described above. Recall that the method for solving second order constant coefficient equations involved the factorization of a differential operator. The factorization of differential operators with nonconstant coefficients is not so easy, because they do *not* multiply like polynomials. In fact, they don't, in general, even commute. For example, $(xD)Dy = xD^2y$, whereas

$$D(xD)y = xD^2y + Dy = (xD^2 + D)y,$$

so $(xD)D = xD^2$ and $D(xD) = xD^2 + D$. In particular,

$$(xD)D \neq D(xD).$$

Nevertheless, if we can factor a differential operator L into a product of first order factors, then we can solve the differential equation $Ly = f$ by solving a sequence of first order equations.

Example 3. Solve $xy'' + y' = \ln x$ on the interval $I_{(0,\infty)}$. In operator notation, this equation becomes

$$(xD^2 + D)y = \ln x, \quad \text{or} \quad (xD + 1)Dy = \ln x.$$

Let $y_1 = Dy$. Then the equation becomes $(xD + 1)y_1 = \ln x$. This first order equation has general solution

$$y_1 = \ln x - 1 + c_1/x.$$

To find y, use the fact that $Dy = y_1$ to get

$$y = \int y_1 \, dx = \int (\ln x - 1 + c_1/x) \, dx$$

or

$$y = x \ln x - 2x + c_1 \ln x + c_2.$$

Note that $x \ln x - 2x$ is a particular solution of the equation $xy'' + y' = \ln x$ and that $\{\ln x, 1\}$ is a basis for the solution space of the homogeneous equation $xy'' + y' = 0$. ∎

Example 4. To solve $(xD - r)(xD - s)y = 0$ on the interval $I_{(0,\infty)}$, where r, $s \in \mathbb{R}$, let $y_1 = (xD - s)y$. Then y_1 satisfies $(xD - r)y_1 = 0$, so $(x^{-r}y_1)' = 0$ and $y_1 = cx^r$. Since

$$(xD - s)y = y_1 = cx^r$$

we find that $(x^{-s}y)' = cx^{r-s-1}$ and hence

$$x^{-s}y = \begin{cases} \dfrac{c}{r - s}x^{r-s} + c_2 & \text{if } r \neq s \\[2mm] c \ln x + c_2 & \text{if } r = s \end{cases}$$

or

$$y = \begin{cases} c_1 x^r + c_2 x^s & \text{if } r \neq s \\[2mm] c_1 x^s \ln x + c_2 x^s & \text{if } r = s. \end{cases} \quad ∎$$

Remark. If we compute the product $(xD - r)(xD - s)$ we will get a second order differential operator of the form $ax^2D^2 + bxD + c$, where $a, b, c \in \mathbb{R}$. The result of Example 4 suggests that solutions of equations of the form

$$(1) \qquad\qquad ax^2y'' + bxy' + cy = 0$$

will often be linear combinations of powers of x. We can verify that this is indeed the case by noting that the substitution $x = e^t$ (so that $x^k = e^{kt}$) converts (1) into an equation with constant coefficients. Denoting $\dfrac{dy}{dt}$ by $\dot{y}$, we have

$$y' = \frac{dy}{dx} = \frac{dy}{dt} \Big/ \frac{dx}{dt} = \frac{1}{x}\dot{y}$$

and

$$y'' = -\frac{1}{x^2}\dot{y} + \frac{1}{x}\frac{d\dot{y}}{dt} \Big/ \frac{dx}{dt} = \frac{1}{x^2}(\ddot{y} - \dot{y})$$

so Equation (1) becomes

$$a\ddot{y} + (b - a)\dot{y} + cy = 0.$$

Thus, if λ_1 and λ_2 are the zeros of the auxiliary polynomial $az^2 + (b - a)z + c$, then

$$y = \begin{cases} c_1 e^{\lambda_1 t} + c_2 e^{\lambda_2 t} & \text{if } \lambda_1 \text{ and } \lambda_2 \text{ are real with } \lambda_1 \neq \lambda_2 \\ c_1 e^{\lambda t} + c_2 t e^{\lambda t} & \text{if } \lambda_1 = \lambda_2 = \lambda \\ c_1 e^{\alpha t} \cos \beta t + c_2 e^{\alpha t} \sin \beta t & \text{if } \lambda_1 = \alpha + \beta i \text{ and } \lambda_2 = \alpha - \beta i, \ \beta \neq 0. \end{cases}$$

Using the fact that $e^t = x$, hence $t = \ln x$, we conclude that

$$y = \begin{cases} c_1 x^{\lambda_1} + c_2 x^{\lambda_2} & \text{if } \lambda_1 \text{ and } \lambda_2 \text{ are real with } \lambda_2 \neq \lambda_2 \\ c_1 x^{\lambda} + c_2 x^{\lambda} \ln x & \text{if } \lambda_1 = \lambda_2 = \lambda \\ c_1 x^{\alpha} \cos(\beta \ln x) + c_2 x^{\alpha} \sin(\beta \ln x) & \text{if } \lambda_1 = \alpha + \beta i \\ & \qquad \text{and } \lambda_2 = \alpha - \beta i, \ \beta \neq 0 \end{cases}$$

is the general solution of the equation $ax^2 y'' + bxy' + cy = 0$, where λ_1 and λ_2 are the zeros of the polynomial $az^2 + (b - a)z + c$.

Example 5.   To solve $x^2 y'' + 4xy' + 2y = 0$, we first find the zeros of the auxiliary polynomial

$$z^2 + (4 - 1)z + 2 = z^2 + 3z + 2 = (z + 1)(z + 2).$$

They are -1 and -2, so the general solution of the differential equation is $y = c_1 x^{-1} + c_2 x^{-2}$. ∎

EXERCISES

Find the general solution of each of the following first order linear differential equations.

1. $xy' + 2y = 1$, on $I_{(0,\infty)}$
2. $(1 + x^2)y' + xy = 0$, on $\mathbb{R}$
3. $y' + e^x y = e^x$, on $\mathbb{R}$

4. $y' + (\tan x)y = \sec x$, on $I_{(-\pi/2, \pi/2)}$

5. $y' + 2xy = 2xe^{-x^2}$, on $\mathbb{R}$

6. $xy' + (1 + x)y = e^x$, on $I_{(0, \infty)}$

7. $(x + 1)y' + 2y = e^x/(x + 1)$, on $I_{(-1, \infty)}$

8. $(x - 1)y' + y = \cos x$, on $I_{(1, \infty)}$

Find the particular solution of the given differential equation that satisfies the stated initial condition.

9. $xy' + 4y = 6x$, $y(1) = 2$

10. $(x^2 + 1)y' + 2xy = x^2$, $y(0) = -2$

11. $xy' + 3y = x^{-2} \sin x$, $y(\pi) = 1$

Write each of the following products of differential operators in the form $a(x)D^2 + b(x)D + c(x)$.

12. $(xD + 1)(D + x)$

13. $(D + 3)(D + 3x)$

14. $(D + 3x)(D + 3)$

15. $(D + \sin x)(xD + \cos x)$

Solve each of the following differential equations.

16. $xy'' + y' = \dfrac{1}{x} \ln x$, on $I_{(0, \infty)}$

17. $xy''' + 2y'' = 0$, on $I_{(0, \infty)}$

18. $(\tan x\, D - 1)(\sin x\, D)y = 0$, on $I_{(-\pi/2, \pi/2)}$

19. $\left(D - \dfrac{1}{x}\right)(D - 1)y = 0$, on $I_{(0, \infty)}$

20. $x^2 y'' - 4xy' + 6y = 0$, on $I_{(0, \infty)}$

21. $x^2 y'' + 3xy' + y = 0$, on $I_{(0, \infty)}$

22. $x^2 y'' + xy' + y = 0$, on $I_{(0, \infty)}$

23. $x^3 y''' + 3x^2 y'' + xy' - y = 0$, on $I_{(0, \infty)}$ [Hint: Let $x = e^t$.]

24. (a) Verify that
$$(D + 1)(xD + 1) = xD^2 + (x + 2)D + 1.$$
    (b) Use the result of (a) to find the general solution of the equation $xy'' + (x + 2)y' + y = 0$ on the interval $I_{(0, \infty)}$.

25. Prove the theorem of this section for $n = 2$. [Hint: Imitate the proof of the theorem in Section 3.1.]

26. Show that if L_1 and L_2 are linear differential operators, then the solution space of the differential equation $L_2 y = 0$ is a subspace of the solution space of the differential equation $L_1 L_2 y = 0$.

27. Verify that the method of undetermined coefficients still gives a valid procedure for finding a particular solution of a linear differential equation $Ly = f$ whenever f satisfies some linear differential equation $L_1 f = 0$ (L and L_1 need not have constant coefficients). That is, show that if $\{y_1, \ldots, y_n\}$ spans the solution space of $Ly = 0$ and $\{y_1, \ldots, y_{n+k}\}$ spans the solution space of $L_1 L y = 0$, then there must be a particular solution of $Ly = f$ of the form
$$y = A_{n+1} y_{n+1} + \cdots + A_{n+k} y_{n+k},$$
$A_{n+1}, \ldots, A_{n+k} \in \mathbb{R}$.

9.2 POWER SERIES SOLUTIONS

In contrast to constant coefficient linear differential equations, variable coefficient equations usually do not have solutions that are expressible in terms of familiar functions like x^n, e^{ax}, $\sin bx$, and $\cos bx$. Often, however, it is possible to describe the solutions of such equations as power series.

In this section we shall discuss the ***power series method*** for solving linear differential equations. This method is especially useful when the coefficients in the differential equation are polynomials. The procedure is similar to the method of undetermined coefficients. We substitute a power series $\sum_{n=0}^{\infty} b_n x^n$ into the differential equation and then solve for the coefficients $b_0, b_1, b_2, \ldots$.

In order to use the power series method we need to know the following facts about power series.

(1) If the power series $\sum_{n=0}^{\infty} b_n x^n$ converges for any nonzero value of x, then either it converges for all values of x, or there exists a real number $R > 0$ (called the ***radius of convergence*** of the series) such that the series converges for all x with $|x| < R$ and diverges for all x with $|x| > R$. The value of R is found from the ratio test to be $R = \lim_{n \to \infty} \left| \dfrac{b_n}{b_{n+1}} \right|$, provided this limit exists. If $\lim_{n \to \infty} \dfrac{b_n}{b_{n+1}} = \infty$, then the series converges for all x.

(2) If $\sum_{n=0}^{\infty} b_n x^n$ converges to $f(x)$ for $|x| < R$, then the series $\sum_{n=1}^{\infty} n b_n x^{n-1}$ converges to $f'(x)$ for $|x| < R$.

(3) If a power series $\sum_{n=0}^{\infty} b_n x^n$ converges to zero on any interval, then all the coefficients b_n ($n \geq 0$) must be equal to zero.

We shall illustrate the power series method by looking at several examples. The first example is chosen to have constant coefficients so that the computations are fairly simple.

Example 1. Let us solve the differential equation

$$y' - 2y = 0$$

by the power series method. We seek a solution of the form $y(x) = \sum_{n=0}^{\infty} b_n x^n$. From (2), we have

$$\sum_{n=1}^{\infty} n b_n x^{n-1} - 2 \sum_{n=0}^{\infty} b_n x^n = 0.$$

In order to make the exponents on x in the two summations match up, we let $n = k$ in the first and $n = k - 1$ in the second. The equation then becomes

$$\sum_{k=1}^{\infty} kb_k x^{k-1} - 2 \sum_{k=1}^{\infty} b_{k-1} x^{k-1} = 0.$$

Notice that the range of the summation on the second sum is $k = 1$ to ∞ because $k = n + 1$, and n runs from 0 to ∞. Now the equation can be rewritten as

$$\sum_{k=1}^{\infty} (kb_k - 2b_{k-1}) x^{k-1} = 0.$$

It follows, using (3), that $kb_k - 2b_{k-1} = 0$, or

$$b_k = \frac{2}{k} b_{k-1} \quad \text{for all } k \geq 1.$$

This **recursion formula** allows us to compute all of the coefficients in terms of b_0. Thus,

$$b_1 = \frac{2}{1} b_0$$

$$b_2 = \frac{2}{2} b_1 = \frac{2 \cdot 2}{2 \cdot 1} b_0,$$

$$b_3 = \frac{2}{3} b_2 = \frac{2 \cdot 2 \cdot 2}{3 \cdot 2 \cdot 1} b_0,$$

and so on, allowing us to conclude that $b_n = \dfrac{2^n}{n!} b_0$, for all $n \geq 1$. Therefore, the function y defined by

$$y = b_0 \sum_{n=0}^{\infty} \frac{2^n}{n!} x^n = b_0 \sum_{n=0}^{\infty} \frac{(2x)^n}{n!}$$

satisfies the equation $y' - 2y = 0$, for each $b_0 \in \mathbb{R}$. You should recognize the series $\displaystyle\sum_{n=0}^{\infty} \frac{(2x)^n}{n!}$ as the Maclaurin series for the function e^{2x}. Since we already know, by the methods of Chapter 3, that the solution space of $y' - 2y = 0$ is spanned by $\{e^{2x}\}$, we have succeeded in finding by the power series method the general solution of $y' - 2y = 0$. ■

Example 2. Let us use the power series method to find solutions of the second order differential equation

$$(1 - x^2)y'' - 2xy' + 2y = 0$$

on some interval about 0. Our trial solution is $y = \sum\limits_{n=0}^{\infty} b_n x^n$. Then

$$y' = \sum_{n=1}^{\infty} nb_n x^{n-1} \quad \text{and} \quad y'' = \sum_{n=2}^{\infty} n(n-1)b_n x^{n-2}.$$

We substitute these series into the differential equation to obtain

$$(1-x^2) \sum_{n=2}^{\infty} n(n-1)b_n x^{n-2} - 2x \sum_{n=1}^{\infty} nb_n x^{n-1} + 2 \sum_{n=0}^{\infty} b_n x^n = 0,$$

or

$$\sum_{n=2}^{\infty} n(n-1)b_n x^{n-2} - \sum_{n=2}^{\infty} n(n-1)b_n x^n - \sum_{n=1}^{\infty} 2nb_n x^n + \sum_{n=0}^{\infty} 2b_n x^n = 0.$$

Now let $n = k$ in the first summation, and change n to $k - 2$ in the others, to get

$$\sum_{k=2}^{\infty} k(k-1)b_k x^{k-2} - \sum_{k=4}^{\infty} (k-2)(k-3)b_{k-2} x^{k-2} - \sum_{k=3}^{\infty} 2(k-2)b_{k-2} x^{k-2}$$

$$+ \sum_{k=2}^{\infty} 2b_{k-2} x^{k-2} = 0.$$

Writing out the $k = 2$ and $k = 3$ terms so that all sums start at $k = 4$, we obtain

$$2(b_2 + b_0) + 3 \cdot 2b_3 x + \sum_{k=4}^{\infty} [k(k-1)b_k - k(k-3)b_{k-2}] x^{k-2} = 0.$$

It follows that

$$2(b_2 + b_0) = 0$$
$$3 \cdot 2b_3 = 0$$

and

$$b_k = \frac{k-3}{k-1} b_{k-2} \quad \text{for all } k \geq 4.$$

Thus b_0 and b_1 are arbitrary and

$$b_2 = -b_0 \qquad\qquad b_3 = 0$$
$$b_4 = \tfrac{1}{3}b_2 = -\tfrac{1}{3}b_0 \qquad b_5 = 0$$
$$b_6 = \tfrac{3}{5}b_4 = -\tfrac{1}{5}b_0 \qquad b_7 = 0$$
$$b_8 = \tfrac{5}{7}b_6 = -\tfrac{1}{7}b_0 \qquad b_9 = 0$$

and so on. Hence we see that

$$y = \sum_{n=0}^{\infty} b_n x^n = b_0 + b_1 x - b_0(x^2 + \tfrac{1}{3}x^4 + \tfrac{1}{5}x^6 + \tfrac{1}{7}x^8 + \cdots)$$

$$= b_0\left(1 - \sum_{n=1}^{\infty} \frac{1}{2n-1}x^{2n}\right) + b_1 x$$

$$= -b_0\left(\sum_{n=0}^{\infty} \frac{1}{2n-1}x^{2n}\right) + b_1 x$$

is a solution of the given differential equation, for each $b_0, b_1 \in \mathbb{R}$.

The series $\sum_{n=0}^{\infty} \dfrac{1}{2n-1}x^{2n}$ has radius of convergence

$$R = \lim_{n\to\infty} \left|\frac{1}{2n-1}\bigg/\frac{1}{2n+1}\right| = 1$$

hence it converges on the interval $-1 < x < 1$. Therefore the functions $f(x) = \sum_{n=0}^{\infty} \dfrac{1}{2n-1}x^{2n}$, $g(x) = x$, and all linear combinations of $f(x)$ and $g(x)$ are solutions of the given differential equation on the interval $I_{(-1,1)}$. ■

In this example we found two functions, $f(x) = \sum_{n=0}^{\infty} \dfrac{1}{2n-1}x^{2n}$ and $g(x) = x$ in $\mathcal{F}(I_{(-1,1)})$, both of which satisfy the second order linear differential equation

$$(1 - x^2)y'' - 2xy' + 2y = 0$$

on the interval $I_{(-1,1)}$. Since g is not a constant multiple of f, the set $\{f, g\}$ is linearly independent. It is reasonable, therefore, to guess that $\{f, g\}$ is a basis for the solution space of the given equation, on the interval $I_{(-1,1)}$. The correctness of this guess is an immediate consequence of the following theorem.

Theorem. *Let $a_n, a_{n-1}, \ldots, a_1, a_0$ be differentiable functions on an open interval I and suppose $a_n(x) \neq 0$ for all $x \in I$. Then the set of solutions of the differential equation*

$$a_n y^{(n)} + a_{n-1} y^{(n-1)} + \cdots + a_1 y' + a_0 y = 0$$

on the interval I is a subspace of $\mathcal{F}(I)$ of dimension n.

We shall not prove this theorem here. You can find a proof in almost any advanced book on differential equations. But, based on your experience, the fact that the solution space has dimension n should come as no surprise. Indeed, we already know this to be true when the coefficients are constant and also when the differential operator

$$a_n D^n + \cdots + a_1 D + a_0$$

factors into a product of first order operators.

Let us apply this theorem to the differential equation in Example 2. Notice that $\left\{ \sum_{n=0}^{\infty} \dfrac{1}{2n-1} x^{2n}, x \right\}$ is a linearly independent subset of the solution space on the interval $I_{(-1,1)}$. Since the dimension of the solution space is 2, this set must be a basis for that space. The general solution is therefore

$$y = c_1 \left(\sum_{n=0}^{\infty} \frac{1}{2n-1} x^{2n} \right) + c_2 x.$$

Example 3.   Here we use the power series method to solve the differential equation

$$y'' + xy' + y = 0.$$

The trial solution is $y = \sum_{n=0}^{\infty} b_n x^n$. Then

$$y' = \sum_{n=1}^{\infty} nb_n x^{n-1} \quad \text{and} \quad y'' = \sum_{n=2}^{\infty} n(n-1) b_n x^{n-2}.$$

We substitute these series into the differential equation to obtain

$$\sum_{n=2}^{\infty} n(n-1) b_n x^{n-2} + x \sum_{n=1}^{\infty} nb_n x^{n-1} + \sum_{n=0}^{\infty} b_n x^n = 0$$

or

$$\sum_{n=2}^{\infty} n(n-1) b_n x^{n-2} + \sum_{n=1}^{\infty} nb_n x^n + \sum_{n=0}^{\infty} b_n x^n = 0.$$

Now we make the exponents on x match up by letting $n = k$ in the first sum and $n = k - 2$ in the other two sums. We get

$$\sum_{k=2}^{\infty} k(k-1) b_k x^{k-2} + \sum_{k=3}^{\infty} (k-2) b_{k-2} x^{k-2} + \sum_{k=2}^{\infty} b_{k-2} x^{k-2} = 0$$

or

$$(2 \cdot 1 b_2 + b_0) + \sum_{k=3}^{\infty} [k(k-1) b_k + (k-1) b_{k-2}] x^{k-2} = 0.$$

It follows that

$$2b_2 + b_0 = 0$$

and that

$$k(k - 1)b_k + (k - 1)b_{k-2} = 0 \quad \text{for all } k \geq 3.$$

We see therefore that b_0 and b_1 can be chosen arbitrarily and that the other coefficients are determined by

$$b_2 = -\tfrac{1}{2}b_0$$

and

$$b_k = -\frac{1}{k}b_{k-2} \quad \text{for } k \geq 3.$$

Thus

$$b_2 = -\tfrac{1}{2}b_0 \qquad\qquad b_3 = -\tfrac{1}{3}b_1$$

$$b_4 = -\tfrac{1}{4}b_2 = \frac{(-1)^2}{4\cdot 2}b_0 \qquad b_5 = -\tfrac{1}{5}b_3 = \frac{(-1)^2}{5\cdot 3}b_1$$

$$b_6 = -\tfrac{1}{6}b_4 = \frac{(-1)^3}{6\cdot 4\cdot 2}b_0 \qquad b_7 = -\tfrac{1}{7}b_5 = \frac{(-1)^3}{7\cdot 5\cdot 3}b_1$$

and so on.

The pattern is clear. We see that the coefficients of the even powers of x are given by

$$b_{2n} = \frac{(-1)^n}{(2n)(2n-2)\cdots 6\cdot 4\cdot 2}b_0 = \frac{(-1)^n}{2^n\cdot n(n-1)\cdots 3\cdot 2\cdot 1}b_0 = \frac{(-1)^n}{2^n n!}b_0$$

whereas the coefficients of the odd powers of x are given by

$$b_{2n+1} = \frac{(-1)^n}{(2n+1)(2n-1)\cdots 7\cdot 5\cdot 3}b_1 = \frac{(-1)^n(2n)(2n-2)\cdots 6\cdot 4\cdot 2}{(2n+1)!}b_1$$

$$= \frac{(-1)^n 2^n n!}{(2n+1)!}b_1.$$

It follows that

$$y = \sum_{n=0}^{\infty} b_n x^n = b_0\left(\sum_{n=0}^{\infty}\frac{(-1)^n}{2^n n!}x^{2n}\right) + b_1\left(\sum_{n=0}^{\infty}\frac{(-1)^n 2^n n!}{(2n+1)!}x^{2n+1}\right)$$

is a solution of the given differential equation, for each $b_0, b_1 \in \mathbb{R}$.

Notice that both of these series converge for all x, since

$$\lim_{n\to\infty}\left|\frac{(-1)^n}{2^n n!}\middle/\frac{(-1)^{n+1}}{2^{n+1}(n+1)!}\right| = \lim_{n\to\infty}\frac{2^{n+1}(n+1)!}{2^n n!} = \lim_{n\to\infty}2(n+1) = \infty$$

and

$$\lim_{n\to\infty}\left|\frac{(-1)^n 2^n n!}{(2n+1)!}\middle/\frac{(-1)^{n+1}2^{n+1}(n+1)!}{(2(n+1)+1)!}\right| = \lim_{n\to\infty}\frac{2^n n!(2n+3)!}{2^{n+1}(n+1)!(2n+1)!}$$

$$= \lim_{n\to\infty}\frac{(2n+3)(2n+2)}{2(n+1)} = \infty.$$

Hence the functions defined by these power series are solutions of the given differential equation on $\mathbb{R}$. Since neither of these functions is a constant multiple of the other, they form a linearly independent set, and hence a basis for the solution space. We conclude that the general solution of the differential equation

$$y'' + xy' + y = 0$$

is

$$y = c_1\left(\sum_{n=0}^{\infty}\frac{(-1)^n}{2^n n!}x^{2n}\right) + c_2\left(\sum_{n=0}^{\infty}\frac{(-1)^n 2^n n!}{(2n+1)!}x^{2n+1}\right). \quad \blacksquare$$

In each of these three examples, we have succeeded in finding power series solutions $y = \sum_{n=0}^{\infty}b_n x^n$ of the linear differential equations under consideration. Sometimes, however, there is no nonzero solution of a given linear differential equation that is valid in an open interval about 0. For example, we know that the general solution of the equation $xy' + y = 0$ on any interval not containing 0 is $y = c\frac{1}{x}$. Clearly we cannot expand $\frac{1}{x}$ in a power series about 0. However, we can expand $\frac{1}{x}$ in a power series about any $x_0 \in \mathbb{R}$ with $x_0 \neq 0$, so it makes sense to try to find a power series solution of this equation using a trial solution of the form $y = \sum_{n=0}^{\infty}b_n(x-x_0)^n$. The properties (1), (2), and (3) of power series stated at the beginning of this section are also valid, with the obvious minor changes, for power series of the form $\sum_{n=0}^{\infty}b_n(x-x_0)^n$. Hence the procedure is the same as before.

Example 4. Let us try to find a solution of the differential equation $xy' + y = 0$ of the form

$$y = \sum_{n=0}^{\infty}b_n(x-1)^n.$$

Then

$$y' = \sum_{n=1}^{\infty} nb_n(x - 1)^{n-1}$$

so in order to satisfy the given equation we must have

$$x \sum_{n=1}^{\infty} nb_n(x - 1)^{n-1} + \sum_{n=0}^{\infty} b_n(x - 1)^n = 0.$$

In order to be able to collect like powers of $(x - 1)$, we need to express the factor x in the form

$$x = 1 + (x - 1)$$

so that the above equation becomes

$$[1 + (x - 1)] \sum_{n=1}^{\infty} nb_n(x - 1)^{n-1} + \sum_{n=0}^{\infty} b_n(x - 1)^n = 0$$

or

$$\sum_{n=1}^{\infty} nb_n(x - 1)^{n-1} + \sum_{n=1}^{\infty} nb_n(x - 1)^n + \sum_{n=0}^{\infty} b_n(x - 1)^n = 0.$$

Let $n = k$ in the first summation, and change n to $k - 1$ in each of the others to get

$$\sum_{k=1}^{\infty} kb_k(x - 1)^{k-1} + \sum_{k=2}^{\infty} (k - 1)b_{k-1}(x - 1)^{k-1} + \sum_{k=1}^{\infty} b_{k-1}(x - 1)^{k-1} = 0$$

or

$$(b_1 + b_0) + \sum_{k=2}^{\infty} (kb_k + kb_{k-1})(x - 1)^{k-1} = 0.$$

It follows that

$$b_1 = -b_0$$

and that

$$b_k = -b_{k-1} \quad \text{for all } k \geq 2;$$

therefore,

$$b_1 = -b_0, \, b_2 = -b_1 = (-1)^2 b_0, \, b_3 = -b_2 = (-1)^3 b_0, \ldots, b_n = (-1)^n b_0.$$

The series solution is therefore

$$y = b_0 \left(\sum_{n=0}^{\infty} (-1)^n (x - 1)^n \right).$$

This series has radius of convergence

$$R = \lim_{n \to \infty} \left| \frac{(-1)^n}{(-1)^{n+1}} \right| = 1$$

and hence it converges for $|x - 1| < 1$.

So the power series method, using the series $\sum_{n=0}^{\infty} b_n (x - 1)^n$, leads us to the

solution $y = c \left(\sum_{n=0}^{\infty} (-1)^n (x - 1)^n \right)$ of the differential equation $xy' + y = 0$ on

the interval $I_{(0, 2)}$. You may recognize this series as the geometric series with ratio $-(x - 1)$ and sum $1/\{1 - [-(x - 1)]\} = 1/x$, so we have succeeded in finding the general solution $y = c(1/x)$ of $xy' + y = 0$ on the interval $I_{(0, 2)}$. ∎

You can expect that the power series method, using series of the form $\sum_{n=0}^{\infty} b_n x^n$,

will fail to lead to the general solution of the nth order linear differential equation

$$a_n y^{(n)} + \cdots + a_1 y' + a_0 y = 0$$

whenever $a_n(0) = 0$. In this case, we can usually find the general solution on an open interval about $x_0 \in \mathbb{R}$, when $a_n(x_0) \neq 0$, by using series of the form

$\sum_{n=0}^{\infty} b_n (x - x_0)^n$. Another method, which will often lead to solutions on an interval

$I_{(0, x_0)}$ for some x_0, is to try a series of the form $y = x^s \sum_{n=0}^{\infty} b_n x^n$, where s is some real

number to be determined. Such a series is called a ***Frobenius series.***

Example 5. Let us try to find a Frobenius series solution to the equation $xy' + y = 0$. Let

$$y = x^s \sum_{n=0}^{\infty} b_n x^n = \sum_{n=0}^{\infty} b_n x^{n+s}.$$

Then

$$y' = \sum_{n=0}^{\infty} (n + s) b_n x^{n+s-1}.$$

Notice that this sum starts at $n = 0$, not at $n = 1$, since the derivative of $b_0 x^s$ is not zero (unless, of course, s turns out to be zero). Now substitute into the differential equation $xy' + y = 0$ to get

$$x \sum_{n=0}^{\infty} (n + s)b_n x^{n+s-1} + \sum_{n=0}^{\infty} b_n x^{n+s} = 0$$

or

$$\sum_{n=0}^{\infty} ((n + s)b_n + b_n)x^{n+s} = 0.$$

Hence

$$(n + s + 1)b_n = 0 \quad \text{for all } n \geq 0.$$

This equation says that $b_n = 0$ unless $n + 1 = -s$. If we take $s = -1$ we find that b_0 is arbitrary and $b_n = 0$ for all $n > 0$. This leads us once again to the solution

$$y = b_0 x^{-1}.$$

But this time we see that our solution is valid on the interval $I_{(0, \infty)}$ rather than only on the interval $I_{(0, 2)}$ as in Example 4. ∎

Finally, we use the method of Frobenius to solve a differential equation that cannot be solved by more elementary methods.

Example 6.   We shall solve *Bessel's equation*

$$x^2 y'' + xy' + (x^2 - \tfrac{1}{9})y = 0$$

on $I_{(0, \infty)}$ using the Frobenius series method. Let

$$y = x^s \sum_{n=0}^{\infty} b_n x^n = \sum_{n=0}^{\infty} b_n x^{n+s}.$$

Then

$$y' = \sum_{n=0}^{\infty} (n + s)b_n x^{n+s-1} \quad \text{and} \quad y'' = \sum_{n=0}^{\infty} (n + s)(n + s - 1)b_n x^{n+s-2}.$$

We substitute these series into the given equation to get

$$x^2 \sum_{n=0}^{\infty} (n + s)(n + s - 1)b_n x^{n+s-2} + x \sum_{n=0}^{\infty} (n + s)b_n x^{n+s-1}$$

$$+ (x^2 - \tfrac{1}{9}) \sum_{n=0}^{\infty} b_n x^{n+s} = 0$$

or

$$\sum_{n=0}^{\infty} (n + s)(n + s - 1)b_n x^{n+s} + \sum_{n=0}^{\infty} (n + s)b_n x^{n+s} + \sum_{n=0}^{\infty} b_n x^{n+s+2}$$

$$- \sum_{n=0}^{\infty} \tfrac{1}{9} b_n x^{n+s} = 0.$$

Now let $n + 2 = k$ in the third sum and replace n by k in each of the other sums to get

$$\sum_{k=0}^{\infty} (k + s)(k + s - 1)b_k x^{k+s} + \sum_{k=0}^{\infty} (k + s)b_k x^{k+s} + \sum_{k=2}^{\infty} b_{k-2} x^{k+s}$$

$$- \sum_{k=0}^{\infty} \tfrac{1}{9} b_k x^{k+s} = 0$$

or

$$(s^2 - \tfrac{1}{9})b_0 x^s + ((1 + s)^2 - \tfrac{1}{9})b_1 x^{1+s} + \sum_{k=2}^{\infty} \{[(k + s)^2 - \tfrac{1}{9}]b_k + b_{k-2}\}x^{k+s} = 0.$$

If we choose $s = \pm\tfrac{1}{3}$ we find that b_0 is arbitrary, $b_1 = 0$, and

$$b_k = -\frac{b_{k-2}}{(k + s)^2 - \tfrac{1}{9}} \quad \text{for all } k \geq 2.$$

In particular, we see that $0 = b_1 = b_3 = b_5 = \cdots$ so all of the odd power coefficients are zero. Since $s^2 = \tfrac{1}{9}$, the recursion formula for b_k can be rewritten as

$$b_k = -\frac{b_{k-2}}{k^2 + 2ks} = -\frac{b_{k-2}}{k(k + 2s)}, \quad k \geq 2.$$

We see that the coefficients of the even powers of x are given by

$$b_2 = -\frac{b_0}{2(2 + 2s)}$$

$$b_4 = -\frac{b_2}{4(4 + 2s)} = (-1)^2 \frac{b_0}{2 \cdot 4(2 + 2s)(4 + 2s)}$$

$$\vdots$$

$$b_{2n} = (-1)^n \frac{b_0}{(2 \cdot 4 \cdot 6 \cdots 2n)(2 + 2s)(4 + 2s) \cdots (2n + 2s)}$$

and so

$$b_{2n} = (-1)^n \frac{b_0}{n!(1 + s)(2 + s) \cdots (n + s) \cdot 2^{2n}}, \quad \text{for } n \geq 1.$$

We therefore obtain two solutions, one corresponding to $s = \frac{1}{3}$ and the other to $s = -\frac{1}{3}$. They are

$$y_1 = x^{\frac{1}{3}}\left(1 + \sum_{n=1}^{\infty} \frac{(-1)^n}{n!(1 + \frac{1}{3})(2 + \frac{1}{3}) \cdots (n + \frac{1}{3})} \left(\frac{x}{2}\right)^{2n}\right)$$

and

$$y_2 = x^{-\frac{1}{3}}\left(1 + \sum_{n=1}^{\infty} \frac{(-1)^n}{n!(1 - \frac{1}{3})(2 - \frac{1}{3}) \cdots (n - \frac{1}{3})} \left(\frac{x}{2}\right)^{2n}\right).$$

It is easy to check that the radius of convergence of both of these power series is ∞, so y_1 and y_2 are solutions on $I_{(0,\infty)}$ of Bessel's equation. Since $\{y_1, y_2\}$ is clearly linearly independent, $\{y_1, y_2\}$ is a basis for the solution space, and the general solution of Bessel's equation on $I_{(0,\infty)}$ is $y = c_1 y_1 + c_2 y_2$. ∎

EXERCISES

1. Use the power series method to find the general solution of each of the following constant coefficient differential equations, and compare your answer with the solution you obtain using the methods of Chapter 3.

   (a) $y' + y = 0$ (c) $y'' + y = 0$
   (b) $y'' - y = 0$ (d) $y'' + 2y' + y = 0$

2. Use the power series method to solve the following variable coefficient linear differential equations.

   (a) $y' + xy = 0$ (d) $y'' - 3xy' - y = 0$
   (b) $y'' - 2xy' + 2y = 0$ (e) $y'' + xy = 0$
   (c) $y'' - xy' - y = 0$ (f) $4y'' + x^2y' - xy = 0$

3. Use a trial solution of the form $y = \sum_{n=0}^{\infty} b_n(x - 1)^n$ to solve the following differential equations on the interval $I_{(0,2)}$. Carry each computation out to the term involving $(x - 1)^4$.

   (a) $x^2y' - y = 0$ (c) $xy'' - y' - 4x^3y = 0$
   (b) $x^2y'' + xy' - y = 0$ (d) $xy'' + 2y' + xy = 0$

4. Use a Frobenius series of the form $y = \sum_{n=0}^{\infty} b_n x^{n+s}$ to solve the following differential equations. Where possible find general formulas for the solutions. Otherwise, find the first four nonzero terms of each series.

   (a) $2xy'' + 3y' - 2xy = 0$
   (b) $2x^2y'' + 3xy' + (2x - 1)y = 0$
   (c) $2x^2y'' + (3x - 2x^2)y' - (x + 1)y = 0$

(d) $16x^2y'' + (3 - x^2)y = 0$

(e) $5x^2y'' + xy' + (x^3 - 1)y = 0$

5. Use the theorem of this section to show that if y_1 and $y_2 \in \mathfrak{F}(I)$ are solutions on an open interval I of a homogeneous second order linear differential equation $ay'' + by' + cy = 0$ with $a(x) \neq 0$ for all $x \in I$, and if

$$y_1(x_0) = 1 \qquad y_2(x_0) = 0$$
$$y_1'(x_0) = 0 \qquad y_2'(x_0) = 1$$

for some $x_0 \in I$, then $\{y_1, y_2\}$ is a basis for the solution space of that equation on I.

6. Generalizing Exercise 5, show that if $y_1, \ldots, y_n \in \mathfrak{F}(I)$ satisfy an nth order homogeneous linear differential equation on an open interval I, with leading coefficient nowhere zero on I, and if

$$\begin{pmatrix} y_1(x_0) & y_2(x_0) & & y_n(x_0) \\ y_1'(x_0) & y_2'(x_0) & & y_n'(x_0) \\ \vdots & \vdots & \vdots & \vdots \\ y_1^{(n-1)}(x_0) & y_2^{(n-1)}(x_0) & & y_n^{(n-1)}(x_0) \end{pmatrix} = \begin{pmatrix} 1 & 0 & & 0 \\ 0 & 1 & & 0 \\ \vdots & \vdots & \vdots & \vdots \\ 0 & 0 & & 1 \end{pmatrix}$$

then $\{y_1, \ldots, y_n\}$ is a basis for the solution space on I of that equation. [Hint: Remember the Wronskian matrix?]

9.3 VARIATION OF PARAMETERS

In Chapter 3 we described the method of undetermined coefficients, a method for finding particular solutions of certain inhomogeneous linear differential equations. Now we shall describe another method, one that is usually more difficult to carry out than the method of undetermined coefficients, but that has the advantage of being more generally applicable. We begin with an example.

Example 1.   Solve $y'' + y = \sec x$ on the interval $I_{(-\pi/2, \pi/2)}$. We know that $\{\cos x, \sin x\}$ is a basis for the solution space of the homogeneous equation $y'' + y = 0$. We shall try to find a particular solution of $y'' + y = \sec x$ of the form

$$y_P = (\cos x)u_1 + (\sin x)u_2$$

where u_1 and u_2 are functions of x. Since we are dealing with two unknown functions, u_1 and u_2, we expect to be able to impose two conditions. One of these conditions is that y_P must satisfy the equation $y'' + y = \sec x$. The other condition will be chosen to simplify the computations. Using the product rule for differentiation, we find

$$y_P' = (-\sin x)u_1 + (\cos x)u_2 + (\cos x)u_1' + (\sin x)u_2'.$$

If we set

(1) $(\cos x)u_1' + (\sin x)u_2' = 0,$

this simplifies to

$$y_P' = (-\sin x)u_1 + (\cos x)u_2.$$

Then differentiation of y_P' yields

$$y_P'' = (-\cos x)u_1 + (-\sin x)u_2 + (-\sin x)u_1' + (\cos x)u_2'.$$

Substituting y_P and y_P'' into the equation $y'' + y = \sec x$ gives

(2) $(-\sin x)u_1' + (\cos x)u_2' = \sec x.$

The conditions (1) and (2) describe a system of two algebraic linear equations in the two unknown functions u_1' and u_2':

$$(\cos x)u_1' + (\sin x)u_2' = 0$$
$$(-\sin x)u_1' + (\cos x)u_2' = \sec x.$$

Notice that the coefficient matrix of this system is the Wronskian matrix $W_x(\cos x, \sin x)$. Solving this system, we find that

$$u_1' = -\sin x \sec x = -\tan x$$
$$u_2' = \cos x \sec x = 1.$$

Finally, integration (any integral will do) yields

$$u_1 = \ln|\cos x|$$
$$u_2 = x$$

so $y_P = (\cos x)(\ln|\cos x|) + (\sin x)x$ is a particular solution of $y'' + y = \sec x$. The general solution is

$$y = (\cos x)(\ln|\cos x|) + x \sin x + c_1 \cos x + c_2 \sin x. \quad \blacksquare$$

The procedure used in Example 1 may seem a bit mysterious, but it always works! Suppose

$$ay'' + by' + cy = f$$

is any second order linear differential equation, with a, b, c, and $f \in \mathcal{C}(I)$ and $a \neq 0$ on I, and suppose $\{y_1, y_2\}$ is any linearly independent pair of solutions of the homogeneous equation

$$ay'' + by' + cy = 0.$$

Set

$$y_P = y_1 u_1 + y_2 u_2.$$

If we require that $y_1 u_1' + y_2 u_2' = 0$, we find that

$$y_P' = y_1' u_1 + y_2' u_2$$

and

$$y_P'' = y_1'' u_1 + y_2'' u_2 + y_1' u_1' + y_2' u_2'$$

so

$$
\begin{aligned}
ay_P'' + by_P' + cy_P &= a(y_1'' u_1 + y_2'' u_2 + y_1' u_1' + y_2' u_2') \\
&\quad + b(y_1' u_1 + y_2' u_2) + c(y_1 u_1 + y_2 u_2) \\
&= u_1(ay_1'' + by_1' + cy_1) + u_2(ay_2'' + by_2' + cy_2) \\
&\quad + a(y_1' u_1' + y_2' u_2') \\
&= a(y_1' u_1' + y_2' u_2').
\end{aligned}
$$

Here we have used the fact that y_1 and y_2 satisfy the equation $ay'' + by' + cy = 0$. Thus $y_P = y_1 u_1 + y_2 u_2$ will satisfy the equation $ay'' + by' + cy = f$ provided that

$$y_1 u_1' + y_2 u_2' = 0$$
$$y_1' u_1' + y_2' u_2' = f/a.$$

The coefficient matrix of this linear system is the Wronskian matrix $W(y_1, y_2)$. The Wronskian determinant $\det W(y_1, y_2)$ is nowhere zero on I (see Exercise 11) and hence this system has a unique solution (u_1', u_2'). Integrating then yields functions u_1 and u_2 such that $y_1 u_1 + y_2 u_2$ is a particular solution of $ay'' + by' + cy = f$.

This procedure is called *variation of parameters* because the general solution $y_H = c_1 y_1 + c_2 y_2$ of the homogeneous equation $ay'' + by' + cy = 0$ depends on the two parameters c_1 and c_2, and we have found a particular solution by "varying these parameters"; that is, by viewing them as functions (u_1 and u_2) rather than as constants.

We now state (without proof, but see Exercise 13) the generalization of this method to equations of higher order.

Method of Variation of Parameters. In order to find a particular solution of

$$a_n y^{(n)} + \cdots + a_1 y' + a_0 y = f:$$

(i) Find a basis $\{y_1, \ldots, y_n\}$ for the solution space of the homogeneous equation

$$a_n y^{(n)} + \cdots + a_1 y' + a_0 y = 0.$$

(ii) Solve the system of algebraic linear equations whose matrix is

$$\begin{pmatrix} y_1 & y_2 & \cdots & y_n & 0 \\ y_1' & y_2' & \cdots & y_n' & 0 \\ \vdots & \vdots & & \vdots & \vdots \\ y_1^{(n-1)} & y_2^{(n-1)} & \cdots & y_n^{(n-1)} & f/a_n \end{pmatrix}$$

Call the solution $(u_1', \ldots, u_n')$.

(iii) For each i $(1 \le i \le n)$, find $u_i = \int u_i'$ (any integral will do). Then

$$y_P = y_1 u_1 + y_2 u_2 + \cdots + y_n u_n$$

is a particular solution.

Example 2. Use variation of parameters to find a particular solution of $y'' - 2y' + y = e^x$. The homogeneous equation $y'' - 2y' + y = 0$ has general solution $y = c_1 e^x + c_2 x e^x$, so we may take $y_1 = e^x$ and $y_2 = x e^x$. Solving the system

$$\begin{cases} y_1 u_1' + y_2 u_2' = 0 \\ y_1' u_1' + y_2' u_2' = f/a_2, \end{cases} \quad \text{or} \quad \begin{cases} e^x u_1' + x e^x u_2' = 0 \\ e^x u_1' + (e^x + x e^x) u_2' = e^x, \end{cases}$$

yields $u_1' = -x$, $u_2' = 1$, so $u_1 = -\frac{1}{2}x^2$, $u_2 = x$, and

$$y_P = (e^x)(-\tfrac{1}{2}x^2) + (x e^x)x = \tfrac{1}{2}x^2 e^x. \quad \blacksquare$$

Example 3. Solve $xy''' + 2y'' = 1/x^2$ on $I_{(0, \infty)}$. Since $xy''' + 2y'' = (xD + 2)D^2 y$, we can use the method described in Section 9.1 to solve the homogeneous equation $xy''' + 2y'' = 0$. Its solution space has basis $\{\ln x, x, 1\}$, so we must solve the linear system whose matrix is

$$\begin{pmatrix} \ln x & x & 1 & 0 \\ 1/x & 1 & 0 & 0 \\ -1/x^2 & 0 & 0 & 1/x^3 \end{pmatrix}.$$

Solving this system yields $(u_1', u_2', u_3') = \left(-\dfrac{1}{x}, \dfrac{1}{x^2}, \dfrac{1}{x} \ln x - \dfrac{1}{x} \right)$. Hence

$$u_1 = \int -\frac{1}{x}\, dx = -\ln x,$$

$$u_2 = \int \frac{1}{x^2}\, dx = -\frac{1}{x}, \text{ and}$$

$$u_3 = \int \left(\frac{1}{x} \ln x - \frac{1}{x} \right) dx = \tfrac{1}{2}(\ln x)^2 - \ln x,$$

so

$$y_P = y_1 u_1 + y_2 u_2 + y_3 u_3$$

$$= (\ln x)(-\ln x) + x\left(-\frac{1}{x} \right) + 1(\tfrac{1}{2}(\ln x)^2 - \ln x)$$

$$= -\tfrac{1}{2}(\ln x)^2 - 1 - \ln x.$$

But 1 and $\ln x$ are solutions of the homogeneous equation $xy''' + 2y'' = 0$, so we can just as well take $y_P = -\tfrac{1}{2}(\ln x)^2$. The general solution is

$$y = -\tfrac{1}{2}(\ln x)^2 + c_1 \ln x + c_2 x + c_3. \quad \blacksquare$$

EXERCISES

Solve the following differential equations.

1. $y'' + y = \tan x$ on $I_{(-\pi/2,\, \pi/2)}$
2. $y'' + y = \csc x$ on $I_{(0,\, \pi)}$
3. $y'' + y = \tan^2 x$ on $I_{(-\pi/2,\, \pi/2)}$
4. $y'' + 9y = \sec 3x$ on $I_{(-\pi/6,\, \pi/6)}$
5. $y'' - 2y' + y = e^x/x$ on $I_{(0,\, \infty)}$
6. $xy''' + 2y'' = 1/x$ on $I_{(0,\, \infty)}$
7. $xy''' + 3y'' = \sqrt{x}$ on $I_{(0,\, \infty)}$
8. $x^2 y'' + xy' - y = \dfrac{1}{x}$ on $I_{(0,\, \infty)}$

9. The method of variation of parameters, applied to the equation $ay'' + by' + cy = f$, yields a particular solution of the form
    $$y_P = y_1 u_1 + y_2 u_2$$
    where $\{y_1, y_2\}$ is a basis for the solution space of the homogeneous equation $ay'' + by' + cy = 0$. Show that the functions u_1 and u_2 can be obtained by integrating the formulas
    $$u_1' = -\left(\frac{y_2}{y_1 y_2' - y_1' y_2} \right) \frac{f}{a}, \quad u_2' = \left(\frac{y_1}{y_1 y_2' - y_1' y_2} \right) \frac{f}{a}.$$

10. If, in the method of variation of parameters, we use $u_i = \int u_i' + c_i$ rather than $u_i = \int u_i'$ (without the arbitrary constant c_i), what do we get for y_p?

11. Show that if $\{y_1, y_2\}$ is a basis for the solution space of the equation $ay'' + by' + cy = 0$ on the open interval I, then the Wronskian determinant

$$\det \begin{pmatrix} y_1 & y_2 \\ y_1' & y_2' \end{pmatrix}$$

is nowhere zero on I. [Hint: Proceed as in Exercise 11 of Section 5.1, but without assuming that a, b, and c are constant.]

12. Verify that the method of variation of parameters is valid for first order equations by showing that if y_1 is a nonzero solution of $ay' + by = 0$, then $y_1 u_1$ is a solution of $ay' + by = f$, where u_1' satisfies the equation $y_1 u_1' = f/a$.

13. Verify that the method of variation of parameters does determine a particular solution of any third order linear equation

$$a_3 y''' + a_2 y'' + a_1 y' + a_0 y = f.$$

[Hint: Imitate the argument that established the validity of the method for second order equations.]

14. A method analogous to variation of parameters can be used to help solve linear differential equations when one solution of the associated homogeneous equation is known. Show that if y_1 is a solution (not the zero solution) of the homogeneous equation

$$ay'' + by' + cy = 0,$$

then the general solution of the equation

$$ay'' + by' + cy = f$$

is of the form $y = y_1 u$, where u is the general solution of the equation

$$(ay_1)y'' + (2ay_1' + by_1)y' = f.$$

[Note that the differential operator occurring in this last equation has an obvious factorization.]

9.4 THE LAPLACE TRANSFORM

In this section we shall discuss the Laplace transform, a linear map from one function space to another. This linear map is the key to a powerful technique for solving linear differential equations subject to specified initial conditions. In contrast to the techniques discussed earlier, the Laplace transform method leads directly to the particular solution of the differential equation that satisfies the prescribed initial conditions. It is not necessary to find the general solution of the differential equation first.

The **Laplace transform** $L(f)$ of a function $f \in \mathcal{F}(I_{[0, \infty)})$ is defined by

$$L(f) = \int_0^\infty e^{-sx} f(x)\, dx,$$

provided that the integral exists. Notice that $L(f)$ is a function of s.

Example 1. Let us compute the Laplace transforms of some elementary functions.

(i) $L(1) = \int_0^\infty e^{-sx}\, dx = \dfrac{e^{-sx}}{-s}\Big]_0^\infty = \dfrac{1}{s}, \quad s > 0.$

(ii) $L(x) = \int_0^\infty e^{-sx}x\, dx = \dfrac{e^{-sx}x}{-s}\Big]_0^\infty - \int_0^\infty \dfrac{e^{-sx}}{-s}\, dx = \dfrac{1}{s^2}, \quad s > 0.$

(iii) $L(e^x) = \int_0^\infty e^{-sx}e^x\, dx = \int_0^\infty e^{-(s-1)x}\, dx = \dfrac{e^{-(s-1)x}}{-(s-1)}\Big]_0^\infty = \dfrac{1}{s-1}, s > 1.$

(iv) $L(e^{-x}) = \int_0^\infty e^{-sx}e^{-x}\, dx = \int_0^\infty e^{-(s+1)x}\, dx = \dfrac{e^{-(s+1)x}}{-(s+1)}\Big]_0^\infty = \dfrac{1}{s+1},$
$$s > -1. \quad \blacksquare$$

In order that the Laplace transform $L(f) = \int_0^\infty e^{-sx} f(x)\, dx$ exist, it is certainly necessary that the function $f \in \mathcal{F}(I_{[0,\infty)})$ be integrable. But this is not enough to guarantee convergence of the integral $\int_0^\infty e^{-sx} f(x)\, dx$. We need to impose another condition on f, one that will guarantee that $e^{-sx} f(x)$ goes to zero sufficiently fast, as $x \to \infty$, so that the integral $\int_0^\infty e^{-sx} f(x)\, dx$ is finite. A simple condition that will guarantee convergence of the integral is that $\lim_{x\to\infty} f(x)/e^{kx} = 0$, for some $k \in \mathbb{R}$.

Theorem 1. *Suppose $f \in \mathcal{F}(I_{[0,\infty)})$ is integrable and $\lim_{x\to\infty} f(x)/e^{kx} = 0$ for some $k \in \mathbb{R}$. Then $L(f) = \int_0^\infty e^{-sx} f(x)\, dx$ exists, for all $s > k$.*

Proof. Since $\lim_{x\to\infty} f(x)/e^{kx} = 0$, there must be a real number b such that $|f(x)/e^{kx}| < 1$ whenever $x > b$. Hence

$$|e^{-sx} f(x)| = e^{-sx}e^{kx}\, |f(x)/e^{kx}| < e^{-(s-k)x}$$

for all $x > b$. Since

$$\int_b^\infty e^{-(s-k)x}\, dx = \dfrac{e^{-(s-k)x}}{-(s-k)}\Big]_b^\infty = \dfrac{e^{-(s-k)b}}{s-k} < \infty$$

for all $s > k$, we can conclude by comparison that the integral $\int_b^\infty |e^{-sx} f(x)|\, dx$ is finite, for all $s > k$. It follows that the integral $\int_0^\infty e^{-sx} f(x)\, dx$ converges absolutely, for all $s > k$. $\blacksquare$

A function $f \in \mathcal{F}(I_{[0,\infty)})$ such that $\lim_{x \to \infty} f(x)/e^{kx} = 0$ for some $k \in \mathbb{R}$ is said to be *of exponential order.* Let $\mathcal{E}$ denote the subset of $\mathcal{F}(I_{[0,\infty)})$ consisting of those functions that are both integrable and of exponential order. It is easy to check that $\mathcal{E}$ is a subspace of $\mathcal{F}(I_{[0,\infty)})$ (see Exercise 16). Moreover, according to Theorem 1, the Laplace transform $L(f)$ exists for each $f \in \mathcal{E}$.

The Laplace transform $L(f)$ of a function $f \in \mathcal{E}$ is a function whose domain contains an interval $I_{(c,\infty)}$ for some $c \in \mathbb{R}$. We shall call an interval of the form $I_{(c,\infty)}$, $c \in \mathbb{R}$, a *neighborhood of* ∞. If we let $\mathcal{F}_\infty$ denote the set of all functions defined on some neighborhood of ∞, then we see that the Laplace transform L maps the vector space $\mathcal{E}$ into $\mathcal{F}_\infty$. We can make $\mathcal{F}_\infty$ into a vector space as follows.

First, we shall treat two functions in $\mathcal{F}_\infty$ as *equal* if they are identical on some neighborhood of ∞. Thus, for f and $g \in \mathcal{F}_\infty$, $f = g$ if and only if $f(s) = g(s)$ for all $s > c$, for some $c \in \mathbb{R}$. For example, as *elements of* $\mathcal{F}_\infty$, the functions s and $|s|$ are the same.

Sums and scalar multiples of functions in $\mathcal{F}_\infty$ are defined as you would expect. If f and g are in $\mathcal{F}_\infty$ then $f + g \in \mathcal{F}_\infty$ is defined by $(f + g)(s) = f(s) + g(s)$. Notice that if f is defined on the interval $I_{(a,\infty)}$ and g is defined on the interval $I_{(b,\infty)}$ then $f + g$ is defined on the interval $I_{(c,\infty)}$ where $c = \max\{a, b\}$. If $f \in \mathcal{F}_\infty$ and $c \in \mathbb{R}$, then $cf \in \mathcal{F}_\infty$ is defined by $(cf)(s) = cf(s)$. Notice that the domain of cf is the same as the domain of f.

With these operations of addition and scalar multiplication, $\mathcal{F}_\infty$ is a vector space. Moreover, the Laplace transform $L: \mathcal{E} \to \mathcal{F}_\infty$ is a linear map from the vector space $\mathcal{E}$ into the vector space $\mathcal{F}_\infty$. The linearity of L is an immediate consequence of the linearity of the integral: for f and $g \in \mathcal{E}$ and $c \in \mathbb{R}$ we have

$$L(f + g) = \int_0^\infty e^{-sx}(f(x) + g(x))\, dx = \int_0^\infty e^{-sx} f(x)\, dx + \int_0^\infty e^{-sx} g(x)\, dx$$

$$= L(f) + L(g)$$

and

$$L(cf) = \int_0^\infty e^{-sx} cf(x)\, dx = c \int_0^\infty e^{-sx} f(x)\, dx = cL(f).$$

Example 2. We can use the linearity of L to help us compute Laplace transforms. From Example 1 we know that $L(e^x) = \dfrac{1}{s - 1}$ and $L(e^{-x}) = \dfrac{1}{s + 1}$. Hence

(i) $$L(\cosh x) = L(\tfrac{1}{2}e^x + \tfrac{1}{2}e^{-x}) = \tfrac{1}{2}L(e^x) + \tfrac{1}{2}L(e^{-x})$$

$$= \tfrac{1}{2} \cdot \frac{1}{s - 1} + \tfrac{1}{2} \cdot \frac{1}{s + 1} = \frac{s}{s^2 - 1},$$

and

(ii) $$L(\sinh x) = L(\tfrac{1}{2}e^x - \tfrac{1}{2}e^{-x}) = \tfrac{1}{2}L(e^x) - \tfrac{1}{2}L(e^{-x})$$

$$= \tfrac{1}{2} \cdot \frac{1}{s - 1} - \tfrac{1}{2} \cdot \frac{1}{s + 1} = \frac{1}{s^2 - 1}. \quad \blacksquare$$

The next theorem tells us why the Laplace transform is useful in solving differential equations.

Theorem 2.   *Let* $y \in \mathcal{E}$ *be differentiable and suppose* $y' \in \mathcal{E}$. *Then*

$$L(y') = sL(y) - y(0).$$

Proof.   Integrate by parts:

$$L(y') = \int_0^\infty e^{-sx} y'(x)\, dx$$

$$= e^{-sx} y(x)\,\Big|_0^\infty - \int_0^\infty - se^{-sx} y(x)\, dx$$

$$= -y(0) + sL(y). \quad \blacksquare$$

Example 3.   Let us use the Laplace transform to solve the initial value problem

$$y' - y = 1, \; y(0) = 1.$$

First, we apply L to the equation $y' - y = 1$ and use linearity to get

$$L(y') - L(y) = L(1) = \frac{1}{s}.$$

Then we use Theorem 2 together with the initial condition $y(0) = 1$ to get

$$sL(y) - 1 - L(y) = \frac{1}{s}$$

or

$$(s - 1)L(y) = 1 + \frac{1}{s} = \frac{s + 1}{s}.$$

Thus

$$L(y) = \frac{s + 1}{s(s - 1)}.$$

Using partial fractions, we can rewrite $L(y)$ as

$$L(y) = -\frac{1}{s} + \frac{2}{s - 1}$$

and we know from Example 1 that

$$\frac{1}{s} = L(1) \quad \text{and} \quad \frac{1}{s - 1} = L(e^x).$$

Hence

$$L(y) = -L(1) + 2L(e^x)$$
$$= L(-1 + 2e^x).$$

Therefore

$$y = -1 + 2e^x$$

is a function whose Laplace transform satisfies the transformed equation. It is easy to check that this y satisfies both $y' - y = 1$ and $y(0) = 1$. ∎

In order to solve higher order initial value problems, we need the following generalization of Theorem 2.

Theorem 3. *Let $y \in \mathcal{E}$ be n-times differentiable and suppose y', y'', ..., $y^{(n)}$ are all in $\mathcal{E}$. Then*

$$L(y^{(n)}) = s^n L(y) - s^{n-1}y(0) - s^{n-2}y'(0) - \cdots - y^{(n-1)}(0).$$

Proof. Apply Theorem 2 repeatedly. Thus

$$L(y'') = sL(y') - y'(0)$$
$$= s[sL(y) - y(0)] - y'(0)$$
$$= s^2 L(y) - sy(0) - y'(0)$$

and

$$L(y''') = sL(y'') - y''(0)$$
$$= s[s^2 L(y) - sy(0) - y'(0)] - y''(0)$$
$$= s^3 L(y) - s^2 y(0) - sy'(0) - y''(0).$$

And so on. ∎

Example 4. Consider the initial value problem

$$y'' - y = 0, \quad y(0) = 1, \quad y'(0) = 3.$$

Let us apply the Laplace transform to the differential equation. We find, using Theorem 3, that

$$s^2 L(y) - sy(0) - y'(0) - L(y) = 0$$

or

$$(s^2 - 1)L(y) = sy(0) + y'(0) = s + 3.$$

Thus

$$L(y) = \frac{s+3}{s^2-1} = \frac{2}{s-1} - \frac{1}{s+1}.$$

Since

$$\frac{1}{s-1} = L(e^x) \quad \text{and} \quad \frac{1}{s+1} = L(e^{-x}),$$

we find that

$$L(y) = 2L(e^x) - L(e^{-x}) = L(2e^x - e^{-x}).$$

Therefore

$$y = 2e^x - e^{-x}$$

is a function whose Laplace transform satisfies the transformed equation. Again it is easy to check that $y = 2e^x - e^{-x}$ satisfies both the given differential equation and the initial conditions. ■

The method for solving initial value problems that was illustrated in Examples 3 and 4 can be summarized as follows.

Laplace transform method.   To solve

$$a_n y^{(n)} + \cdots + a_1 y' + a_0 y = f,$$

where $a_0, a_1, \ldots, a_n \in \mathbb{R}$ and $f \in \mathcal{E}$, subject to the initial conditions

$$y(0) = b_1, y'(0) = b_2, \ldots, y^{(n-1)}(0) = b_n,$$

proceed as follows:

(1)   Apply the Laplace transform L to both sides of the given equation.
(2)   Use linearity together with Theorem 3 (or Theorem 4, below) to write the transformed equation in the form

$$h(s)L(y) = g(s).$$

(3)   Solve for $L(y)$: $L(y) = g(s)/h(s)$.
(4)   Find $y \in \mathcal{E}$ so that $L(y) = g(s)/h(s)$.

Remark 1.   If f is continuous, the Laplace transform method will yield a function y that is n-times continuously differentiable and that satisfies the given differential equation at every point of $I_{[0, \infty)}$. The case in which f is not continuous is somewhat more subtle. We will discuss this case in detail shortly.

Remark 2. The success of the Laplace transform method depends on the following two facts:

(1)   If $f \in \mathcal{E}$ is continuous and $y \in \mathcal{F}(I_{[0, \infty)})$ is any solution of the constant coefficient linear differential equation

$$a_n y^{(n)} + \cdots + a_1 y' + a_0 y = f,$$

then y and its first n derivatives $y', \ldots, y^{(n)}$ are all in $\mathcal{E}$.

(2)   If y_1 and y_2 are two continuous functions in $\mathcal{E}$ with $L(y_1) = L(y_2)$, then $y_1 = y_2$.

Fact (1) guarantees that the function y we seek does have a Laplace transform and that its transform can be found by applying Steps 1–3 of the Laplace transform method. Fact (2) guarantees that if $y \in \mathcal{E}$ is continuous and $L(y)$ satisfies the transformed equation, then y does satisfy the given initial value problem. For if y_1 is a solution of the given initial value problem, then $L(y_1)$ must also satisfy the transformed equation, so $L(y) = L(y_1)$ and hence $y = y_1$.

We will not prove these two facts here. You will find these matters discussed in more advanced books on the Laplace transform. See, for example, R. V. Churchill, *Operational Mathematics* (3rd edition), McGraw-Hill, 1958.

Remark 3.   In order to apply the Laplace transform method successfully to solve a given initial value problem

$$a_n y^{(n)} + \cdots + a_1 y' + a_0 y = f, \quad y(0) = b_1, \ldots, y^{(n-1)}(0) = b_n,$$

you must first find the Laplace transform of f. Then, eventually, you must find the solution y from its Laplace transform. For both of these steps, it is helpful to have available a table of transforms. A short table is provided on page 309. More extensive tables may be found elsewhere, e.g., in the book by Churchill referred to above.

Remark 4.   You will recall from calculus that the method of partial fractions enables you to express any quotient $g(s)/h(s)$, where $g(s)$ and $h(s)$ are polynomials, as a sum of simple terms. You will often find this procedure helpful in carrying out the final step in the Laplace transform method. If you have forgotten how to do partial fraction decompositions, then this is a good time to take out your calculus book and review the method.

Example 5.   Let us use the Laplace transform to solve the initial value problem

$$y'' - 6y' + 9y = 18, \quad y(0) = y'(0) = 0.$$

A SHORT TABLE OF LAPLACE TRANSFORMS

	FUNCTION	TRANSFORM
(i)	1	$\dfrac{1}{s}$
(ii)	x	$\dfrac{1}{s^2}$
(iii)	x^n	$\dfrac{n!}{s^{n+1}}$
(iv)	$e^{\alpha x}$	$\dfrac{1}{s - \alpha}$
(v)	$xe^{\alpha x}$	$\dfrac{1}{(s - \alpha)^2}$
(vi)	$\cos \beta x$	$\dfrac{s}{s^2 + \beta^2}$
(vii)	$\sin \beta x$	$\dfrac{\beta}{s^2 + \beta^2}$
(viii)	$e^{\alpha x} \cos \beta x$	$\dfrac{s - \alpha}{(s - \alpha)^2 + \beta^2}$
(ix)	$e^{\alpha x} \sin \beta x$	$\dfrac{\beta}{(s - \alpha)^2 + \beta^2}$
(x)	$u_a(x) = \begin{cases} 0 & \text{if } x < a \\ 1 & \text{if } x \geq a \end{cases} (a > 0)$	$\dfrac{e^{-as}}{s}$
(xi)*	$u_a(x)f(x - a)$	$e^{-as}L(f)$
(xii)*	$x^n f(x)$	$(-1)^n \dfrac{d^n}{ds^n} L(f)$
(xiii)*	$e^{\alpha x}f(x)$	$g(s - \alpha)$, where $g(s) = L(f)$
(xiv)*	$\displaystyle\int_0^x f(t)dt$	$\dfrac{1}{s} L(f)$
(xv)*	$\displaystyle\int_0^x f(x - t)g(t)dt$	$L(f)L(g)$
(xvi)*	y'	$sL(y) - y(0)$
(xvii)*	$y^{(n)}$	$s^n L(y) - s^{n-1}y(0) - s^{n-2}y'(0) - \cdots - y^{(n-1)}(0)$

*It is assumed in formulas (xi)–(xiv) that $f \in \mathcal{E}$, in (xv) that both f and $g \in \mathcal{E}$, in (xvi) that $y' \in \mathcal{E}$, and in (xvii) that $y^{(n)} \in \mathcal{E}$.

First apply L to the equation to get

$$s^2 L(y) - 6s L(y) + 9L(y) = 18L(1)$$

or

$$(s^2 - 6s + 9)L(y) = \frac{18}{s}.$$

Hence

$$L(y) = \frac{18}{s(s^2 - 6s + 9)} = \frac{18}{s(s-3)^2}.$$

Referring now to the table of transforms, we see that

$$L(y) = \frac{18}{s} \cdot \frac{1}{(s-3)^2} \overset{\text{(v)}}{=} \frac{18}{s} L(xe^{3x}) \overset{\text{(xiv)}}{=} 18L\left(\int_0^x te^{3t}\,dt\right)$$

$$= 18L(\tfrac{1}{3}xe^{3x} - \tfrac{1}{9}e^{3x} + \tfrac{1}{9})$$

$$= L(6xe^{3x} - 2e^{3x} + 2).$$

Hence

$$y = 6xe^{3x} - 2e^{3x} + 2$$

is the solution to the given initial value problem.

Notice that we could also have found y using partial fractions. Since

$$L(y) = \frac{18}{s(s-3)^2} = \frac{-2}{s-3} + \frac{6}{(s-3)^2} + \frac{2}{s}$$

we can see that

$$y = -2e^{3x} + 6xe^{3x} + 2. \quad \blacksquare$$

Now let us consider the case in which the function f in the differential equation

$$a_n y^{(n)} + \cdots + a_1 y' + a_0 y = f$$

is not continuous. In this case we must relax slightly our concept of what constitutes a solution. If, say, $f \in \mathcal{E}$ is not required to be continuous at $a \in I_{[0,\infty)}$, then we may alter the value of f at a to obtain a new function

$$g(x) = \begin{cases} f(x) & \text{if } x \neq a \\ b & \text{if } x = a \end{cases}$$

where $b \neq f(a)$. Then $g \in \mathcal{E}$ whenever $f \in \mathcal{E}$, and $L(g) = L(f)$. Hence the transformed equations of

$$a_n y^{(n)} + \cdots + a_1 y' + a_0 y = f$$

and of

$$a_n y^{(n)} + \cdots + a_1 y' + a_0 y = g$$

are identical. Therefore, it is not possible that the Laplace transform method can lead to a function y that satisfies both of these equations at a. What usually happens is that the function y found by the method is not even n-times differentiable at a. So we must modify our definition of solution, and we must relax the hypotheses in Theorem 3 to allow functions y that are not n-times differentiable everywhere.

To keep the discussion reasonably simple, we shall restrict our attention to the case in which the function f has discontinuities only on a discrete subset of $I_{[0, \infty)}$. This case includes most of the functions that arise in applications.

A subset S of $I_{[0, \infty)}$ is **discrete** if for each $c \in I_{[0, \infty)}$ the set $\{x \in S \mid x \leq c\}$ is either finite or empty. Finite subsets of $I_{[0, \infty)}$ are discrete, as is the set of positive integers.

A function f is said to be **piecewise continuous** on the interval $I_{[0, \infty)}$ if

(i) f is defined and continuous on the interval $I_{[0, \infty)}$ except possibly on a discrete subset S, and

(ii) for each $a \in S$, the right hand limit $\lim_{x \to a_+} f(x)$ exists and, if $a \neq 0$, the left hand limit $\lim_{x \to a_-} f(x)$ also exists.

Example 6. For each $a > 0$, the **unit step function** u_a defined by

$$u_a(x) = \begin{cases} 0 & \text{if } 0 \leq x < a \\ 1 & \text{if } x \geq a \end{cases}$$

is piecewise continuous (see Figure 9.1). ■

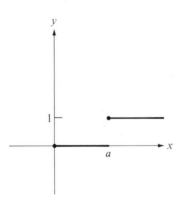

FIGURE 9.1

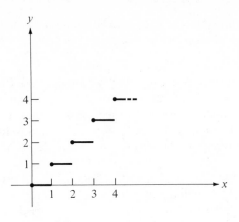

FIGURE 9.2

Example 7.   The *staircase function*

$$f(x) = [x]$$

is piecewise continuous (recall that $[x]$ means the greatest integer less than or equal to x; see Figure 9.2). ■

Example 8.   The function

$$f(x) = \frac{1}{x - 1}$$

is *not* piecewise continuous because, although it has only one discontinuity, at $x = 1$, the limits $\lim\limits_{x \to 1_-} f(x)$ and $\lim\limits_{x \to 1_+} f(x)$ do not exist (see Figure 9.3). ■

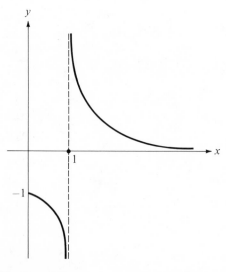

FIGURE 9.3

Example 9. The function $f(x) = |\sin x|$ is continuous on $I_{[0,\infty)}$ but fails to be differentiable on the discrete subset $\{\pi, 2\pi, 3\pi, \ldots\}$ of $I_{[0,\infty)}$. Nevertheless, its derivative

$$f'(x) = \begin{cases} \cos x & \text{if } 0 < x < \pi \\ -\cos x & \text{if } \pi < x < 2\pi \\ \cos x & \text{if } 2\pi < x < 3\pi \\ -\cos x & \text{if } 3\pi < x < 4\pi \\ \cdots \end{cases}$$

is piecewise continuous on $I_{[0,\infty)}$ (see Figure 9.4a, b). The formula for $f'(x)$ can be written more compactly as $f'(x) = (-1)^{[x/\pi]} \cos x$, for x not an integer multple of π. ∎

Often it is convenient to associate with a piecewise continuous function f on $I_{[0,\infty)}$ a function $\overline{f}$ that is defined on the whole interval $I_{[0,\infty)}$ and that agrees with f wherever f is defined. A convenient way of doing this is to set

$$\overline{f}(x) = \begin{cases} f(x) & \text{if } x \text{ is in the domain of } f \\ \lim_{t \to x_+} f(t) & \text{otherwise} \end{cases}.$$

If the function $\overline{f}$ is in $\mathcal{E}$, then we shall say that f **extends to a function in** $\mathcal{E}$. Notice that if f is piecewise continuous on $I_{[0,\infty)}$ and extends to a function in $\mathcal{E}$, then the Laplace transform $L(f)$ of f exists, and in fact $L(f) = L(\overline{f})$.

Example 10. Consider the piecewise continuous function $f'(x) = (-1)^{[x/\pi]} \cos x$, where x is not an integer multiple of π, from Example 9. The function $\overline{f}'$ is given by $\overline{f}'(x) = (-1)^{[x/\pi]} \cos x$ for *all* $x \in I_{[0,\infty)}$ (see Figure 9.4c). Since $\overline{f}' \in \mathcal{E}$, we see that this f' does extend to a function in $\mathcal{E}$. ∎

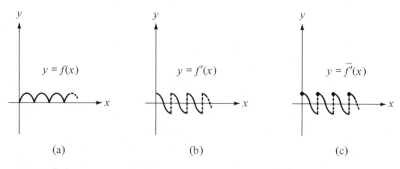

(a) (b) (c)

FIGURE 9.4

We are now ready to state the generalization of Theorem 3 that will allow us to apply the Laplace transform method to initial value problems

$$a_n y^{(n)} + \cdots + a_1 y' + a_0 y = f, \quad y(0) = b_1, \ldots, y^{(n-1)}(0) = b_n$$

in which the function f is only piecewise continuous.

Theorem 4. *Let $y \in \mathcal{E}$ be $(n-1)$-times continuously differentiable and suppose that $y^{(n-1)}$ has a piecewise continuous derivative $y^{(n)}$ on $I_{[0,\infty)}$. Assume that $y', \ldots, y^{(n-1)}$ are all in $\mathcal{E}$ and that $y^{(n)}$ extends to a function in $\mathcal{E}$. Then*

$$L(y^{(n)}) = s^n L(y) - s^{n-1} y(0) - s^{n-2} y'(0) - \cdots - y^{(n-1)}(0).$$

Proof. The proofs of Theorems 2 and 3 carry over to this case without change. ■

Now let $a_0, \ldots, a_n \in \mathbb{R}$. Suppose f is a piecewise continuous function on the interval $I_{[0,\infty)}$. We shall say that $y \in \mathcal{F}(I_{[0,\infty)})$ is a **solution** of the differential equation

$$a_0 y^{(n)} + \cdots + a_1 y' + a_0 y = f$$

provided that

(i) y is continuous and $(n-1)$-times differentiable on $I_{[0,\infty)}$,

(ii) The nth derivative $y^{(n)}$ exists on $I_{[0,\infty)}$ except possibly on the discrete subset S where f is discontinuous, and

(iii) y satisfies the given equation at each $x \in I_{[0,\infty)}$, $x \notin S$.

With this definition of solution, and with the help of Theorem 4, the Laplace transform method does yield the unique solution of the initial value problem

$$a_n y^{(n)} + \cdots + a_1 y' + a_0 y = f, \quad y(0) = b_1, \ldots, y^{(n-1)}(0) = b_n,$$

for each $a_0, \ldots, a_n, b_1, \ldots, b_n \in \mathbb{R}$ and each piecewise continuous function f that extends to a function in $\mathcal{E}$.

Example 11.   Consider the initial value problem

$$y'' + y = \begin{cases} 0 & \text{if } x < \pi, \\ 1 & \text{if } x \geq \pi, \end{cases} \quad y(0) = 0, \, y'(0) = 1.$$

The equation is $y'' + y = u_\pi$. Its Laplace transform is

$$s^2 L(y) - 1 + L(y) \stackrel{(x)}{=} \frac{e^{-\pi s}}{s}$$

or

$$(s^2 + 1)L(y) = 1 + \frac{e^{-\pi s}}{s}$$

so

$$L(y) = \frac{1}{s^2 + 1} + \frac{e^{-\pi s}}{(s^2 + 1)s} \overset{\text{(vii)}}{\underset{\text{(x)}}{=}} L(\sin x) + L(\sin x)L(u_\pi).$$

But

$$L(\sin x)L(u_\pi) \overset{\text{(xv)}}{=} L\left(\int_0^x \sin(x - t)u_\pi(t)\, dt \right).$$

If $x < \pi$, then

$$\int_0^x \sin(x - t)u_\pi(t)\, dt = 0.$$

If $x \geq \pi$, then

$$\int_0^x \sin(x - t)u_\pi(t)\, dt = \int_\pi^x \sin(x - t)\, dt = \cos(x - t) \Big|_{t=\pi}^{t=x}$$

$$= 1 - \cos(x - \pi) = 1 + \cos x.$$

Hence

$$L(\sin x)L(u_\pi) = L((1 + \cos x)u_\pi(x))$$

and

$$L(y) = L(\sin x) + L((1 + \cos x)u_\pi(x))$$
$$= L(\sin x + (1 + \cos x)u_\pi(x)).$$

We can conclude, therefore, that

$$y = \sin x + (1 + \cos x)u_\pi(x)$$

or

$$y = \begin{cases} \sin x & \text{if } x < \pi \\ 1 + \sin x + \cos x = 1 + \sqrt{2}\sin\left(x + \dfrac{\pi}{4}\right) & \text{if } x \geq \pi \end{cases}$$

(see Figure 9.5) is the solution of the given initial value problem. ∎

We conclude this section by verifying formulas (xi)–(xv) in the short table of Laplace transforms. You will be asked in the exercises to verify the remaining formulas.

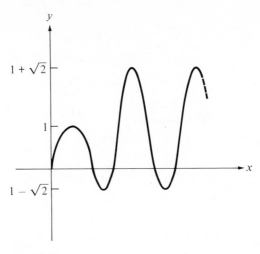

FIGURE 9.5

Formula (xi) $L(u_a(x)f(x-a)) = e^{-as}L(f)$.

Proof. $L(u_a(x)f(x-a)) = \displaystyle\int_0^\infty e^{-sx}u_a(x)f(x-a)\,dx$

$$= \int_a^\infty e^{-sx}f(x-a)\,dx$$

$$= \int_0^\infty e^{-s(v+a)}f(v)\,dv \quad \text{where } v = x - a$$

$$= e^{-as}\int_0^\infty e^{-sv}f(v)\,dv$$

$$= e^{-as}L(f). \quad \blacksquare$$

Formula (xii) $L(x^n f(x)) = (-1)^n \dfrac{d^n}{ds^n}L(f)$.

Proof.   We have

$$\frac{d}{ds}L(f(x)) = \frac{d}{ds}\int_0^\infty e^{-sx}f(x)\,dx = \int_0^\infty \frac{\partial}{\partial s}e^{-sx}f(x)\,dx$$

$$= -\int_0^\infty e^{-sx}xf(x)\,dx = -L(xf(x))$$

and hence

$$L(xf(x)) = -\frac{d}{ds}L(f(x)).$$

This is Formula (xii) when $n = 1$. Using this formula, we get the case when $n = 2$:

$$L(x^2 f(x)) = -\frac{d}{ds}L(xf(x)) = (-1)^2 \frac{d^2}{ds^2}L(f(x)).$$

This process can be repeated n times. ■

Formula (xiii) $L(e^{\alpha x} f(x)) = g(s - \alpha)$, where $g(s) = L(f)$.

Proof. $L(e^{\alpha x} f(x)) = \int_0^\infty e^{-sx} e^{\alpha x} f(x)\, dx = \int_0^\infty e^{-(s-\alpha)x} f(x)\, dx$

$$= g(s - \alpha), \text{ where } g(s) = L(f) = \int_0^\infty e^{-sx} f(x)\, dx. \quad ■$$

Formula (xiv) $L\left(\int_0^x f(t)\, dt\right) = \frac{1}{s}L(f).$

Proof. Let $F(x) = \int_0^x f(t)\, dt$. Then $F' = f$. Hence, by Theorem 1,

$$L(f) = L(F') = sL(F) - F(0) = sL(F),$$

so

$$L\left(\int_0^x f(t)\, dt\right) = L(F) = \frac{1}{s}L(f). \quad ■$$

Formula (xv) $L\left(\int_0^x f(x - t)g(t)\, dt\right) = L(f)L(g).$

Proof. This proof uses double integrals. If you have not yet studied double integrals, you should not attempt to follow the proof. We have

$$L\left(\int_0^x f(x - t)g(t)\, dt\right) = \int_0^\infty e^{-sx}\left(\int_0^x f(x - t)g(t)\, dt\right)dx$$

$$= \int_0^\infty \int_0^x e^{-sx} f(x - t)g(t)\, dt\, dx.$$

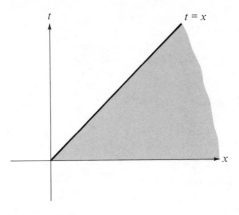

FIGURE 9.6

The region of integration for this double integral is the triangular region $\{(x, t)|0 \leq t \leq x, 0 \leq x < \infty\}$ (see Figure 9.6). This region may also be described as $\{(x, t)|t \leq x < \infty, 0 \leq t < \infty\}$ so, if we interchange the order of integration, we get

$$L\left(\int_0^x f(x - t)g(t)\, dt\right) = \int_0^\infty \int_t^\infty e^{-sx} f(x - t)g(t)\, dx\, dt$$

$$= \int_0^\infty g(t)\left(\int_t^\infty e^{-sx} f(x - t)\, dx\right) dt$$

$$= \int_0^\infty g(t)\left(\int_0^\infty e^{-s(u+t)} f(u)\, du\right) dt \quad \text{(where } u = x - t\text{)}$$

$$= \int_0^\infty e^{-st} g(t)\left(\int_0^\infty e^{-su} f(u)\, du\right) dt$$

$$= \left(\int_0^\infty e^{-su} f(u)\, du\right)\left(\int_0^\infty e^{-st} g(t)\, dt\right)$$

$$= L(f)L(g). \quad ■$$

Remark. The function $h(x) = \int_0^x f(x - t)g(t)\, dt$ appearing in formula (xv) is called the **convolution** of f and g.

EXERCISES

1. Sketch the graph and calculate the Laplace transform of each of the following functions.

   (a) $f(x) = \begin{cases} 1 & \text{if } x < 1 \\ 0 & \text{if } x \geq 1 \end{cases}$

(b) $f(x) = \begin{cases} x & \text{if } 0 \le x < 1 \\ 1 & \text{if } x \ge 1 \end{cases}$

(c) $f(x) = \begin{cases} x & \text{if } x < 1 \\ x - 1 & \text{if } 1 \le x < 2 \\ 0 & \text{if } x \ge 2 \end{cases}$

(d) $f(x) = \begin{cases} 1 & \text{if } 4 \le x < 5 \\ 0 & \text{otherwise} \end{cases}$

(e) $f(x) = \begin{cases} 0 & \text{if } [x] \text{ is odd} \\ 1 & \text{if } [x] \text{ is even} \end{cases}$

(f) $f(x) = [x]$

[Hint: In (e) and (f), use the formula $\sum_{n=0}^{\infty} ar^n = \dfrac{a}{1 - r}$ $(|r| < 1)$ for the sum of a geometric series.]

2. Verify the following by direct calculation.

(a) $L(e^{\alpha x}) = \dfrac{1}{s - \alpha}$ (c) $L(x^2) = \dfrac{2}{s^3}$

(b) $L(xe^{\alpha x}) = \dfrac{1}{(s - \alpha)^2}$ (d) $L(u_a(x)) = \dfrac{e^{-as}}{s}$, where

$$u_a(x) = \begin{cases} 0 & \text{if } x < a \\ 1 & \text{if } x \ge a \end{cases} \quad (a > 0)$$

3. Verify that each of the following functions is of exponential order.

(a) $e^{\alpha x}$ (e) $e^{\alpha x} \sin \beta x$

(b) $\cos \beta x$ (f) x^n

(c) $\sin \beta x$ (g) $x^n e^{\alpha x} \cos \beta x$

(d) $e^{\alpha x} \cos \beta x$ (h) $x^n e^{\alpha x} \sin \beta x$

4. Show that $f(x) = e^{x^2}$ is not of exponential order.

5. Use the short table of Laplace transforms, especially Formula (xii), to help you find the Laplace transforms of

(a) $x^2 e^x$ (d) $xe^{\alpha x} \cos \beta x$

(b) $x \cos x$ (e) $xe^{\alpha x} \sin \beta x$

(c) $x \sin x$

6. Use trigonometric identities together with the linearity of the Laplace transform to help you find the Laplace transforms of

(a) $\cos^2 x$ (c) $\sin x \cos x$

(b) $\sin^2 x$

7. Let $f(x) = (-1)^{[x]}$ and let $g(x) = \int_0^x f(t)\, dt$.

(a) Sketch the graphs of f and g.

(b) Find $L(f)$. [Hint: You'll need to sum a geometric series.]

(c) Find $L(g)$.

8. Use the short table of Laplace transforms to help you find $f \in \mathcal{E}$ such that $L(f) =$

(a) $\dfrac{1}{s^2 - 1}$ (d) $\dfrac{e^{-2s}}{s^2 + 1}$

(b) $\dfrac{1}{s(s^2 - 1)}$ (e) $\dfrac{e^{-2s}}{s(s + 1)}$

(c) $\dfrac{1}{s(s - 1)^2}$

9. Formulas (i), (ii), and (xi)–(xvii) in the short table of Laplace transforms are verified in the text. You were asked to verify Formulas (iv), (v), and (x) in Exercise 2. Verify the remaining formulas in the table as follows.

   (iii) $L(x^n) = n!/s^{n+1}$ [Use (i) and (xvii)]

   (vi) $L(\cos \beta x) = s/(s^2 + \beta^2)$ [Use (xvii) together with the fact that if $y = \cos \beta x$, then $y'' + y = 0$, $y(0) = 1$, $y'(0) = 0$]

   (vii) $L(\sin \beta x) = \beta/(s^2 + \beta^2)$ [Use (vi) and (xiv)]

   (viii) $L(e^{\alpha x} \cos \beta x) = (s - \alpha)/((s - \alpha)^2 + \beta^2)$ [Use (vi) and (xiii)]

   (ix) $L(e^{\alpha x} \sin \beta x) = \beta/((s - \alpha)^2 + \beta^2)$ [Use (vii) and (xiii)]

10. Sketch the graphs of the following functions and decide which of them are piecewise continuous on $I_{[0, \infty)}$.

   (a) $f(x) = \begin{cases} 1 & \text{if } x < 2 \\ x & \text{if } x > 2 \end{cases}$

   (b) $f(x) = \begin{cases} x & \text{if } [x] \text{ is even} \\ 0 & \text{if } [x] \text{ is odd} \end{cases}$

   (c) $f(x) = x - [x]$

   (d) $f(x) = \sec^2 x$

   (e) $f(x) = 1/[x + 1]$

   (f) $f(x) = \begin{cases} (-1)^{[1/x]} & \text{if } x \neq 0 \\ 1 & \text{if } x = 0 \end{cases}$

11. Using the Laplace transform, solve the initial value problems.

   (a) $y' + 3y = x$, $y(0) = 0$

   (b) $y' - 2y = \cos x$, $y(0) = 1$

   (c) $y' + 5y = \begin{cases} 0 & \text{if } x < 1 \\ x - 1 & \text{if } x \geq 1 \end{cases}$, $y(0) = 0$

   [Hint: Notice that the right hand side in (c) is just $u_1(x)f(x - 1)$, where $f(x) = x$.]

12. Using the Laplace transform, solve

   (a) $y'' - 3y' + 2y = 0$, $y(0) = 1$, $y'(0) = 0$

   (b) $y'' - 3y' + 2y = 0$, $y(0) = 0$, $y'(0) = 1$

   (c) $y'' - 6y' + 9y = e^{-x}$, $y(0) = 5$, $y'(0) = 12$

   (d) $y'' + 5y' + 6y = e^{-2x}$, $y(0) = 1$, $y'(0) = 4$

   (e) $y'' + 2y' + 5y = 1$, $y(0) = y'(0) = 0$

13. Solve

   (a) $y'' + 4y = \begin{cases} 0 & \text{if } x < \pi \\ 1 & \text{if } x \geq \pi \end{cases}$, $y(0) = y'(0) = 0$

   (b) $y'' + y = \begin{cases} 0 & \text{if } x < 1 \\ x - 1 & \text{if } x \geq 1 \end{cases}$, $y(0) = 0$, $y'(0) = 1$

   (c) $y'' - 4y = \begin{cases} 0 & \text{if } x < \ln 2 \\ e^{2x} & \text{if } x \geq \ln 2 \end{cases}$, $y(0) = 2$, $y'(0) = -8$

14. Suppose an object is attached to a wall by a spring as in Exercise 25 of Section

3.2, but now assume that there acts an additional driving force f. Assume no viscosity, so that the equation of motion becomes

$$mx'' + kx = f(t).$$

Find $x(t)$ so that $x(0) = 0$, $x'(0) = 0$, when

(a) $f(t) = \sin \omega t$, $\omega \neq \sqrt{k/m}$

(b) $f(t) = \sin \omega t$, $\omega = \sqrt{k/m}$

(c) $f(t) = \begin{cases} 1/h & \text{if } 1 \leq t < 1 + h \\ 0 & \text{otherwise} \end{cases}$

(d) What is the limiting behavior of $x(t)$ in part (c) as $h \to 0$?

Remark. The limiting case, as $h \to 0$, corresponds physically to striking the object a sharp blow with a hammer at time $t = 1$.

15. Show that $f \in \mathcal{F}(I_{[0, \infty)})$ is of exponential order if and only if there exist positive real numbers a, b, and c such that $|f(x)| \leq ae^{bx}$ for all $x \geq c$.

16. Show that the set $\mathcal{E}$ of all functions in $\mathcal{F}(I_{[0, \infty)})$ that are integrable and of exponential order is a subspace of $\mathcal{F}(I_{[0, \infty)})$.

17. Suppose $f \in \mathcal{E}$ is such that $\lim\limits_{x \to 0_+} f(x)/x$ exists. Show that if we define $f(x)/x$ to take on the value $\lim\limits_{x \to 0_+} f(x)/x$ when $x = 0$, then $f(x)/x \in \mathcal{E}$ and

$$L(f(x)/x) = \int_s^\infty L(f).$$

18. Suppose $f \in \mathcal{E}$ is periodic with period a, so that $f(x + a) = f(x)$ for all $x \in I_{[0, \infty)}$. Show that

$$L(f) = \int_0^a e^{-sx} f(x)\, dx \Big/ (1 - e^{-as}).$$

19. Sketch the graph of $u_a(x) f(x - a)$, where

(a) $f(x) = \sin x$, $a = \pi$

(b) $f(x) = e^{-x}$, $a = 1$

(c) $f(x) = e^{-x}$, $a = 2$

(d) In general, describe the graph of $u_a(x) f(x - a)$ $(a > 0)$ in terms of the graph of f.

20. (a) Show that if y is the solution to the initial value problem

$$a_n y^{(n)} + \cdots + a_1 y' + a_0 y = f, \quad y(0) = \cdots = y^{(n-1)}(0) = 0$$

where $a_0, \ldots, a_n \in \mathbb{R}$ and $f \in \mathcal{E}$, then

$$L(y) = \frac{L(f)}{p(s)}$$

where $p(z)$ is the auxiliary polynomial of the given differential equation.

(b) Show that the solution of the initial value problem in (a) can be expressed as the integral

$$y = \int_0^x g(x - t) f(t)\, dt$$

where $g \in \mathcal{E}$ is the function such that $L(g) = 1/p(s)$.

Remark. The function $G(x, t) = g(x - t)$ is called the **Green's function** of this initial value problem.

Answers

SECTION 1.1 (page 6)

1. (a) $(0, 1)$
   (b) $(1 - c, c)$, c arbitrary
   (c) This system has no solution

2. (a) $\begin{pmatrix} 1 & 0 & 0 \\ 0 & 1 & 0 \\ 0 & 0 & 1 \end{pmatrix}$ (b) $\begin{pmatrix} 1 & 0 & 0 \\ 0 & 1 & 0 \\ 0 & 0 & 1 \\ 0 & 0 & 0 \end{pmatrix}$ (c) $\begin{pmatrix} 1 & 1 & 0 & 0 & -\frac{2}{5} \\ 0 & 0 & 1 & 0 & -3 \\ 0 & 0 & 0 & 1 & \frac{3}{5} \end{pmatrix}$

3. $(\frac{13}{5} + \frac{4}{5}c, \frac{4}{5} - \frac{3}{5}c, c)$, c arbitrary

4. $(3 - 2c, -1 - c, c)$, c arbitrary

5. $(11 + c, 5 + c, 3 + c, c)$, c arbitrary

6. $(-\frac{2}{5} - \frac{4}{5}c, \frac{11}{10} + \frac{17}{10}c, c)$, c arbitrary

7. $(\frac{15}{8} - \frac{3}{8}c_1 + \frac{3}{8}c_2, \frac{11}{8} + \frac{17}{8}c_1 - \frac{25}{8}c_2, c_1, c_2)$, c_1, c_2 arbitrary

8. $(\frac{12}{5} + \frac{7}{5}c, \frac{6}{5} + \frac{2}{5}c, \frac{4}{5} + \frac{3}{5}c, c)$ c arbitrary

9. $(1, \frac{1}{2}, -1, \frac{3}{2}, -\frac{1}{2})$

SECTION 1.2 (page 11)

1. (a) $(0, 0, 0, 0, 0)$ (b) $(1, 6, 0, 14)$ (c) $(2, 2, -2)$
   (d) $(-1, 0, 0, 1)$ (e) $(-1, 2, 9, 24)$ (f) $(0, 4, 1, 1)$

2. (a) $\mathbf{x} = (-1, 2, 1, 3)$ (b) $\mathbf{x} = (1, -\frac{1}{3}, -\frac{1}{3})$ (c) $\mathbf{x} = (1, 0), \mathbf{y} = (0, 1)$

3. #3: $(\frac{13}{5}, \frac{4}{5}, 0) + c(\frac{4}{5}, -\frac{3}{5}, 1)$
   #4: $(3, -1, 0) + c(-2, -1, 1)$
   #5: $(11, 5, 3, 0) + c(1, 1, 1, 1)$
   #6: $(-\frac{2}{5}, \frac{11}{10}, 0) + c(-\frac{4}{5}, \frac{17}{10}, 1)$
   #7: $(\frac{15}{8}, \frac{11}{8}, 0, 0) + c_1(-\frac{3}{8}, \frac{17}{8}, 1, 0) + c_2(\frac{3}{8}, -\frac{25}{8}, 0, 1)$
   #8: $(\frac{12}{5}, \frac{6}{5}, \frac{4}{5}, 0) + c(\frac{7}{5}, \frac{2}{5}, \frac{3}{5}, 1)$
   #9: $(1, \frac{1}{2}, -1, \frac{3}{2}, -\frac{1}{2})$

4. $(1, 0, 0) + c_1(-2, 1, 0) + c_2(1, 0, 1)$

5. $(5, -8)$

6. $(\frac{1}{5}, 0, \frac{6}{5}, -\frac{1}{5}, 0, 0) + c_1(2, 1, 0, 0, 0, 0) + c_2(-\frac{3}{5}, 0, \frac{2}{5}, \frac{3}{5}, 1, 0) + c_3(\frac{3}{5}, 0, \frac{3}{5}, \frac{2}{5}, 0, 1)$

SECTION 1.3 (page 21)

1.

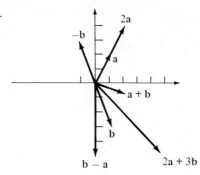

2. (a) $\mathbf{x} = (-3, 3) + t(2, 1)$, slope $\frac{1}{2}$, y-intercept $\frac{9}{2}$.
   (b) $\mathbf{x} = (-3, 3) + t(4, 2)$, slope $\frac{1}{2}$, y-intercept $\frac{9}{2}$.
   (c) $\mathbf{x} = (-6, 6) + t(2, 1)$, slope $\frac{1}{2}$, y-intercept 9.
   (d) $\mathbf{x} = (-3, 3) + t(5, -2)$, slope $-\frac{2}{5}$, y-intercept $\frac{9}{5}$.

3. $\mathbf{x} = (\frac{7}{5}, 0) + t(-\frac{7}{5}, \frac{7}{2})$ (Note: There are many correct answers.)

4. (a) $\sqrt{10}$ (b) $\sqrt{97}$ (c) $3\sqrt{17}$ (d) $\sqrt{61}$ (e) $(\frac{3}{2}, 6)$
   (f) $(\frac{7}{3}, 7)$ (g) $(\frac{1}{2}, \frac{24}{5})$

8.

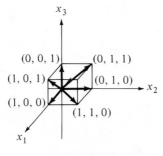

(a)–(f)

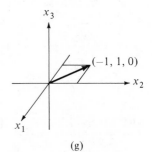

(g)

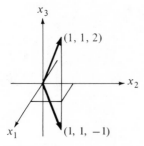

(h)–(i)

9. (a) $\mathbf{x} = (1, 2, 1) + t(-3, 1, 4)$ (b) $\mathbf{x} = (-1, 1, -1) + t(2, 1, 2)$
   (c) $\mathbf{x} = (1, -2, 3) + t(-4, 1, 1)$ (d) $\mathbf{x} = (1, 2, 1) + t(-2, -1, -2)$
   (e) $\mathbf{x} = (1, -2, 3) + t(-5, 3, -2)$

10. (a) $\mathbf{x} = (1, 1, -1, 0) + t(1, 2, 3, 4)$
    (b) $\mathbf{x} = (1, 0, 1, 0, 1, 0) + t(-2, -1, 0, 1, 2, 3)$
    (c) $\mathbf{x} = (1, 1, -1, 0) + t(0, 1, 4, 4)$
    (d) $\mathbf{x} = (1, 2, 3, 4, 5) + t(4, 2, 0, -2, -4)$
    (e) $\mathbf{x} = (1, 2, \ldots, n) + t(-1, -1, \ldots, -1)$

11. (a) $(0, 0, 0)$ (b) $(0, \frac{1}{2}, \frac{5}{2})$ (c) $(5, 1, 0, 4)$
    (d) $\left(\dfrac{n+1}{2}, \dfrac{n+1}{2}, \ldots, \dfrac{n+1}{2} \right)$

12. (a) $(0, -1, 2)$ and $(1, -4, 5)$
    (b) $(-\frac{1}{4}, -\frac{1}{4}, \frac{5}{4})$ (c) $(\frac{4}{5}, -\frac{17}{5}, \frac{22}{5})$

13. $\mathbf{x} = (1, -1, 0) + t(\frac{3}{2}, \frac{3}{2}, 1)$

14. (a) $\mathbf{x} = (1, 0, 0) + t(-1, \frac{1}{2}, \frac{1}{2})$
        $\mathbf{x} = (0, 1, 0) + t(\frac{1}{2}, -1, \frac{1}{2})$
        $\mathbf{x} = (0, 0, 1) + t(\frac{1}{2}, \frac{1}{2}, -1)$
    (b) $(\frac{1}{3}, \frac{1}{3}, \frac{1}{3})$

15. $\mathbf{x} = (1, 3, 5) + t(-1, -1, 1)$

16. $\mathbf{x} = t(-\pi, -\pi, 2\pi)$

17. (a) yes   (b) no   (c) no   (d) yes

18. (a) intersect   (b) neither   (c) parallel   (d) both (these two lines are identical)

19. no (their trajectories intersect but they do not collide)

SECTION 1.4 (page 31)

1. (a) $\mathbf{a} \cdot \mathbf{b} = 4$, $\cos \theta = 4/\sqrt{65}$, θ is acute
   (b) $\mathbf{a} \cdot \mathbf{b} = 0$, $\cos \theta = 0$, $\theta = \pi/2$
   (c) $\mathbf{a} \cdot \mathbf{b} = 4$, $\cos \theta = 1$, $\theta = 0$
   (d) $\mathbf{a} \cdot \mathbf{b} = -10$, $\cos \theta = -1$, $\theta = \pi$
   (e) $\mathbf{a} \cdot \mathbf{b} = -5$, $\cos \theta = -1/\sqrt{5}$, θ is obtuse
   (f) $\mathbf{a} \cdot \mathbf{b} = 2$, $\cos \theta = \sqrt{2}/3$, θ is acute
   (g) $\mathbf{a} \cdot \mathbf{b} = 4$, $\cos \theta = 4/\sqrt{66}$, θ is acute
   (h) $\mathbf{a} \cdot \mathbf{b} = -11$, $\cos \theta = -11/7\sqrt{6}$, θ is obtuse
   (i) $\mathbf{a} \cdot \mathbf{b} = 0$, $\cos \theta = 0$, $\theta = \pi/2$
   (j) $\mathbf{a} \cdot \mathbf{b} = 8$, $\cos \theta = \frac{4}{5}$, θ is acute

2. (b) (i) $(2/\sqrt{13}, 3/\sqrt{13})$ (ii) $(-1/\sqrt{2}, 1/\sqrt{2})$ (iii) $(\frac{2}{3}, \frac{1}{3}, \frac{2}{3})$

3. $x - y = 5$, $\mathbf{x} = (3, -2) + t(1, 1)$

4. (a) $-3x_1 + x_2 + 4x_3 = 3$ (b) $2x_1 + x_2 + 2x_3 = -3$
   (c) $-4x_1 + x_2 + x_3 = -3$

5. $2x_1 + 3x_2 - x_3 = 6$

6. (a) $\mathbf{x} = (-2, 3, 1) + t(3, -2, 1)$ (b) $(\frac{10}{7}, \frac{5}{7}, \frac{15}{7})$

7. (a) $2x_1 - x_2 + 3x_3 + 4x_4 = -3$ (b) $-2x_1 - x_2 + x_4 + 2x_5 = 0$
    (c) $2x_1 + 4x_2 + \cdots + (2n)x_n = 2n$

8. $x_1 + x_4 = 5$

9. $\mathbf{x} = t(-3, 1, 2, -4)$

10. (a) $\mathbf{x} = (3, 2, 0) + t(-1, 0, 1)$
    (b) $\mathbf{x} = (0, -\frac{1}{4}, 0) + t(-\frac{1}{2}, \frac{3}{8}, 1)$
    (c) $\mathbf{x} = t(0, 0, 1)$

11. $\mathbf{x} = (\frac{5}{4}, \frac{1}{2}, -\frac{3}{4}, 0) + t(-\frac{1}{2}, -1, \frac{1}{2}, 1)$

19. (a) $7/\sqrt{13}$ (b) $8/\sqrt{13}$ (c) $10/\sqrt{11}$ (d) 0

SECTION 1.5 (page 43)

1. (a) $(-3, -1, 5)$ (b) $(-14, 2, -25)$ (c) $(0, 0, 0)$
    (d) $(4, -10, 8)$ (e) $(-2, -2, 0)$ (f) $(2, -5, 6)$

2. (a) $x_1 + x_2 - x_3 = 0$ (b) $x_1 = 1$ (c) $4x_1 + 2x_2 + x_3 = 9$
    (d) $x_1 + x_2 + x_3 = 1$ (e) $6x_1 - 4x_2 - 2x_3 = 0$
    (f) $2x_1 - 3x_2 - 11x_3 = -33$

3. (a) $\sqrt{14}$ (b) $4\sqrt{6}$ (c) 1   (d) $7\sqrt{3}$ (e) $\sqrt{89}$ (f) $\sqrt{419}$

4. (a) $3\sqrt{3}$ (b) $\sqrt{6}$ (c) $5\sqrt{74}$ (d) 1   (e) 3

5. (a) $\sqrt{3}/2$ (b) $\frac{5}{2}$ (c) $\sqrt{14}$ (d) $\frac{3}{2}$ (e) $\sqrt{21}$

6. (a) 18   (b) 1   (c) 9   (d) 2   (e) 3

7. (a) $\frac{1}{6}$ (b) 1   (c) 8   (d) 0

SECTION 2.1 (page 53)

3. $(x + 2)^2 = x^2 + 4(x + 1)$

4. $1 = \sec^2 x + (-1)\tan^2 x$

5. $\cos^2 x = \frac{1}{2} \cdot 1 + \frac{1}{2}\cos 2x$

6. (a) it is the line through $\dfrac{x^3}{3(1 + x^2)}$ in the direction of $\dfrac{1}{1 + x^2}$
    (b) it is the line through xe^{-x} in the direction of e^{-x}
    (c) it is the line through $\frac{1}{2}x^2 e^x$ in the direction of e^x
    (d) it is the line through $e^{-p(x)} \displaystyle\int_0^t q(t)\, dt$ in the direction of $e^{-p(x)}$

7. (a) $(1, 3, 5, 7, 9, \ldots)$ (b) $(1, 1, 1, 1, 1, \ldots)$
    (c) $(2, 4, 6, 8, 10, \ldots)$ (d) $(\frac{1}{2}, -\frac{1}{2}, -\frac{3}{2}, -\frac{5}{2}, -\frac{7}{2}, \ldots)$
    (e) $(0, \pi, 2\pi, 3\pi, 4\pi, \ldots)$

8. (a) $\begin{pmatrix} 0 & 1 \\ 0 & 4 \\ 2 & 6 \end{pmatrix}$ (b) $\begin{pmatrix} -2 & -1 \\ 2 & 2 \\ -2 & -2 \end{pmatrix}$ (c) $\begin{pmatrix} \frac{1}{2} & \frac{1}{2} \\ -\frac{1}{2} & \frac{1}{2} \\ 1 & 2 \end{pmatrix}$

    (d) $\begin{pmatrix} -3 & 2 \\ 3 & 17 \\ 4 & 18 \end{pmatrix}$ (e) $\begin{pmatrix} -2\pi & 0 \\ 2\pi & 3\pi \\ 0 & 2\pi \end{pmatrix}$

9. $(1, 0) = \frac{1}{3}(1, -1) + \frac{1}{3}(2, 1)$

10. (a) yes   (b) no   (c) no   (d) yes

11. (a) $(a_1, a_2) = a_1(1, 0) + a_2(0, 1)$
    (b) $(a_1, a_2, a_3) = a_1(1, 0, 0) + a_2(0, 1, 0) + a_3(0, 0, 1)$

12. $\cos 2x = \cos^2 x + (-1)\sin^2 x$

13. $2x^2 + 3x - 1 = 4 \cdot 1 + 7(x - 1) + 2(x - 1)^2$

14. $e^{-x} = (-1)e^x + 2 \cosh x$

20. all properties except A_4 are satisfied; V is not a vector space

21. yes

SECTION 2.2 (page 58)

1. (a) yes   (b) yes   (c) no   (d) no   (e) yes
    (f) no   (g) yes   (h) yes   (i) no

2. (a) yes   (b) no   (c) yes   (d) no

3. (a) yes   (b) no   (c) no   (d) yes

4. (a) yes   (b) no   (c) yes   (d) yes

5. (a) no   (b) yes   (c) yes

    (d) yes   (e) yes   (f) no

6. $\{0\}$ and $\mathbb{R}^1$

SECTION 2.3 (page 64)

1. (a) $x_3 = 0$ (b) $x_1 + x_2 + x_3 = 0$ (c) $3x_1 + 3x_2 + x_3 = 0$
    (d) $x_1 - 2x_2 + x_3 = 0$ (e) $-4x_1 + 7x_2 - 13x_3 = 0$

2. (a) yes   (b) yes   (c) no   (d) yes

3. (a) no   (b) yes   (c) no

4. (a) yes   (b) no   (c) yes

5. (a) no   (b) yes   (c) yes   (d) yes   (e) no   (f) yes

7. (a) yes   (b) yes   (c) no

8. $\{1, x, x^2, x^3\}$

9. $\{1, x, x^2, \ldots, x^{n-1}\}$

10. $\{1, 0, 0, \ldots, 0), (0, 1, 0, \ldots, 0), \ldots, (0, 0, 0, \ldots, 1)\}$

12. $\{1, x\}$

SECTION 2.4 (page 70)

1. (a) 2, $\{(2, -3, 1)\}$
   (b) 2, $\{(-6, 1, 0, 0), (-11, 0, 8, 1)\}$
   (c) 3, the spanning set is the empty set
   (d) 3, $\{(-3, 6, -2, 1, 0), (-1, 1, -5, 0, 1)\}$

2. (a) $\mathbf{x} = (-12, 0, 5, 0) + c_1(-6, 1, 0, 0) + c_2(-11, 0, 8, 1)$
   (b) $\mathbf{x} = (19, -8, -11, 0, 0) + c_1(-3, 6, -2, 1, 0) + c_2(-1, 1, -5, 0, 1)$

3. (a) $\{(-1, 1, 0), (-1, 0, 1)\}$ (b) $\{(1, 0, 0), (0, -1, 1)\}$
   (c) $\{(\frac{3}{2}, 1, 0), (-2, 0, 1)\}$ (d) $\{(1, 0, 0), (0, 1, 0)\}$

4. $\{(-3, 1, 0, 0, 0), (1, 0, 1, 0, 0), (2, 0, 0, 1, 0), (-4, 0, 0, 0, 1)\}$

SECTION 3.1 (page 78)

1. $y = 2 + ce^{-x}$

2. $y = x^2 + ce^{-x}$

3. $y = 5x + c_1 e^{-x} + c_2$

4. $y = \sin x + c_1 e^{-x} + c_2$

5. $y = -e^{-x} + c_1 x^2 + c_2 x + c_3$

6. $y = xe^x + ce^x$

7. $y = e^x + ce^{-3x}$

8. $y = \frac{1}{2}x - \frac{1}{4} + ce^{-2x}$

9. $y = -\frac{1}{3}e^{-x} + ce^{2x}$

10. $y = \frac{1}{2}\sin x - \frac{1}{2}\cos x + ce^{-x}$

11. $y = \dfrac{1}{k-1}e^{kx} + ce^x$ if $k \neq 1$; $y = xe^x + ce^x$ if $k = 1$

12. $y = \dfrac{1}{1+\pi^2}\cos \pi x + \dfrac{\pi}{1+\pi^2}\sin \pi x + e^{-\pi x}$

13. $y = x^2 - 4x + 8 + ce^{-x/2}$

14. $y = -c_1 e^{2x} + c_2 e^{3x}$

15. $y = c_1 x e^{3x} + c_2 e^{3x}$

16. (a) $N = ce^{kt}$ (b) $k = \ln \frac{5}{4}$

17. (a) $N = \dfrac{r}{k} + ce^{kt}$

(b) Notice that $N = \dfrac{r}{k} + \left(N_0 - \dfrac{r}{k}\right)e^{kt}$. If $N_0 = r/k$, the population stays constant: $N = N_0$ for all t. If $N_0 < r/k$, the population decreases and eventually dies out.

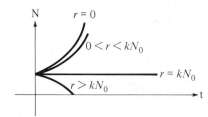

18. (a) $T = K + ce^{-kt}$ (b) $\dfrac{5 \ln \frac{1}{5}}{\ln \frac{4}{5}} - 5 \approx 31$ hours

SECTION 3.2 (page 86)

1. $y = c_1 e^{3x} + c_2 x e^{3x}$

2. $y = c_1 e^{3x} + c_2 e^{-3x}$

3. $y = c_1 e^{2x} + c_2 e^{-3x}$

4. $y = c_1 e^{x} + c_2 e^{-3x}$

5. $y = c_1 \cos 3x + c_2 \sin 3x$

6. $y = c_1 e^{-x} \cos \sqrt{2}x + c_2 e^{-x} \sqrt{2}x$

7. $y = c_1 e^{(-1+\sqrt{2})x} + c_2 e^{(-1-\sqrt{2})x}$

8. $y = c_1 e^{\frac{1}{3}x} + c_2 e^{-2x}$

9. $y = c_1 e^{\frac{1}{7}x} + c_2 e^{-3x}$

10. $y = c_1 e^{-x} \cos x + c_2 e^{-x} \sin x$

11. $y = c_1 \cos \sqrt{7}x + c_2 \sin \sqrt{7}x$

12. $y = c_1 e^{\sqrt{2}x} + c_2 x e^{\sqrt{2}x}$

13. $y = c_1 + c_2 e^{-4x}$

14. $y = c_1 e^{(1+\sqrt{5})x} + c_2 e^{(1-\sqrt{5})x}$

15. $y = c_1 e^{5x} + c_2 e^{\frac{1}{4}x}$

16. $y = c_1 e^{-\frac{1}{4}x} \cos \dfrac{\sqrt{7}}{4} x + c_2 e^{-\frac{1}{4}x} \sin \dfrac{\sqrt{7}}{4} x$

17. $y = \cos 3x - \frac{2}{3} \sin 3x$

18. $y = \frac{1}{6} e^{4x} - \frac{1}{6} e^{-2x}$

19. $y = 5 e^x \cos x - e^x \sin x$

20. $y = \frac{1}{4} e^{4x} - \frac{9}{4} e^{-4x}$

25. (a) $x(t) = (1/\beta) \sin \beta t$, where $\beta = \sqrt{k/m}$

    (b) $x(t) = \begin{cases} (1/\beta) e^{\alpha t} \sin \beta t & \text{if } \nu^2 < 4km \\ (1/\beta) e^{\alpha t} \sinh \beta t & \text{if } \nu^2 > 4km \\ t e^{\alpha t} & \text{if } \nu^2 = 4km. \end{cases}$

    where $\alpha = -\nu/2m$ and $\beta = |(k/m) - (\nu/2m)^2|^{\frac{1}{2}}$.

SECTION 3.3 (page 95)

1. $y = c_1 e^{-5x} + c_2 e^{2x} + c_3 e^{3x}$

2. $y = c_1 e^x + c_2 x e^x + c_3 e^{-3x}$

3. $y = c_1 + c_2 e^{2x} + c_3 x e^{2x}$

4. $y = c_1 e^{4x} + c_2 x e^{4x} + c_3 x^2 e^{4x} + c_4 x^3 e^{4x}$

5. $y = c_1 + c_2 x + c_3 x^2 + c_4 x^2$

6. $y = c_1 \cos 3x + c_2 \sin 3x + c_3 x \cos 3x + c_4 x \sin 3x$

7. $y = c_1 e^x + c_2 e^{-x} + c_3 \cos x + c_4 \sin x$

8. $y = c_1 e^x + c_2 x e^x + c_3 e^{-2x} + c_4 x e^{-2x}$
$$+ c_5 e^{-x} \cos x + c_6 e^{-x} \sin x + c_7 x e^{-x} \cos x + c_8 x e^{-x} \sin x$$

9. $y = c_1 e^{x/\sqrt{2}} \cos(x/\sqrt{2}) + c_2 e^{x/\sqrt{2}} \sin(x\sqrt{2})$
$$+ c_3 e^{-x/\sqrt{2}} \cos(x/\sqrt{2}) + c_4 e^{-x/\sqrt{2}} \sin(x/\sqrt{2})$$

10. $y = c_1 e^{2x} + c_2 e^{-2x} + c_3 \cos 2x + c_4 \sin 2x$

11. $y = c_1 e^x + c_2 e^{-x} + c_3 e^{\frac{1}{2}x} \cos \dfrac{\sqrt{3}}{2} x + c_4 e^{\frac{1}{2}x} \sin \dfrac{\sqrt{3}}{2} x$
$$+ c_5 e^{-\frac{1}{2}x} \cos \dfrac{\sqrt{3}}{2} x + c_6 e^{-\frac{1}{2}x} \sin \dfrac{\sqrt{3}}{2} x$$

12. $y = \frac{1}{2} + \frac{1}{2} e^{2x}$

13. $y = \frac{1}{2} e^x + e^{-x} - \frac{1}{2} \cos x - \frac{1}{2} \sin x$

SECTION 3.4 (page 102)

1. $y = -\frac{1}{2} \sin x + c_1 e^x + c_2 e^{-x}$

2. $y = \frac{1}{4} x^2 e^x - \frac{1}{4} x e^x + c_1 e^x + c_2 e^{-x}$

3. $y = \frac{1}{4}e^x + c_1 e^{-x} + c_2 x e^{-x}$

4. $y = \frac{1}{2}x^2 e^{-x} + c_1 e^{-x} + c_2 x e^{-x}$

5. $y = -\frac{1}{2}x \cos x + c_1 \cos x + c_2 \sin x$

6. $y = x^3 - 3x^2 + 6 + c_1 e^{-\frac{1}{2}x} \cos \dfrac{\sqrt{3}}{2} x + c_2 e^{-\frac{1}{2}x} \sin \dfrac{\sqrt{3}}{2} x$

7. $y = -\frac{1}{5}e^x \cos x + \frac{2}{5}e^x \sin x + c_1 e^x + c_2 e^{-x}$

8. $y = -\frac{1}{5}e^x \cos x + \frac{2}{5}e^x \sin x - \frac{1}{2}xe^{-x} + c_1 e^x + c_2 e^{-x}$

9. $y = \frac{1}{2}x^2 - x - xe^{-x} + c_1 + c_2 e^{-x}$

10. $y = e^{-x} \cos x - e^{-x} \sin x + c_1 e^{-\frac{1}{2}x} \cos \dfrac{\sqrt{3}}{2} x + c_2 e^{-\frac{1}{2}x} \sin \dfrac{\sqrt{3}}{2} x$

11. $y = \frac{1}{10}e^{3x} + \frac{9}{10} \cos x - \frac{3}{10} \sin x$

12. $y = \frac{1}{3}e^{-2x} + \frac{2}{3}e^x - e^{-x}$

13. $y = -xe^{2x} + e^{2x}$

14. $y = \frac{1}{3}xe^x - \frac{1}{3}e^x + \frac{1}{3}e^{-\frac{1}{2}x} \cos \dfrac{\sqrt{3}}{2} x + \dfrac{\sqrt{3}}{9} e^{-\frac{1}{2}x} \sin \dfrac{\sqrt{3}}{2} x$

15. $y = \frac{1}{27}e^{2x} + \frac{26}{27}e^{-x} + \frac{26}{9} xe^{-x} + \frac{11}{6} x^2 e^{-x}$

16. If $\omega \neq \sqrt{k/m}$ then

$$x(t) = \frac{1}{k - m\omega^2}\left(\cos \omega t - \cos \sqrt{\frac{k}{m}}\, t \right)$$

or, equivalently,

$$x(t) = \frac{2}{k - m\omega^2} \sin\left[\frac{1}{2}\left(\sqrt{\frac{k}{m}} + \omega\right)t\right] \sin\left[\frac{1}{2}\left(\sqrt{\frac{k}{m}} - \omega\right)t\right].$$

If $\omega = \sqrt{k/m}$ then

$$x(t) = \frac{1}{2m\omega} t \sin \omega t.$$

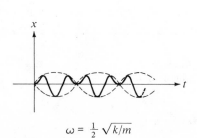

$\omega = \frac{1}{2}\sqrt{k/m}$

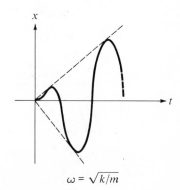

$\omega = \sqrt{k/m}$

SECTION 4.1 (page 110)

1. (a) independent   (b) dependent   (c) dependent
   (d) independent   (e) independent   (f) dependent
   (g) dependent   (h) dependent   (i) independent
2. (a) independent   (b) independent   (c) dependent
   (d) independent   (e) independent   (f) dependent
   (g) independent   (h) independent   (i) independent

SECTION 4.2 (page 116)

1. (a) no   (b) yes   (c) no   (d) no   (e) no

2. (a) yes   (b) yes   (c) no   (d) no   (e) yes   (f) no   (g) yes

3. (a) yes   (b) no   (c) yes   (d) no   (e) yes

4. (a) $\{(-1, 1, 0), (-1, 0, 1)\}$ (b) $\{(2, 1, 0, 0), (-3, 0, 1, 0), (-1, 0, 0, 1)\}$
   (c) $\{(\frac{2}{3}, 1, 0, 0), (0, 0, 1, 0), (\frac{1}{3}, 0, 0, 1)\}$ (d) $\{(\frac{29}{12}, -\frac{11}{12}, 1)\}$
   (e) $\{(\frac{3}{7}, -\frac{1}{7}, 1, 0), (-\frac{3}{7}, \frac{8}{7}, 0, 1)\}$

5. (a) $\{(1, 0, \frac{5}{7}), (0, 1, \frac{11}{7})\}$ (b) $\{(1, 0, 0), (0, 1, 0), (0, 0, 1)\}$
   (c) $\{1, \sin 2x, \cos 2x\}$ (d) $\{x^2 + x + 1, x^2 - x + 1\}$

6. (a) $\{e^{16x}\}$ (b) $\{e^{4x}, e^{-4x}\}$ (c) $\{e^{2x}, e^{-2x}, \cos 2x, \sin 2x\}$
   (d) $\{e^x, \cos x, \sin x\}$ (e) $\{e^{-\frac{1}{2}x}, xe^{-\frac{1}{2}x}\}$

SECTION 4.3 (page 121)

1. (a) $(-1, 2)$ (b) $(-1, -2)$ (c) $(-\frac{1}{5}, -\frac{7}{5})$ (d) $(3, -2)$

2. $a, \frac{1}{2}a$

3. (a) $(1, 0)$ (b) $(-12, 1, 3)$ (c) $x - 5$ (d) $3e^x - 2e^{-x}$

4. (a) $(1, -3, 5)$ (b) $(1, 1, \frac{2}{3})$ (c) $(-2, 1, 3)$
   (d) $(\frac{1}{2}, -\frac{1}{2})$ (e) $(\frac{1}{2}, 0, \frac{1}{2})$

5. (b) $(p(a), p'(a), \frac{1}{2}p''(a), \ldots, \frac{1}{(n-1)!}p^{(n-1)}(a))$

SECTION 4.4 (page 127)

1. (a) independent   (b) dependent   (c) independent
   (d) independent   (e) dependent   (f) independent

2. (a) $\{(1, 1, 2), (1, 0, 0), (0, 1, 0)\}$
   (b) $\{(1, 0, 1), (1, 2, 1), (1, 0, 0)\}$
   (c) $\{(1, 1, 1), (1, 1, 0), (0, 1, 1)\}$

3. (a) $\{1, x - 1, x^2\}$
    (b) $\{x^2 - 1, x^2 + 2x + 1, 1\}$
    (c) $\{2, x^2 + 2, x\}$

4. (a) 3   (b) n (c) mn (d) 3   (e) 2   (f) $n - k$

6. (a) Example: $V = \mathcal{P}^5$, $S = \{1, x, x^2, x^3, x^4, 1 + x\}$
    (b) Example: $V = \mathcal{P}^5$, $S = \{x, x + 1, x + 2, x + 3, x + 4, x + 5\}$

7. $\{a\}$, where a is any nonzero real number

SECTION 5.1 (page 137)

1. (a) -10 (b) 5   (c) 6   (d) -1 (e) -8 (f) 1

2. (a) 1   (b) $b - a$ (c) 1

3. (a) -28 (b) -5 (c) 7   (d) 1   (e) 120   (f) 0

4. (a) 1   (b) $(a - b)(b - c)(c - a)$

5. (a) -72 (b) 1   (c) 64   (d) -210 (e) 12

7. 1 and 2

SECTION 5.2 (page 148)

1. (a) 58   (b) 5   (c) 69   (d) 60   (e) 10   (f) -76

2. (a) -1 (b) $\lambda^3 - 4\lambda^2 - 4\lambda + 16$
    (c) $(a_{11}a_{22} - a_{12}a_{21})(b_{11}b_{22} - b_{12}b_{21})$

3. (a) 0   (b) $2abc$ (c) $(af + cd - be)^2$

5. (b) $\det \begin{pmatrix} 1 & x_1 & x_2 & x_3 \\ 1 & a_1 & a_2 & a_3 \\ 1 & b_1 & b_2 & b_3 \\ 1 & c_1 & c_2 & c_3 \end{pmatrix} = 0$

SECTION 5.3 (page 158)

1. Permutations with sign $+1$: $(1, 2, 3, 4)$, $(1, 3, 4, 2)$, $(1, 4, 2, 3)$, $(2, 3, 1, 4)$, $(2, 1, 4, 3)$, $(2, 4, 3, 1)$, $(3, 1, 2, 4)$, $(3, 2, 4, 1)$, $(3, 4, 1, 2)$, $(4, 1, 3, 2)$, $(4, 2, 1, 3)$, $(4, 3, 2, 1)$
    Permutations with sign -1: $(1, 2, 4, 3)$, $(1, 3, 2, 4)$, $(1, 4, 3, 2)$, $(2, 3, 4, 1)$, $(2, 1, 3, 4)$, $(2, 4, 1, 3)$, $(3, 1, 4, 2)$, $(3, 2, 1, 4)$, $(3, 4, 2, 1)$, $(4, 1, 2, 3)$, $(4, 3, 1, 2)$, $(4, 2, 3, 1)$.

2. (a) 242   (b) -143 (c) -55 (d) -253

3. (a) $x^2 - 3x + 2$ (b) λ^3 (c) $-a - b - c - d$

SECTION 6.1 (page 168)

1. (b), (c), (d), and (e) are orthogonal
   (c) and (d) are orthonormal

2. (a) $(0, 1, 0)$ (b) $(1, 0, \sqrt{2})$ (c) $\left(\frac{2}{3}, \frac{\sqrt{2}}{2}, \frac{\sqrt{2}}{6}\right)$

   (d) $\left(\frac{2}{3}, \sqrt{2}, -\frac{4\sqrt{2}}{3}\right)$ (e) $(0, 1, 1)$

3. (a) $\left\{\frac{1}{\sqrt{2}}(1, 1, 0), \frac{1}{\sqrt{2}}(-1, 1, 0)\right\}$

   (b) $\left\{\frac{1}{\sqrt{3}}(1, 1, 1), \frac{1}{\sqrt{6}}(-2, 1, 1)\right\}$

   (c) $\left\{\frac{1}{2}(1, 1, 1, 1), \frac{1}{\sqrt{10}}(-2, -1, 1, 2), \frac{1}{7\sqrt{10}}(7, -4, -16, 13)\right\}$

   (d) $\{(0, 1, 0, \ldots, 0), (0, 0, 1, 0, \ldots, 0), \ldots, (0, 0, \ldots, 0, 1)\}$

4. (a) $\left\{\frac{1}{\sqrt{10}}(-1, 3, 0), \frac{1}{\sqrt{110}}(3, 1, 10)\right\}$

   (b) $\left\{\frac{1}{\sqrt{5}}(1, 2, 0, 0), \frac{1}{\sqrt{70}}(-6, 3, 5, 0), \frac{1}{\sqrt{210}}(2, -1, 3, 14)\right\}$

   (c) $\left\{\frac{1}{\sqrt{6}}(-1, -1, 2, 0), \frac{1}{\sqrt{66}}(1, -5, -2, 6)\right\}$

6. (b) $(\frac{42}{50}, -\frac{47}{50}, -\frac{47}{50}, \frac{103}{50}, \frac{103}{50})$ (c) $(\frac{8}{50}, \frac{47}{50}, -\frac{53}{50}, -\frac{53}{50}, \frac{47}{50})$

7. (a) $\left\{\frac{1}{\sqrt{3}}(1, -1, 1), \frac{1}{\sqrt{2}}(1, 1, 0)\right\}$

   (b) $(\frac{13}{3}, \frac{23}{3}, -\frac{5}{3})$ (c) $(\frac{2}{3}, -\frac{2}{3}, -\frac{4}{3})$

SECTION 6.2 (page 179)

1. (a) $y = x + \frac{4}{3}$ (b) $y = -\frac{5}{2}x + \frac{23}{3}$ (c) $y = -\frac{1}{10}x - \frac{1}{5}$
   (d) $y = 2x - 1$ (e) $y = -\frac{17}{10}x + \frac{11}{10}$ (f) $y = -\frac{1}{5}x + \frac{11}{2}$
   (g) $y = \frac{39}{35}x - \frac{22}{7}$

2. (a) $y = 2.97x^{1.92}$ (b) $y = 2.09x^{-1.23}$ (c) $y = 100x^{-5.06}$
   (d) $y = 2.98x^{0.46}$ (e) $9.9x^{2.96}$

3. (a) $y = 4.13e^{0.54t}$ (b) $y = 10e^{1.61t}$ (c) $y = 0.91e^{0.55t}$
   (d) $y = 1.53e^{-5.1t}$ (e) $y = 7.42e^{1.08t}$

4. $N = 50.17e^{1.94t}$. When $t = 3$, N should be approximately 16,900.

5. $a = 795.5$, $b = -0.49$. Approximately 83 worker-hours should be required to assemble the 100th automobile.

6. (a) $\mathbf{x} = (\frac{7}{6}, \frac{10}{11})$ (b) $\mathbf{x} = (\frac{17}{14}, \frac{11}{3})$ (c) $\mathbf{x} = (-\frac{1}{2}, 1, -\frac{1}{2})$

7. $m = \mathbf{x} \cdot \mathbf{y}/\mathbf{x} \cdot \mathbf{x}$, $b = \mathbf{1} \cdot \mathbf{y}/\mathbf{1} \cdot \mathbf{1}$

SECTION 6.3 (page 188)

1. (a) no   (b) no   (c) no   (d) no   (e) yes   (f) no

3. (a) yes   (b) no   (c) yes   (d) no   (e) yes

4. (a) $\left(\frac{11}{6}, \frac{\sqrt{3}}{3}, \frac{\sqrt{5}}{30}\right)$ (b) $\left(\frac{1}{2}, \frac{\sqrt{3}}{6}, 0\right)$ (c) $\left(-\frac{1}{2}, \frac{\sqrt{3}}{6}, 0\right)$

   (d) $\left(\frac{1}{3}, -\frac{\sqrt{3}}{6}, \frac{\sqrt{5}}{30}\right)$

5. (a) $\{\frac{1}{2}\sqrt{2}, \frac{1}{2}\sqrt{6}x, \frac{3}{4}\sqrt{10}(x^2 - \frac{1}{3})\}$
   (b) $\{\frac{1}{2}\sqrt{2}, \frac{1}{2}\sqrt{6}x, \frac{3}{4}\sqrt{10}(x^2 - \frac{1}{3}), \frac{5}{4}\sqrt{14}(x^3 - \frac{3}{5}x)\}$

6. $\{\sqrt{3}x, 4\sqrt{5}(x^2 - \frac{3}{4}x)\}$

7. $\left\{\dfrac{1}{\sqrt{2\pi}}, \dfrac{1}{\sqrt{\pi}}\cos x, \dfrac{1}{\sqrt{\pi}}\cos 2x, \dfrac{1}{\sqrt{\pi}}\sin x, \dfrac{1}{\sqrt{\pi}}\sin 2x\right\}$

8. (a) $1/\sqrt{1 - a^2}$ (b) $\{(1, \frac{1}{2}, \frac{1}{4}, \frac{1}{8}, \ldots), (\frac{1}{10}, -\frac{7}{60}, -\frac{41}{360}, -\frac{163}{2160}, \ldots)\}$

10. (a) $\dfrac{4\pi^2}{3} + \displaystyle\sum_{k=1}^{\infty}\left(\dfrac{4\pi}{k^2}\cos kx - \dfrac{4\pi}{k}\sin kx\right)$ (b) $\frac{1}{2} - \frac{1}{2}\cos 2x$

   (c) $\frac{1}{2} + \frac{1}{2}\cos 2x$ (d) $\displaystyle\sum_{k=1}^{\infty}\dfrac{2}{k}\sin kx$

11. (b) $a_k = \dfrac{1}{\pi}\displaystyle\int_0^l f(x)\cos\dfrac{2k\pi x}{l}\,dx, \; b_k = \dfrac{1}{\pi}\int_0^l f(x)\sin\dfrac{2k\pi x}{l}\,dx$

SECTION 7.1 (page 200)

1. (a) is a linear map if and only if $b = 0$.
   (b), (f), and (g) are linear maps; (c), (d), and (e) are not.

2. (a), (b), (c), (e), and (f) are linear maps.
   (d) is not a linear map.

3. (a), (b), and (d) are linear maps.
   (c) and (e) are not.

4. (a), (c), and (d) are linear maps
   (b), (e), and (f) are not.

5. (a) $(29, 10, 39)$ (b) $(-5, 4, 10)$ (c) $(1, 6, 33)$
   (d) $(2, -1, 1)$ (e) $(-1, 7, 4)$ (f) $(3, 2, 9)$

6. (a) $(3x_1 - x_2 + 2x_3, 2x_1 + 4x_2 + 3x_3)$
   (b) $3x_1 + 2x_2 + x_3$
   (c) $(x, 2x, 3x)$
   (d) $\left(\dfrac{x_1}{\sqrt{2}} - \dfrac{x_2}{\sqrt{2}}, \dfrac{x_1}{\sqrt{2}} + \dfrac{x_2}{\sqrt{2}}\right)$

7. (a) $\begin{pmatrix} 2 & -3 \\ 5 & 4 \end{pmatrix}$ (b) $\begin{pmatrix} 1 & 1 \\ 1 & -1 \\ 2 & 3 \end{pmatrix}$ (c) $\begin{pmatrix} 1 & -1 & 1 \\ 1 & 2 & 1 \end{pmatrix}$

     (d) $\begin{pmatrix} 2 & 0 & -1 \\ 0 & 3 & 0 \\ 0 & 0 & -1 \end{pmatrix}$ (e) $\begin{pmatrix} 0 & 1 & 0 \\ 0 & 0 & 1 \\ 1 & 0 & 0 \end{pmatrix}$ (f) $(0 \ \ 1 \ \ 0)$ (g) $\begin{pmatrix} 1 \\ -1 \\ 0 \end{pmatrix}$

8. (a) $\begin{pmatrix} 0 & -1 \\ 1 & 0 \end{pmatrix}$ (b) $\begin{pmatrix} -1 & 0 \\ 0 & 1 \end{pmatrix}$ (c) $\begin{pmatrix} 0 & 1 \\ 1 & 0 \end{pmatrix}$

     (d) $\begin{pmatrix} 0 & 0 & 0 \\ 0 & 1 & 0 \\ 0 & 0 & 0 \end{pmatrix}$ (e) $\begin{pmatrix} a_1^2 & a_1a_2 & a_1a_3 \\ a_1a_2 & a_2^2 & a_2a_3 \\ a_1a_3 & a_2a_3 & a_3^2 \end{pmatrix}$ (f) $\begin{pmatrix} 0 & -a_3 & a_2 \\ a_3 & 0 & -a_1 \\ -a_2 & a_1 & 0 \end{pmatrix}$

9. (a) $(2, 1)$ (b) $(2, 4)$ (c) $(1, 13)$

10. (a) $(3, -3)$ (b) $(1, 2)$ (c) $(4, 1)$

11. (a) 1 (b) $\cos 2x$ (c) 0 (d) -1 (e) 1

SECTION 7.2 (page 208)

1. (a) $\begin{pmatrix} 2 & 1 \\ 6 & -7 \end{pmatrix}$ (b) $\begin{pmatrix} -1 \\ 7 \end{pmatrix}$ (c) $\begin{pmatrix} -3 & 4 & 1 \\ 5 & -5 & 0 \\ -9 & 12 & 3 \end{pmatrix}$

     (d) $(5 \ \ -6)$ (e) (5) (f) $\begin{pmatrix} 2 & -2 & 2 \\ -1 & 1 & -1 \\ 2 & -2 & 2 \end{pmatrix}$

3. (a) $\begin{pmatrix} 4 & -10 \\ 33 & 15 \end{pmatrix}$ (b) $\begin{pmatrix} 5 \\ -10 \end{pmatrix}$ (c) $\begin{pmatrix} 9 \\ 2 \\ 18 \end{pmatrix}$

     (d) $\begin{pmatrix} -6 & -13 & -3 \\ -17 & 6 & 2 \\ -1 & 4 & 0 \end{pmatrix}$ (e) $(6 \ \ 11 \ \ 17)$ (f) $\begin{pmatrix} 1 & 0 & 0 \\ 0 & 1 & 0 \\ 0 & 0 & 1 \end{pmatrix}$

4. (a) $A^2 = \begin{pmatrix} 0 & 0 & ac \\ 0 & 0 & 0 \\ 0 & 0 & 0 \end{pmatrix}, A^3 = \begin{pmatrix} 0 & 0 & 0 \\ 0 & 0 & 0 \\ 0 & 0 & 0 \end{pmatrix}$

     (b) $A^2 = \begin{pmatrix} 1 & 0 & 2a \\ 0 & 1 & 0 \\ 0 & 0 & 1 \end{pmatrix}, A^3 = \begin{pmatrix} 1 & 0 & 3a \\ 0 & 1 & 0 \\ 0 & 0 & 1 \end{pmatrix}$

5. (a) $A^2 = \begin{pmatrix} 1 & 2 \\ 0 & 1 \end{pmatrix}, A^3 = \begin{pmatrix} 1 & 3 \\ 0 & 1 \end{pmatrix}, A^4 = \begin{pmatrix} 1 & 4 \\ 0 & 1 \end{pmatrix}$

     (b) $A^k = \begin{pmatrix} 1 & k \\ 0 & 1 \end{pmatrix}$ (c) $B = \begin{pmatrix} 1 & \frac{1}{2} \\ 0 & 1 \end{pmatrix}$

6. $A^2 = \begin{pmatrix} 0 & 0 & 1 & 0 \\ 0 & 0 & 0 & 1 \\ 0 & 0 & 0 & 0 \\ 0 & 0 & 0 & 0 \end{pmatrix}, A^3 = \begin{pmatrix} 0 & 0 & 0 & 1 \\ 0 & 0 & 0 & 0 \\ 0 & 0 & 0 & 0 \\ 0 & 0 & 0 & 0 \end{pmatrix}, A^4 = \begin{pmatrix} 0 & 0 & 0 & 0 \\ 0 & 0 & 0 & 0 \\ 0 & 0 & 0 & 0 \\ 0 & 0 & 0 & 0 \end{pmatrix}$

7. (a) $A = \begin{pmatrix} -3 & 1 \\ 1 & 1 \end{pmatrix}$, $B = \begin{pmatrix} 2 & -1 \\ 3 & 4 \end{pmatrix}$, $AB = \begin{pmatrix} 3 & 7 \\ 5 & 3 \end{pmatrix}$

$\psi \circ \phi(x_1, x_2) = (3x_1 + 7x_2, 5x_1 + 3x_2)$

(b) $A = \begin{pmatrix} 0 & 0 & 1 \\ 0 & 1 & 0 \\ 1 & 0 & 0 \end{pmatrix}$, $B = \begin{pmatrix} 1 & 1 \\ 1 & -1 \\ 2 & 0 \end{pmatrix}$, $AB = \begin{pmatrix} 2 & 0 \\ 1 & -1 \\ 1 & 1 \end{pmatrix}$

$\psi \circ \phi(x_1, x_2) = (2x_1, x_1 - x_2, x_1 + x_2)$

(c) $A = \begin{pmatrix} 1 & -1 & 0 \\ 0 & 1 & -1 \\ -1 & 0 & 1 \end{pmatrix}$, $B = \begin{pmatrix} 1 & 1 & 1 \\ 1 & 1 & 0 \\ 1 & 0 & 0 \end{pmatrix}$, $AB = \begin{pmatrix} 0 & 0 & 1 \\ 0 & 1 & 0 \\ 0 & -1 & -1 \end{pmatrix}$

$\psi \circ \phi(x_1, x_2, x_3) = (x_3, x_2, - x_2 - x_3)$

(d) $A = (1 \quad -1)$, $B = \begin{pmatrix} 1 & 1 & 0 \\ 0 & 1 & 1 \end{pmatrix}$, $AB = (1 \quad 0 \quad -1)$

$\psi \circ \phi(x_1, x_2, x_3) = x_1 - x_3$

(e) $A = (1 \quad 1 \quad 1 \quad 1)$, $B = \begin{pmatrix} 1 \\ -1 \\ 2 \\ 3 \end{pmatrix}$, $AB = (5)$

$\psi \circ \phi(x) = 5x$

SECTION 7.3 (page 219)

1. (a) $\phi^{-1}(\mathbf{v}) = \frac{1}{2}\mathbf{v}$ (b) $\phi^{-1}(\mathbf{v}) = -\mathbf{v}$ (c) $\phi^{-1} = \phi$
   (d) $\phi^{-1}(x_1, x_2) = (\frac{1}{2}x, \frac{1}{3}x_2)$ (e) $\phi^{-1}(a + bx + cx^2) = (a, b, c)$
   (f) $\phi^{-1}(a, b) = a + (b - a)x$

2. Only (b) is invertible.

3. (a) $\begin{pmatrix} \frac{1}{2} & \frac{1}{2} \\ \frac{1}{2} & -\frac{1}{2} \end{pmatrix}$ (b) $\begin{pmatrix} \frac{1}{4} & -\frac{3}{8} \\ \frac{1}{4} & \frac{1}{8} \end{pmatrix}$ (c) Not invertible

   (d) $\begin{pmatrix} -1 & 0 & 1 \\ 1 & 1 & 0 \\ 3 & 1 & -1 \end{pmatrix}$ (e) $\begin{pmatrix} 3 & -\frac{5}{2} & \frac{1}{2} \\ -3 & 4 & -1 \\ 1 & -\frac{3}{2} & \frac{1}{2} \end{pmatrix}$ (f) $\begin{pmatrix} -7 & 5 & 3 \\ 3 & -2 & -2 \\ 3 & -2 & -1 \end{pmatrix}$

4. (a) $\begin{pmatrix} 1 & -a \\ 0 & 1 \end{pmatrix}$ (b) $\begin{pmatrix} 1/a & 0 \\ 0 & 1/b \end{pmatrix}$ (c) $\begin{pmatrix} 1 & 0 & -a \\ 0 & 1 & 0 \\ 0 & 0 & 1 \end{pmatrix}$

   (d) $\begin{pmatrix} 1 & -a & ac - b \\ 0 & 1 & -c \\ 0 & 0 & 1 \end{pmatrix}$ (e) $\begin{pmatrix} 1 & 0 & 0 & -a \\ 0 & 1 & 0 & 0 \\ 0 & 0 & 1 & 0 \\ 0 & 0 & 0 & 1 \end{pmatrix}$

5. (a) $\begin{pmatrix} 1 & -2 & 9 & -52 \\ 0 & 1 & -4 & 22 \\ 0 & 0 & 1 & -5 \\ 0 & 0 & 0 & 1 \end{pmatrix}$ (b) $\begin{pmatrix} -42 & 63 & -5 & -30 \\ 10 & -16 & 1 & 8 \\ 9 & -13 & 1 & 6 \\ -1 & 2 & 0 & -1 \end{pmatrix}$

6. $\det A \neq 0$ if and only if A has rank n so Theorem 4 applies.

8. (a) $\phi^{-1}(x_1, x_2) = (x_2, x_1 - x_2)$
   (b) $\phi^{-1}(x_1, x_2) = (x_1 - 3x_2, -x_1 + 4x_2)$
   (c) $\phi^{-1}(x_1, x_2, x_3) = \left(x_3, \dfrac{\sqrt{3}}{4}x_1 - \tfrac{1}{4}x_2, \tfrac{1}{4}x_1 + \dfrac{\sqrt{3}}{4}x_2\right)$
   (d) $\phi^{-1}(x_1, \ldots, x_n) = (x_1, -x_1 + x_2, -x_2 + x_3, \ldots, -x_{n-1} + x_n)$

9. $A^{-1} = \begin{pmatrix} 23 & -4 \\ -17 & 3 \end{pmatrix}$;   (a) $\begin{pmatrix} 42 \\ -31 \end{pmatrix}$ (b) $\begin{pmatrix} -89 \\ 66 \end{pmatrix}$ (c) $\begin{pmatrix} 23\pi \\ -17\pi \end{pmatrix}$

   (d) $\begin{pmatrix} -4 \\ 3 \end{pmatrix}$

10. $A^{-1} = \begin{pmatrix} \tfrac{1}{4} & \tfrac{1}{4} & -\tfrac{1}{4} \\ -\tfrac{37}{4} & -\tfrac{13}{4} & \tfrac{17}{4} \\ 11 & 4 & -5 \end{pmatrix}$;

   (a) $\begin{pmatrix} 0 \\ -6 \\ 7 \end{pmatrix}$ (b) $\begin{pmatrix} 1 \\ -53 \\ 64 \end{pmatrix}$ (c) $\begin{pmatrix} -\tfrac{3}{2} \\ \tfrac{77}{2} \\ -45 \end{pmatrix}$ (d) $\begin{pmatrix} -8 \\ 40 \\ -48 \end{pmatrix}$

16. (a) $\begin{pmatrix} 2 & -3 \\ -1 & 2 \end{pmatrix}$ (b) $\begin{pmatrix} \tfrac{1}{2} & 0 \\ -\tfrac{1}{6} & \tfrac{1}{3} \end{pmatrix}$ (c) $\begin{pmatrix} 1 & -a \\ 0 & 1 \end{pmatrix}$

   (d) $\begin{pmatrix} 1 & -a & ac - b \\ 0 & 1 & -c \\ 0 & 0 & 1 \end{pmatrix}$ (e) $\begin{pmatrix} \tfrac{2}{5} & \tfrac{2}{5} & -\tfrac{1}{5} \\ -\tfrac{28}{5} & -\tfrac{13}{5} & \tfrac{19}{5} \\ -\tfrac{17}{5} & -\tfrac{7}{5} & \tfrac{11}{5} \end{pmatrix}$

18. (a) $(\tfrac{9}{7}, -\tfrac{5}{7})$ (b) $(\tfrac{7}{26}, \tfrac{27}{26})$ (c) $(-\tfrac{7}{5}, -\tfrac{6}{5}, \tfrac{4}{5})$ (d) $(-\tfrac{1}{5}, \tfrac{14}{5}, \tfrac{6}{5})$

19. (a) $(\tfrac{19}{62}, \tfrac{13}{31})$ (b) $(\tfrac{2}{3}, \tfrac{2}{3})$ (c) $(\tfrac{9}{14}, -\tfrac{1}{7})$
    (d) $(\tfrac{11}{6}, 0) + c(\tfrac{4}{3}, 1)$ (e) $(\tfrac{1}{7}, \tfrac{1}{7}, \tfrac{1}{7})$

20. (a) $\sqrt{62}/62$ (b) $\sqrt{3}/3$ (c) $\sqrt{14}/14$ (d) $\sqrt{2}/2$ (e) $2\sqrt{7}/7$

SECTION 7.4 (p. 228)

1. (a) $\begin{pmatrix} 1 & 1 & 1 \\ 1 & 1 & 0 \end{pmatrix}$ (b) $\begin{pmatrix} 3 & 2 & 1 \\ 0 & 1 & 0 \end{pmatrix}$ (c) $\begin{pmatrix} 4 & 5 & 2 \\ 5 & 0 & -1 \end{pmatrix}$

2. (a) $\begin{pmatrix} 2 & 1 \\ -1 & 3 \end{pmatrix}$ (b) $\begin{pmatrix} 1 & 0 \\ 0 & 1 \end{pmatrix}$ (c) $\begin{pmatrix} 3 & -\tfrac{7}{2} \\ -2 & -\tfrac{7}{2} \end{pmatrix}$ (d) $\begin{pmatrix} -2 & -\tfrac{7}{2} \\ 3 & -\tfrac{7}{2} \end{pmatrix}$

3. (a) $\begin{pmatrix} 0 & 1 & 0 & 0 & 0 \\ 0 & 0 & 2 & 0 & 0 \\ 0 & 0 & 0 & 3 & 0 \\ 0 & 0 & 0 & 0 & 4 \end{pmatrix}$ (b) $4x^3 + 6x^2 - 4x + 1$

4. (a) $\begin{pmatrix} 1 & 1 & 0 & 0 \\ 0 & 1 & 2 & 0 \\ 0 & 0 & 1 & 3 \\ 0 & 0 & 0 & 1 \end{pmatrix}$ (b) $\begin{pmatrix} 1 & 0 & 0 & 0 \\ 3 & 1 & 0 & 0 \\ 0 & 2 & 1 & 0 \\ 0 & 0 & 1 & 1 \end{pmatrix}$ (c) $\begin{pmatrix} 1 & 1 & -1 & -1 \\ 0 & 1 & 2 & -1 \\ 0 & 0 & 1 & 3 \\ 0 & 0 & 0 & 1 \end{pmatrix}$

5. (a) $\begin{pmatrix} 1 & 0 & 0 \\ 4 & 1 & 0 \\ 4 & 2 & 1 \end{pmatrix}$ (b) $\begin{pmatrix} 4 & 2 & 1 \\ 4 & 1 & 0 \\ 1 & 0 & 0 \end{pmatrix}$ (c) $\begin{pmatrix} 0 & 0 & 1 \\ 0 & 1 & 4 \\ 1 & 2 & 4 \end{pmatrix}$

   (d) $\begin{pmatrix} 1 & 0 & 0 \\ 0 & 1 & 0 \\ 0 & 0 & 1 \end{pmatrix}$

6. (a) $\begin{pmatrix} 1 & 0 & 0 \\ 0 & 2 & 0 \\ 0 & 0 & 3 \end{pmatrix}$ (b) $\begin{pmatrix} 0 & 0 & 0 \\ 0 & 1 & 0 \\ 0 & 0 & 2 \end{pmatrix}$ (c) $\begin{pmatrix} 0 & 0 & 0 \\ 0 & 0 & 0 \\ 0 & 0 & 2 \end{pmatrix}$ (d) $\begin{pmatrix} 0 & 0 & 0 \\ 0 & 0 & 0 \\ 0 & 0 & 0 \end{pmatrix}$

7. (a) $-x + 4$ (b) $2x + 1$ (c) $9x - 8$ (d) $5x - 6$

8. (a) $\phi \circ \phi(p(x)) = p(x + 2)$

   (b) $\begin{pmatrix} 1 & 0 & 0 & 0 \\ 6 & 1 & 0 & 0 \\ 12 & 4 & 1 & 0 \\ 8 & 4 & 2 & 1 \end{pmatrix}$

9. (a) $M_1 = \begin{pmatrix} 1 & 0 \\ 0 & -1 \end{pmatrix}$, $M_2 = \begin{pmatrix} 0 & 1 \\ 1 & 0 \end{pmatrix}$, $P = \begin{pmatrix} \frac{1}{2} & \frac{1}{2} \\ \frac{1}{2} & -\frac{1}{2} \end{pmatrix}$

   (b) $M_1 = \begin{pmatrix} 0 & 0 \\ 0 & -2 \end{pmatrix}$, $M_2 = \begin{pmatrix} -1 & 1 \\ 1 & -1 \end{pmatrix}$, $P = \begin{pmatrix} \frac{1}{2} & \frac{1}{2} \\ \frac{1}{2} & -\frac{1}{2} \end{pmatrix}$

10. (a) $\begin{pmatrix} 2 & 0 \\ 0 & 3 \end{pmatrix}$ (b) $\begin{pmatrix} \frac{5}{2} & \frac{1}{2} \\ \frac{1}{2} & \frac{5}{2} \end{pmatrix}$ (c) $\begin{pmatrix} -1 & 1 \\ 1 & 1 \end{pmatrix}$

11. (a) $\begin{pmatrix} 1 & 3 \\ 3 & 1 \end{pmatrix}$ (b) $\begin{pmatrix} -2 & 0 \\ 0 & 4 \end{pmatrix}$ (c) $\begin{pmatrix} -1 & 1 \\ 1 & 1 \end{pmatrix}$

SECTION 7.5 (page 239)

1. $\mathcal{P}^2 = \mathcal{L}(1, x)$

2. (a) $\mathbf{x} = t(1, 1, 1)$ (b) $x_1 + x_2 + x_3 = 0$

3. (a) $\{\cos x, \sin x\}$ (b) $\{e^x, e^{-x}\}$ (c) $\{e^x, xe^x, x^2 e^x\}$
   (d) $\{1, x, e^{2x}, e^{-2x}, \cos x, \sin x\}$

4. (a) $\ker \phi_A = \{0\}$; there is no basis
   (b) $\{(-1, 1)\}$ (c) $\ker \phi_A = \{0\}$, no basis
   (d) $\{(-3, -1, 1)\}$ (e) $\ker \phi_A = \{0\}$, no basis (f) $\{(-2, -4, 3, 1)\}$

5. (a) $\{(1, 0), (0, 1)\}$ (b) $\{(1, -1)\}$ (c) $\{(1, 0, 0), (0, 1, 0), (0, 0, 1)\}$
   (d) $\{(1, 0, 2), (0, 1, 1)\}$ (e) $\{(1, 0, 0, 2), (0, 1, 0, 4), (0, 0, 1, -3)\}$
   (f) $\{(1, 0, 0), (0, 1, 0), (0, 0, 1)\}$

6. (a) 2 (b) 1 (c) 3 (d) 2 (e) 3 (f) 3

7. (a) $\ker \phi = \{0\}$; ϕ *is* one-to-one
   (b) $\ker \phi = \{0\}$; ϕ *is* one-to-one
   (c) $\ker \phi$ is the line $\mathbf{x} = t(1, 1, 1)$; ϕ is *not* one-to-one
   (d) $\ker \phi$ is the plane $x_1 + x_2 + x_3 = 0$; ϕ is *not* one-to-one
   (e) $\ker \phi$ is the line $\mathbf{x} = t(1, 0, 0, \ldots)$; ϕ is *not* one-to-one

8. (a) im $\phi = \mathcal{L}((1, 0, 1), (0, 1, -1))$ is the plane $-x_1 + x_2 + x_3 = 0$; ϕ is *not* onto
   (b) im $\phi = \mathbb{R}^n$; ϕ *is* onto
   (c) im $\phi = \mathcal{L}((1, 0, -1), (0, 1, -1))$ is the plane $x_1 + x_2 + x_3 = 0$; ϕ is *not* onto
   (d) im ϕ is the line $\mathbf{x} = t(1, 1)$; ϕ is *not* onto
   (e) im $\phi = \mathbb{R}^\infty$; ϕ *is* onto.

10. (b) $a \neq 1, 2$, or 3   (c) $a \neq 1, 2$, or 3

11. (b) $\{x, x^2, \ldots, x^{n-1}\}$ (c) $\mathbb{R}$

12. (b) ker $\phi = $ those polynomials in $\mathcal{P}^n$ that have a as a zero.
    (c) $\mathbb{R}$

13. (a) ker $\phi = \{\mathbf{0}\}$; ϕ *is* one-to-one
    (b) im $\phi = $ polynomials with constant term equal to zero; ϕ is *not* one-to-one

14. (d) $\phi^{-1}(p(x)) = p(x + 1)$

SECTION 8.1 (page 252)

1. (a) no eigenvalues or eigenvectors
   (b) eigenvalue 3: eigenvectors $\{c(1, 1)|c \neq 0\}$
       eigenvalue 7: eigenvectors $\{c(-1, 1)|c \neq 0\}$
   (c) no eigenvalues or eigenvectors
   (d) eigenvalue -2: eigenvectors $\{c(\frac{2}{7}, 1)|c \neq 0\}$
       eigenvalue 3: eigenvectors $\{c(\frac{1}{3}, 1)|c \neq 0\}$
   (e) eigenvalue -4: eigenvectors $\{c(-2, 1)|c \neq 0\}$
       eigenvalue 3: eigenvectors $\{c(-1, 1)|c \neq 0\}$
   (f) eigenvalue -2: eigenvectors $\{c(1, 1)|c \neq 0\}$
       eigenvalue 2: eigenvectors $\{c(-1, 1)|c \neq 0\}$
   (g) eigenvalue -1: eigenvectors $\{c(\frac{2}{7}, 1)|c \neq 0\}$
       eigenvalue 1: eigenvectors $\{c(\frac{1}{3}, 1)|c \neq 0\}$
   (h) no eigenvalues or eigenvectors

2. (a) eigenvalue 0: eigenvectors $\{c(-\frac{3}{2}, -\frac{1}{2}, 1)|c \neq 0\}$
       eigenvalue 2: eigenvectors $\{c(-\frac{1}{2}, \frac{3}{2}, 1)|c \neq 0\}$
       eigenvalue 3: eigenvectors $\{c(0, 1, 1)|c \neq 0\}$
   (b) eigenvalue 2: eigenvectors $\{c(0, -1, 1)|c \neq 0\}$
   (c) eigenvalue -1: eigenvectors $\{c(\frac{5}{2}, \frac{3}{2}, 1)|c \neq 0\}$
       eigenvalue 1: eigenvectors $\{c_1(\frac{10}{3}, 1, 0) + c_2(-\frac{7}{3}, 0, 1)|c_1$ and c_2 not both zero$\}$
   (d) eigenvalue 4: eigenvectors $\{c_1(3, 1, 0) + c_2(5, 0, 1)|c_1$ and c_2 not both zero$\}$
       eigenvalue -8: eigenvectors $\{c(1, 0, 1)|c \neq 0\}$
   (e) eigenvalue 1: eigenvectors $\{c(0, 1, 0)|c \neq 0\}$
       eigenvalue $\sqrt{2}$: eigenvectors $\{c(1 + \sqrt{2}, 0, 1)|c \neq 0\}$
       eigenvalue $-\sqrt{2}$: eigenvectors $\{c(1 - \sqrt{2}, 0, 1)|c \neq 0\}$

3. (a) eigenvalue $\dfrac{1 + \sqrt{41}}{2}$: eigenvectors $\left\{c\left(\dfrac{-5 + \sqrt{41}}{4}, 1\right)|c \neq 0\right\}$
       eigenvalue $\dfrac{1 - \sqrt{41}}{2}$: eigenvectors $\left\{c\left(\dfrac{-5 - \sqrt{41}}{4}, 1\right)|c \neq 0\right\}$

(b) Every real number λ is an eigenvalue.
For $\lambda > 0$, the λ-eigenspace is $\mathcal{L}(e^{\sqrt{\lambda}x}, e^{-\sqrt{\lambda}x})$.
For $\lambda = 0$, the λ-eigenspace is $\mathcal{L}(1, x)$.
For $\lambda < 0$, the λ-eigenspace is $\mathcal{L}(\cos(\sqrt{-\lambda}x, \sin\sqrt{-\lambda}x))$.

(c) Every real number λ is an eigenvalue.
For $\lambda > 1$, the λ-eigenspace is $\mathcal{L}(e^{\sqrt{\lambda-1}x}, e^{-\sqrt{\lambda-1}x})$.
For $\lambda = 1$, the λ-eigenspace is $\mathcal{L}(1, x)$.
For $\lambda < 1$, the λ-eigenspace is $\mathcal{L}(\cos\sqrt{1-\lambda}x, \sin\sqrt{1-\lambda}x)$.

(d) Every real number λ is an eigenvalue.
For $\lambda > -\frac{1}{4}$, the λ-eigenspace is $\mathcal{L}(e^{(-1+\sqrt{1+4\lambda})x/2}, e^{(-1-\sqrt{1+4\lambda})x/2})$.
For $\lambda = -\frac{1}{4}$, the λ-eigenspace is $\mathcal{L}(e^{-x/2}, xe^{-x/2})$.
For $\lambda < -\frac{1}{4}$, the λ-eigenspace is $\mathcal{L}\left(e^{-x/2}\cos\dfrac{\sqrt{-1-4\lambda}}{2}x, e^{-x/2}\sin\dfrac{\sqrt{-1-4\lambda}}{2}x\right)$.

(e) no eigenvalues or eigenvectors

4. (a) eigenvalue -1, eigenspace $\mathcal{L}(x)$
eigenvalue $+1$, eigenspace $\mathcal{L}(1)$

(b) eigenvalue -1, eigenspace: all polynomials with only odd power terms
eigenvalue $+1$, eigenspace: all polynomials with only even power terms

(c) The eigenvalues are the nonnegative integers $0, 1, 2, 3, \ldots$
The eigenspace corresponding to eigenvalue n is $\mathcal{L}(x^n)$, $n = 0, 1, \ldots$.

(d) The eigenvalues are the nonnegative real numbers.
The λ-eigenspace, where $\lambda \geq 0$, is $\mathcal{L}(x^\lambda)$.
[Note: $x^\lambda \notin \mathcal{D}^\infty$ if $\lambda < 0$.]

5. (b) If θ is an odd multiple of π, the only eigenvalue is -1 and every nonzero vector is an eigenvector.
If θ is an even multiple of π, then the only eigenvalue is $+1$ and every nonzero vector is an eigenvector.

6. (a) λ^2

7. The eigenvalues of ϕ^{-1} are the reciprocals of the eigenvalues of ϕ.

SECTION 8.2 (page 256)

1. (a) and (d)

2. (a) $P = \begin{pmatrix} -1 & 1 \\ 1 & 1 \end{pmatrix}$, $D = \begin{pmatrix} -1 & 0 \\ 0 & 5 \end{pmatrix}$

(d) $P = \begin{pmatrix} -1 & -1 & \frac{1}{2} \\ 1 & 0 & \frac{1}{2} \\ 0 & 1 & 1 \end{pmatrix}$, $D = \begin{pmatrix} 1 & 0 & 0 \\ 0 & 1 & 0 \\ 0 & 0 & 5 \end{pmatrix}$

4. (a) $P = \begin{pmatrix} 2 & 1 \\ 1 & 1 \end{pmatrix}$, $D = \begin{pmatrix} -1 & 0 \\ 0 & 1 \end{pmatrix}$

(b) $P = \begin{pmatrix} 1 & \frac{5}{4} \\ 1 & 1 \end{pmatrix}$, $D = \begin{pmatrix} -3 & 0 \\ 0 & 7 \end{pmatrix}$

(c) $P = \begin{pmatrix} -1 & -\frac{1}{2} \\ 1 & 1 \end{pmatrix}$, $D = \begin{pmatrix} 1 & 0 \\ 0 & 10 \end{pmatrix}$

(d) $P = \begin{pmatrix} 1 - \sqrt{2} & 1 + \sqrt{2} \\ 1 & 1 \end{pmatrix}$, $D = \begin{pmatrix} -\sqrt{2} & 0 \\ 0 & \sqrt{2} \end{pmatrix}$

(e) $P = \begin{pmatrix} -\frac{3}{2} & -\frac{7}{5} \\ 1 & 1 \end{pmatrix}$, $D = \begin{pmatrix} 3 & 0 \\ 0 & 4 \end{pmatrix}$

5. (a) $P = \begin{pmatrix} \frac{1}{3} & -\frac{1}{3} & \frac{2}{3} \\ \frac{2}{3} & 1 & 0 \\ 1 & 0 & 1 \end{pmatrix}$, $D = \begin{pmatrix} 1 & 0 & 0 \\ 0 & 2 & 0 \\ 0 & 0 & 2 \end{pmatrix}$

(b) $P = \begin{pmatrix} -\frac{1}{5} & -\frac{1}{4} & -\frac{1}{4} \\ -\frac{3}{5} & -\frac{7}{12} & -\frac{5}{8} \\ 1 & 1 & 1 \end{pmatrix}$, $D = \begin{pmatrix} -1 & 0 & 0 \\ 0 & 1 & 0 \\ 0 & 0 & 2 \end{pmatrix}$

(c) $P = \begin{pmatrix} -2 & 1 & \frac{1}{3} \\ -1 & 1 & \frac{2}{3} \\ 1 & 1 & 1 \end{pmatrix}$, $D = \begin{pmatrix} 0 & 0 & 0 \\ 0 & 1 & 0 \\ 0 & 0 & 2 \end{pmatrix}$

(d) $P = \begin{pmatrix} \frac{3}{2} & -1 & 1 \\ \frac{3}{2} & 3 & 2 \\ 1 & 1 & 1 \end{pmatrix}$, $D = \begin{pmatrix} 1 & 0 & 0 \\ 0 & 2 & 0 \\ 0 & 0 & 3 \end{pmatrix}$

7. (a) $\begin{pmatrix} 121 & 122 \\ 122 & 121 \end{pmatrix}$ (b) $\begin{pmatrix} -857 & 1650 \\ -550 & 1068 \end{pmatrix}$

(c) $\begin{pmatrix} 13 & -4 \\ 42 & -13 \end{pmatrix}$ (d) $\begin{pmatrix} 1893 & -550 \\ 5775 & -1682 \end{pmatrix}$

SECTION 8.3 (page 265)

1. (a) $\begin{aligned} x_1 &= -c_1 e^t - 2c_2 e^t + c_3 e^{5t} \\ x_2 &= c_1 e^t + c_3 e^{5t} \\ x_3 &= c_2 e^t + c_3 e^{5t} \end{aligned}$

(b) $\begin{aligned} x_1 &= c_1 e^{2t} + c_2 e^{-2t} \\ x_2 &= \phantom{c_1 e^{2t} +} c_2 e^{-2t} + c_3 e^{4t} \\ x_3 &= c_1 e^{2t} \phantom{+ c_2 e^{-2t}} + c_3 e^{4t} \end{aligned}$

2. (a) $\mathbf{x} = c_1 e^{-t} \begin{pmatrix} -1 \\ 1 \end{pmatrix} + c_2 e^{5t} \begin{pmatrix} 1 \\ 1 \end{pmatrix}$

(d) $\mathbf{x} = c_1 e^t \begin{pmatrix} -1 \\ 1 \\ 0 \end{pmatrix} + c_2 e^t \begin{pmatrix} -1 \\ 0 \\ 1 \end{pmatrix} + c_3 e^{5t} \begin{pmatrix} \frac{1}{2} \\ \frac{1}{2} \\ 1 \end{pmatrix}$

3. (b) $\mathbf{x} = \begin{pmatrix} c_1 t + c_2 \\ c_1 \end{pmatrix}$ (c) $\mathbf{x} = \begin{pmatrix} c_1 t e^t + c_2 e^t \\ c_1 e^t \end{pmatrix}$

4. (a) $\mathbf{x} = c_1 e^{-t} \begin{pmatrix} 2 \\ 1 \end{pmatrix} + c_2 e^t \begin{pmatrix} 1 \\ 1 \end{pmatrix}$ (b) $\mathbf{x} = c_1 e^{-3t} \begin{pmatrix} 1 \\ 1 \end{pmatrix} + c_2 e^{7t} \begin{pmatrix} \frac{5}{4} \\ 1 \end{pmatrix}$

(c) $\mathbf{x} = c_1 e^t \begin{pmatrix} -1 \\ 1 \end{pmatrix} + c_2 e^{10t} \begin{pmatrix} -\frac{1}{2} \\ 1 \end{pmatrix}$

(d) $\mathbf{x} = c_1 e^{-\sqrt{2}t} \begin{pmatrix} 1 - \sqrt{2} \\ 1 \end{pmatrix} + c_2 e^{\sqrt{2}t} \begin{pmatrix} 1 + \sqrt{2} \\ 1 \end{pmatrix}$

(e) $\mathbf{x} = c_1 e^{3t} \begin{pmatrix} -\frac{3}{2} \\ 1 \end{pmatrix} + c_2 e^{4t} \begin{pmatrix} -\frac{7}{5} \\ 1 \end{pmatrix}$

5. (a) $\mathbf{x} = c_1 e^t \begin{pmatrix} \frac{1}{3} \\ \frac{2}{3} \\ 1 \end{pmatrix} + c_2 e^{2t} \begin{pmatrix} -\frac{1}{3} \\ 1 \\ 0 \end{pmatrix} + c_3 e^{2t} \begin{pmatrix} \frac{2}{3} \\ 0 \\ 1 \end{pmatrix}$

(b) $\mathbf{x} = c_1 e^{-t} \begin{pmatrix} -\frac{1}{5} \\ -\frac{3}{5} \\ 1 \end{pmatrix} + c_2 e^t \begin{pmatrix} -\frac{1}{4} \\ -\frac{7}{12} \\ 1 \end{pmatrix} + c_3 e^{2t} \begin{pmatrix} -\frac{1}{4} \\ -\frac{5}{8} \\ 1 \end{pmatrix}$

(c) $\mathbf{x} = c_1 \begin{pmatrix} -2 \\ -1 \\ 1 \end{pmatrix} + c_2 e^t \begin{pmatrix} 1 \\ 1 \\ 1 \end{pmatrix} + c_3 e^{2t} \begin{pmatrix} \frac{1}{3} \\ \frac{2}{3} \\ 1 \end{pmatrix}$

(d) $\mathbf{x} = c_1 e^t \begin{pmatrix} \frac{3}{2} \\ \frac{3}{2} \\ 1 \end{pmatrix} + c_2 e^{2t} \begin{pmatrix} -1 \\ 3 \\ 1 \end{pmatrix} + c_3 e^{3t} \begin{pmatrix} 1 \\ 2 \\ 1 \end{pmatrix}$

6. (a) $\begin{pmatrix} e^2 & 0 \\ 0 & e^{-3} \end{pmatrix}$ (b) $\begin{pmatrix} -1 & 0 \\ 0 & -1 \end{pmatrix}$ (c) $\begin{pmatrix} 0 & -1 \\ 1 & 0 \end{pmatrix}$

(d) $\begin{pmatrix} 1 & 1 \\ 0 & 1 \end{pmatrix}$ (e) $\begin{pmatrix} e & e \\ 0 & e \end{pmatrix}$

7. (a) $\begin{pmatrix} e^{-1} & 0 & 0 \\ 0 & e & 0 \\ 0 & 0 & 1 \end{pmatrix}$ (b) $\begin{pmatrix} 1 & 1 & \frac{1}{2} \\ 0 & 1 & 1 \\ 0 & 0 & 1 \end{pmatrix}$ (c) $\begin{pmatrix} 1 & 1 & \frac{3}{2} \\ 0 & 1 & 1 \\ 0 & 0 & 1 \end{pmatrix}$

(d) $\begin{pmatrix} 1 & a & b + \frac{1}{2}ac \\ 0 & 1 & c \\ 0 & 0 & 1 \end{pmatrix}$ (e) $\begin{pmatrix} e^a & 0 & 0 \\ 0 & \cos b & -\sin b \\ 0 & \sin b & \cos b \end{pmatrix}$

SECTION 8.4 (page 276)

1. (a) $\mathbf{x} = c_1(1, 1, 1, \ldots) + c_2(1, \frac{1}{5}, \frac{1}{25}, \ldots)$
    or $x_n = c_1 + c_2(\frac{1}{5})^{n-1}$
    (b) $\mathbf{x} = c_1(1, 1, 1, \ldots) + c_2(1, 5, 25, \ldots)$
    or $x_n = c_1 + c_2(5)^{n-1}$
    (c) $\mathbf{x} = c_1(1, -2, 4, -8, \ldots) + c_2(1, -4, 16, -64, \ldots)$
    or $x_n = c_1(-2)^{n-1} + c_2(-4)^{n-1}$
    (d) $\mathbf{x} = c_1(1, 1 + \sqrt{3}, 4 + 2\sqrt{3}, \ldots) + c_2(1, 1 - \sqrt{3}, 4 - 2\sqrt{3}, \ldots)$
    or $x_n = c_1(1 + \sqrt{3})^{n-1} + c_2(1 - \sqrt{3})^{n-1}$
    (e) $\mathbf{x} = c_1(1, -2, 4, -8, \ldots) + c_2(0, 1, -4, 12, \ldots)$
    or $x_n = c_1(-2)^{n-1} + c_2(n - 1)(-2)^{n-2}$
    (f) $\mathbf{x} = c_1(1, -\frac{1}{2}, \frac{1}{4}, -\frac{1}{8}, \ldots) + c_2(0, 1, -1, \frac{3}{4}, \ldots)$
    or $x_n = c_1(-\frac{1}{2})^{n-1} + c_2(n - 1)(-\frac{1}{2})^{n-2}$

(g) $\mathbf{x} = c_1(1, -3, 9, -27, \ldots) + c_2(0, 1, -6, 27, \ldots)$

or $x_n = c_1(-3)^{n-1} + c_2(n-1)(-3)^{n-2}$

(h) $\mathbf{x} = c_1(1, -\frac{1}{2}, -\frac{1}{2}, 1, \ldots) + c_2\left(0, \dfrac{\sqrt{3}}{2}, -\dfrac{\sqrt{3}}{2}, 0 \ldots\right)$

or $x_n = c_1 \cos\left[(n-1)\dfrac{2\pi}{3}\right] + c_2 \sin\left[(n-1)\dfrac{2\pi}{3}\right]$

(i) $\mathbf{x} = c_1(1, -1, 0, 2, \ldots) + c_2(0, 1, -2, 2, \ldots)$

or $x_n = c_1(\sqrt{2})^{n-1}\cos\left[(n-1)\dfrac{3\pi}{4}\right] + c_2(\sqrt{2})^{n-1}\sin\left[(n-1)\dfrac{3\pi}{4}\right]$

(j) $\mathbf{x} = c_1(1, 0, -1, 0, \ldots) + c_2(0, 1, 0, -1, \ldots)$

or $x_n = c_1 \cos\left[(n-1)\dfrac{\pi}{2}\right] + c_2 \sin\left[(n-1)\dfrac{\pi}{2}\right]$

2. $x_n = \dfrac{1}{2\sqrt{2}}[(1 + \sqrt{2})^{n-1} - (1 - \sqrt{2})^{n-1}]$

3. (a) $\mathbf{x} = (0, 1, 1, 0, -1, \ldots);\ x_n = \dfrac{2}{\sqrt{3}} \sin\left[(n-1)\dfrac{\pi}{3}\right]$

(b) $\mathbf{x} = (-1, 1, 2, -2, -4, \ldots);$

$x_n = -(\sqrt{2})^{n-1}\cos\left[(n-1)\dfrac{\pi}{2}\right] + (\sqrt{2})^{n-2}\sin\left[(n-1)\dfrac{\pi}{2}\right]$

(c) $\mathbf{x} = (0, 1, 2, 2, 0, \ldots);\ x_n = (\sqrt{2})^{n-1}\sin\left[(n-1)\dfrac{\pi}{4}\right]$

(d) $\mathbf{x} = (0, 1, 4, 12, 32, \ldots);\ x_n = (n-1)2^{n-2}$

(e) $\mathbf{x} = (0, 1, 5, 20, 75, \ldots);$

$x_n = \dfrac{1}{\sqrt{5}}\left[\left(\dfrac{5 + \sqrt{5}}{2}\right)^{n-1} - \left(\dfrac{5 - \sqrt{5}}{2}\right)^{n-1}\right]$

4. (b) $P_n = P_0(1.06)^n$

5. (b) $\Delta^2\mathbf{x} = (x_3 - 2x_2 + x_1, x_4 - 2x_3 + x_2, x_5 - 2x_4 + x_3, \ldots)$

(c) $p = a,\ q = 2a + b,\ r = a + b + c$

9. (b)   If the zeros of $\displaystyle\sum_{i=0}^{k} a_i\lambda^{n-i}$ are $\lambda_1, \ldots, \lambda_k$, then

$\{(1, \lambda_1, \lambda_1^2, \ldots), (1, \lambda_2, \lambda_2^2, \ldots), \ldots, (1, \lambda_k, \lambda_k^2, \ldots)\}$ is a basis.

SECTION 9.1 (page 283)

1. $y = \frac{1}{2} + c\dfrac{1}{x^2}$

2. $y = c(1 + x^2)^{-1/2}$

3. $y = 1 + ce^{-e^x}$

4. $y = \sin x + c \cos x$

5. $y = x^2 e^{-x^2} + ce^{-x^2}$

6. $y = \dfrac{e^x}{2x} + c\dfrac{e^{-x}}{x}$

7. $y = \dfrac{e^x}{(x+1)^2} + c\dfrac{1}{(x+1)^2}$

8. $y = \dfrac{\sin x}{x-1} + c\dfrac{1}{x-1}$

9. $y = \frac{6}{5}x + \frac{4}{5}\cdot\dfrac{1}{x^4}$

10. $y = \dfrac{x^3 - 6}{3(x^2 + 1)}$

11. $y = (\pi^3 - 1 - \cos x)/x^3$

12. $xD^2 + (x^2 + 1)D + 2x$

13. $D^2 + (3x + 3)D + (9x + 3)$

14. $D^2 + (3x + 3)D + 9x$

15. $xD^2 + (1 + x\sin x + \cos x)D + (-1 + \cos x)\sin x$

16. $y = \frac{1}{6}(\ln x)^3 + c_1 \ln x + c_2$

17. $y = c_1 \ln x + c_2 x + c_3$

18. $y = c_1 x + c_2$

19. $y = c_1(x + 1) + c_2 e^x$

20. $y = c_1 x^2 + c_2 x^3$

21. $y = c_1 \dfrac{1}{x} + c_2 \dfrac{\ln x}{x}$

22. $y = c_1 \cos(\ln x) + c_2 \sin(\ln x)$

23. $y = c_1 x + c_2 x^{-\frac{1}{2}} \cos\left(\dfrac{\sqrt{3}}{2}\ln x\right) + c_3 x^{-\frac{1}{2}} \sin\left(\dfrac{\sqrt{3}}{2}\ln x\right)$

24. $y = c_1 \dfrac{e^{-x}}{x} + c_2 \dfrac{1}{x}$

SECTION 9.2 (page 296)

1. (a) $y = c\left(\displaystyle\sum_{n=0}^{\infty} \dfrac{(-x)^n}{n!}\right) = ce^{-x}$

    (b) $y = c_1\left(\displaystyle\sum_{n=0}^{\infty} \dfrac{x^{2n}}{(2n)!}\right) + c_2\left(\displaystyle\sum_{n=0}^{\infty} \dfrac{x^{2n+1}}{(2n+1)!}\right) = c_1 \cosh x + c_2 \sinh x$

    (c) $y = c_1\left(\displaystyle\sum_{n=0}^{\infty} (-1)^n \dfrac{x^{2n}}{(2n)!}\right) + c_2\left(\displaystyle\sum_{n=0}^{\infty} (-1)^n \dfrac{x^{2n+1}}{(2n+1)!}\right) = c_1 \cos x + c_2 \sin x$

    (d) $y = c_1(1 - \frac{1}{2}x^2 + \frac{1}{3}x^3 - \frac{1}{8}x^4 + \cdots) + c_2(x - x^2 + \frac{1}{2}x^3 - \frac{1}{6}x^4 + \cdots)$
    $= c_1(e^{-x} + xe^{-x}) + c_2 xe^{-x}$

2. (a) $y = c\left[\displaystyle\sum_{n=0}^{\infty} \dfrac{(-1)^n}{n!2^n}x^{2n}\right] = ce^{-x^2/2}$

(b) $y = c_1 \left[1 - \sum_{n=1}^{\infty} \frac{2^n \cdot 1 \cdot 3 \cdot 5 \cdots (2n-3)}{(2n)!} x^{2n} \right] + c_2 x$

(c) $y = c_1 \left[\sum_{n=0}^{\infty} \frac{x^{2n}}{2 \cdot 4 \cdot 6 \cdots (2n)} \right] + c_2 \sum_{n=0}^{\infty} \frac{x^{2n+1}}{1 \cdot 3 \cdot 5 \cdots (2n+1)}$

(d) $y = c_1 \left[1 + \sum_{n=1}^{\infty} \frac{1 \cdot 7 \cdot 13 \cdots (6n-5)}{(2n)!} x^{2n} \right]$

$\qquad + c_2 \left[x + \sum_{n=1}^{\infty} \frac{4 \cdot 10 \cdot 16 \cdots (6n-2)}{(2n+1)!} x^{2n+1} \right]$

(e) $y = c_1 \left[1 + \sum_{n=1}^{\infty} (-1)^n \frac{1 \cdot 4 \cdot 7 \cdots (3n-2)}{(3n)!} x^{3n} \right]$

$\qquad + c_2 \left[x + \sum_{n=1}^{\infty} (-1)^n \frac{2 \cdot 5 \cdot 8 \cdots (3n-1)}{(3n+1)!} x^{3n+1} \right]$

(f) $y = c_1 \left[1 - \sum \frac{2 \cdot 5 \cdot 8 \cdots (3n-4)}{(2 \cdot 3)(5 \cdot 6) \cdots (3n-1)(3n)4^n} x^{3n} \right] + c_2 x$

3. (a) $y = c[1 + (x-1) - \frac{1}{2}(x-1)^2 + \frac{1}{6}(x-1)^3 + \frac{1}{24}(x-1)^4 + \cdots] = ce^{-1/x}$

(b) $y = c_1[1 + \frac{1}{2}(x-1)^2 - \frac{1}{2}(x-1)^3 + \frac{1}{2}(x-1)^4 - \cdots]$
$\qquad + c_2[(x-1) - \frac{1}{2}(x-1)^2 + \frac{1}{2}(x-1)^3 - \frac{1}{2}(x-1)^4 + \cdots]$
$\qquad = c_1 \left(\frac{x}{2} + \frac{1}{2x} \right) + c_2 \left(\frac{x}{2} - \frac{1}{2x} \right) = C_1 x + C_2 \frac{1}{x}$

(c) $y = c_1[1 + 2(x-1)^2 + 2(x-1)^3 + \frac{7}{6}(x-1)^4 + \cdots]$
$\qquad + c_2[(x-1) + \frac{1}{2}(x-1)^2 + \frac{2}{3}(x-1)^3 + (x-1)^4 + \cdots]$

(d) $y = c_1[1 - \frac{1}{2}(x-1)^2 + \frac{1}{3}(x-1)^3 - \frac{7}{24}(x-1)^4 + \cdots]$
$\qquad + c_2[(x-1) - (x-1)^2 + \frac{5}{6}(x-1)^3 - \frac{5}{4}(x-1)^4 + \cdots]$

4. (a) $y = c_1 \left[\sum_{n=0}^{\infty} \frac{x^{2n}}{n! \, 1 \cdot 5 \cdot 9 \cdots (4n+1)} \right] + c_2 x^{-\frac{1}{2}} \left[1 + \sum_{n=1}^{\infty} \frac{x^{2n}}{n! \, 3 \cdot 7 \cdot 11 \cdots (4n-1)} \right]$

$\qquad = c_1(1 + \frac{1}{5}x^2 + \frac{1}{90}x^4 + \frac{1}{3510}x^6 + \cdots) + c_2 x^{-\frac{1}{2}}(1 + \frac{1}{3}x^2 + \frac{1}{42}x^4 + \frac{1}{1386}x^6 + \cdots$

(b) $y = c_1 x^{-1} \left[1 + \sum_{n=1}^{\infty} \frac{(-1)^{n+1} 4^n}{2n(2n-2)!} x^n \right]$

$\qquad + c_2 x^{\frac{1}{2}} \left[\sum_{n+0}^{\infty} \frac{(-1)^n 3 \cdot 4^n}{(2n+3)(2n+1)!} x^n \right]$

$\qquad = c_1 x^{-1}(1 + 2x - 2x^2 + \frac{4}{9}x^3 - \cdots) + c_2 x^{\frac{1}{2}}(1 - \frac{2}{5}x + \frac{2}{35}x^2 - \frac{4}{945}x^3 + \cdots)$

(c) $y = c_1 x^{\frac{1}{2}} \left[\sum_{n=0}^{\infty} \frac{3 \cdot 2^{2n+1}(n+1)!}{(2n+3)!} x^n \right] + c_2 x^{-1} \left[\sum_{n=0}^{\infty} \frac{x^n}{n!} \right]$

$\qquad = c_1 x^{\frac{1}{2}}(1 + \frac{2}{5}x + \frac{4}{35}x^2 + \frac{8}{315}x^3 + \cdots) + c_2 x^{-1}(1 + x + \frac{1}{2}x^2 + \frac{1}{6}x^3 + \cdots)$

(d) $y = c_1 x^{\frac{1}{4}} \left[1 + \sum_{n=1}^{\infty} \frac{1}{n! \, 3 \cdot 7 \cdot 11 \cdots (4n-1)} \left(\frac{x}{4} \right)^{2n} \right]$

$\qquad + c_2 x^{3/4} \left[\sum_{n=0}^{\infty} \frac{1}{n! \, 5 \cdot 9 \cdot 13 \cdots (4n+1)} \left(\frac{x}{4} \right)^{2n} \right]$

$\qquad = c_1 x^{\frac{1}{4}} \left(1 + \frac{x^2}{48} + \frac{x^4}{10,752} + \frac{x^6}{5,677,056} + \cdots \right)$

$\qquad + c_2 x^{\frac{3}{4}} \left(1 + \frac{x^2}{80} + \frac{x^4}{23,040} + \frac{x^6}{14,376,960} + \cdots \right)$

(e) $y = c_1 x^{-\frac{1}{5}}\left[1 + \sum_{n=1}^{\infty} \frac{(-1)^n x^{3n}}{3^n n! \, 9 \cdot 24 \cdot 39 \cdots (15n - 6)}\right]$

$\qquad + c_2 \left[\sum_{n=0}^{\infty} \frac{(-1)^n 6 x^{3n+1}}{3^n n! \, 6 \cdot 21 \cdot 36 \cdots (15n + 6)}\right]$

$\qquad = c_1 x^{-\frac{1}{5}}\left(1 - \frac{x^3}{27} + \frac{x^6}{3888} - \frac{x^9}{1,364,688} + \cdots\right)$

$\qquad + c_2 \left(x - \frac{x^4}{63} + \frac{x^7}{13,608} - \frac{x^{10}}{6,246,072} + \cdots\right)$

SECTION 9.3 (page 301)

1. $y = (-\cos x)\ln(\tan x + \sec x) + c_1 \cos x + c_2 \sin x$

2. $y = -x \cos x + (\sin x)\ln(\sin x) + c_1 \cos x + c_2 \sin x$

3. $y = -2 + (\sin x)\ln(\tan x + \sec x) + c_1 \cos x + c_2 \sin x$

4. $y = \frac{1}{9}(\cos 3x)\ln(\cos 3x) + \frac{1}{3}x \sin 3x + c_1 \cos 3x + c_2 \sin 3x$

5. $y = x e^x \ln x + c_1 e^x + c_2 x e^x$

6. $y = \ln x + c_1 \ln x + c_2 x + c_3$

7. $y = \frac{8}{105}x^{\frac{5}{2}} + c_1 + c_2 x + c_3 \frac{1}{x}$

8. $y = -\frac{1}{4x} - \frac{\ln x}{2x} + c_1 x + c_2 \frac{1}{x}$

10. The general solution of the given equation.

SECTION 9.4 (page 318)

1. (a) $(1 - e^{-s})/s$ (b) $(1 - e^{-s})/s^2$ (c) $[1 - (se^{-s} + e^{-2s} + se^{-2s})]/s^2$
   (d) $(e^{-4s} - e^{-5s})/s$ (e) $1/s(1 + e^{-s})$ (f) $1/s(e^s - 1)$

5. (a) $2/(s - 1)^3$ (b) $(s^2 - 1)/(s^2 + 1)^2$ (c) $2s/(s^2 + 1)^2$
   (d) $[(s - \alpha)^2 - \beta^2]/[(s - \alpha)^2 + \beta^2]^2$ (e) $2\beta(s - \alpha)/[(s - \alpha)^2 + \beta^2]^2$

6. (a) $\dfrac{s^2 + 2}{s(s^2 + 4)}$ (b) $2/s(s^2 + 4)$ (c) $1/(s^2 + 4)$

7. (b) $(1 - e^{-s})/s(1 + e^{-s})$ (c) $(1 - e^{-s})/s^2(1 + e^{-s})$

8. (a) $\frac{1}{2}e^x - \frac{1}{2}e^{-x}$ (b) $\frac{1}{2}e^x + \frac{1}{2}e^{-x} - 1$ (c) $x e^x - e^x + 1$
   (d) $u_2(x) \sin(x - 2)$ (e) $u_2(x)(1 - e^{-(x-2)})$

10. (a), (b), (c), and (e) are piecewise continuous; (d) and (f) are not.

11. (a) $y = \frac{1}{3}x - \frac{1}{9} + \frac{1}{9}e^{-3x}$ (b) $y = \frac{1}{5}\sin x - \frac{2}{5}\cos x + \frac{7}{5}e^{2x}$

(c) $y = \begin{cases} 0 & \text{if } x < 1 \\ \frac{1}{25}e^{-5(x-1)} + \frac{1}{5}(x-1) - \frac{1}{25} & \text{if } x \geq 1 \end{cases}$

12. (a) $y = 2e^x - e^{2x}$ (b) $y = e^{2x} - e^x$

(c) $y = \frac{1}{16}e^{-x} + \frac{79}{16}e^{3x} - \frac{11}{4}xe^{3x}$

(d) $y = -5e^{-3x} + 6e^{-2x} + xe^{-2x}$

(e) $y = \frac{1}{5} - \frac{1}{5}e^{-x}\cos 2x - \frac{1}{10}e^{-x}\sin 2x$

13. (a) $y = \begin{cases} 0 & \text{if } x < \pi \\ \frac{1}{4} - \frac{1}{4}\cos[2(x - \pi)] & \text{if } x \geq \pi \end{cases}$

(b) $y = \begin{cases} \sin x & \text{if } x < 1 \\ \sin x + x - 1 - \sin(x - 1) & \text{if } x \geq 1 \end{cases}$

(c) $y = \begin{cases} 3e^{-2x} - e^{2x} & \text{if } x < \ln 2 \\ 4e^{-2x} - \left(\frac{17}{16} + \frac{\ln 2}{4}\right)e^{2x} + \frac{1}{4}xe^{2x} & \text{if } x \geq \ln 2 \end{cases}$

14. (a) $x = \dfrac{1}{k - m\omega^2}\left[\sin \omega t - \dfrac{\omega}{\sqrt{k/m}}\sin\sqrt{\dfrac{k}{m}}\,t\right]$

(b) $x = \dfrac{1}{2m\omega^2}(\sin \omega t - \omega t \cos \omega t)$

(c) $x = \dfrac{1}{hk/m}\left\{u_1(t)\left[1 - \cos\sqrt{\dfrac{k}{m}}(t - 1)\right] - u_{1+h}(t)\left[1 - \cos\sqrt{\dfrac{k}{m}}(t - 1 - h)\right]\right\}$

$= \begin{cases} 0 & \text{if } 0 \leq t < 1 \\ \dfrac{1}{hk}\left[1 - \cos\sqrt{\dfrac{k}{m}}(t - 1)\right] & \text{if } 1 \leq t < 1 + h \\ \dfrac{1}{hk}\left[-\cos\sqrt{\dfrac{k}{m}}(t - 1) + \cos\sqrt{\dfrac{k}{m}}(t - 1 - h)\right] & \text{if } t \geq 1 + h \end{cases}$

(d) $x = \dfrac{1}{\sqrt{km}}u_1(t)\sin\sqrt{\dfrac{k}{m}}(t - 1)$

$= \begin{cases} 0 & \text{if } 0 \leq t < 1 \\ \dfrac{1}{\sqrt{km}}\sin\sqrt{\dfrac{k}{m}}(t - 1) & \text{if } t \geq 1 \end{cases}$

19. (d) For $x < a$, the graph is the graph of the zero function; for $x \geq a$, the graph is equal to the graph of f shifted a units to the right.

Index